Coding, Shaping, Making

Coding, Shaping, Making combines inspiration from architecture, mathematics, biology, chemistry, physics and computation to look towards the future of architecture, design and art. It presents ongoing experiments in the search for fundamental principles of form and form-making in nature so that we can better inform our own built environment.

In the coming decades, matter will become encoded with shape information so that it shapes itself, as happens in biology. Physical objects, shaped by forces as well, will begin to design themselves based on information encoded in matter they are made of. This knowledge will be scaled and trickled up to architecture. Consequently, architecture will begin to design itself and the role of the architect will need redefining. This heavily illustrated book highlights Haresh Lalvani's efforts towards this speculative future through experiments in form and form-making, including his work in developing a new approach to shape coding, exploring higher dimensional geometry for designing physical structures and organizing form in higher-dimensional diagrams. Taking an in-depth look at Lalvani's pioneering experiments of mass customization in industrial products in architecture, combined with his idea of a form continuum, this book argues for the need for integration of coding, shaping and making in future technologies into one seamless process.

Drawing together decades of research, this book will be a thought-provoking read for architecture professionals and students, especially those interested in the future of the discipline as it relates to mathematics, science, technology and art. It will also interest those in the latter fields for its broader implications.

Haresh Lalvani is a tenured professor of Architecture at Pratt Institute. Known worldwide for his morphological, structural, and design innovations, Lalvani holds a Ph.D. in Architecture from the University of Pennsylvania. He has worked at NASA-Langley Research Center, Computer Graphics Laboratory, NYIT, Tata Institute of Fundamental Research (Mumbai) and was an artist-in-residence at the Cathedral of St. John the Divine, New York. His work is in the permanent design collection of the Museum of Modern Art, New York, and his sculptures are installed in New York City.

CODING SHAPING MAKING

Experiments in Form and Form-Making

Haresh Lalvani

NEW YORK AND LONDON

First published 2023
by Routledge
605 Third Avenue, New York, NY 10158

and by Routledge
2 Park Square, Milton Park, Abingdon, Oxon, OX14 4RN

Routledge is an imprint of the Taylor & Francis Group, an informa business

© 2023 Haresh Lalvani

The right of Haresh Lalvani to be identified as author of this work has been asserted in accordance with sections 77 and 78 of the Copyright, Designs and Patents Act 1988.

All rights reserved. No part of this book may be reprinted or reproduced or utilised in any form or by any electronic, mechanical, or other means, now known or hereafter invented, including photocopying and recording, or in any information storage or retrieval system, without permission in writing from the publishers.

Trademark notice: Product or corporate names may be trademarks or registered trademarks, and are used only for identification and explanation without intent to infringe.

Routledge does not guarantee the stability or longevity of the electronic information embedded within the QR codes inside the book. Neither the publisher nor the author is responsible for loss, corruption or access to this information. Please advise the publisher of any difficulty in accessing such information so it can be corrected in future editions.

Every effort has been made to contact copyright-holders. Please advise the publisher of any errors or omissions, and these will be corrected in subsequent editions.

ISBN: 9780367638801 (hbk)
ISBN: 9780367638795 (pbk)
ISBN: 9781003121145 (ebk)

DOI: 10.4324/9781003121145

Typeset in Ergizio, Fieldwork, LoRes, Swiss 721 and Times New Roman

To Asha and Zaran

CONTENTS

	Introduction	1
1	**Meta Architecture** Postscript 1	5 17
2	**Genomic Architecture** Postscript 2	28 44
3	**The Milgo Experiment**: An Interview with Haresh Lalvani Postscript 3	63 81
4	**The Pattern Continuum**: Mass Customization of Emergent Designs	96
5	**Form Follows Force**: The Epigenetic Continuum Postscript 5	130 159
6	**Self Shaping**: Form is Process	162
7	**Hyper Architecture**: Hyperstructures, Hyperspaces, Hypersurfaces	212
8	**Morphoverse**: And The Universal Morph Tool Kit	270
9	**Abiogenesis and the Future of Architecture**	340
	Acknowledgements	395
	Index	399

VIDEOS

The 39 videos below can be accessed through QR-codes which appear throughout the book. The videos are grouped by concepts within the chapters. Credits are inside the videos.

1-3	Video 1.	*Meta-Morphing Squares* (1989, 2003)	50-51
	Video 2.	*Algorhythm Column* (1997, 2003)	58-59
	Video 3.	*Algorhythm Wave Column* (1997, 2003)	58-59
	Video 4.	*Algorhythm Umbrella* (1999, 2003)	56-57
	Video 5.	*Algorhythm Double Umbrella* (2003)	56-57
	Video 6.	*Algorhythm Beam* (2003)	52-53
	Video 7.	*Algorhythm Canopy* (2003)	52-53
	Video 8.	*Interripples* (2003)	54-55
	Video 9.	*Hyperwall* (1999, 2004)	70-71
4	Video 10.	*Lautomaton* (2010)	99
	Video 11.	*Morphing Sphere 1D Series* (2010)	98
	Video 12.	*Morphing Platters 1D Series* (2011)	100-101
5-6	Video 13.	*Gravitational Rotational Forming* (2007)	149
	Video 14.	*PERI-TUBE* (2017)	153
	Video 15.	*Magic Carpet* (2016)	154
	Video 16.	*Pulsating Sphere* (2017)	154
	Video 17.	*GR FLORA 24 100 2* (2012)	178
	Video 18.	*GR FLORA, Real-time Forming* (2012)	179
	Video 19.	*X-TOWER 88.2* (2014)	188
	Video 20.	*X-POD Prototype* (2014)	194
	Video 21.	*Double Mushroom X-Columns* (2017)	200
	Video 22.	*Zipcode Arches* (2016)	207
7-8	Video 23.	*Morphoverse* (2004)	271
	Video 24.	*Triangle with 2 Vertices* (2017)	280
	Video 25.	*Triangle Continuum* (2010)	320
	Video 26.	*Decagon Continuum* (2010)	220
	Video 27.	*Decagon Continuum 2* (2010)	303
	Video 28.	*Explosions-Implosions* (Lalvani, McDermott, Hanrahan) (1984, NYIT)	289
	Video 29.	*5D Cube of …33 43 53… Families* (2005)	289
	Video 30.	*33 Family, Interactive 4D Cube* (2019)	292
	Video 31.	*4D Cube of 33 Family in AR* (2022)	292
	Video 32.	*Morphing 4D Cube* (2004)	315
	Video 33.	*Fruit Sections 1* (2020)	322
	Video 34.	*Fruit Sections 2* (2020)	322
	Video 35.	*Gaudi 4D Cube* (2004)	329
	Video 36.	*Gaudi's Portico Parc Guell* (2004)	330
	Video 37.	*A Snip of Form: Face-Becoming-Cell* (2017)	384
9	Video 38.	*Morpho-Space of Physical Phenomena* (2006)	341
	Video 39.	*4D Periodic Table of Elements, 5 versions* (2019)	342

IMAGE CREDITS

p.115 Sea shells

Volutoconus bednalli
Credit: Carles Dorado, www.allspira.com

p.121 Sea shells

1	2	3	4
	5		6
7			8
	9	10	11

1. *Phalium aeriola*
   Credit: Joop Trausel and Frans Slieker, Natural History Museum Rotterdam (WoRMS* image)
   CC BY-NC-SA 4.0
2. *Phalium areola*
   Credit: H. Zell
   CC BY-SA 3.0
3. *Phalium bandatum*
   Credit: H. Zell
   CC BY-SA 3.0
4. *Phalium bandatum*
   Credit: Joop Trausel and Frans Slieker (see #1)
   CC BY-NC-SA 4.0
5. *Phalium bandatum*
   Credit: Naturalis Biodiversity Center
   CC0 1.0
6. *Phalium bandatum*
   Credit: Naturalis Biodiversity Center
   CC0 1.0
7. *Phalium decussatum*
   Credit: Naturalis Biodiversity Center
   CC0 1.0
8. *Phalium flammiferum*
   Credit: H. Zell
   CC BY-SA 3.0
9. *Semicassis bisculata booleyi*
   Credit: H. Zell
   CC BY-SA 3.0
10. *Semicassis bisculata*
    Credit: Joop Trausel and Frans Slieker (see left column, #1)
    CC BY-NC-SA 4.0
11. *Phalium muangmani*
    Credit: Joop Trausel and Frans Slieker (see left column, #1)
    CC BY-NC-SA 4.0

pp.126-127 Sea shells

1	2	3	4	5		7	8
9	10		12	13	14	15	16

1. *Ovula ovum*
   Credit: James St. John
   CC BY 2.0
2. *Conus eburneus*
   Credit: Joop Trausel and Frans Slieker (see left column, #1)
   CC BY-NC-SA 4.0
3. *Amoria canaliculata*
   Credit: Indopacific Seashells
4. *Scaphella dohrni*
   Credit: Smithsonian Museum of Natural History
   CC0
5. *Scaphella junonia curryi*
   Credit: Edward J. Petuch
7. *Opeatostoma pseudodon*
   Credit: Joop Trausel and Frans Slieker (see left column, #1)
   CC BY-NC-SA 4.0
8. *Amoria zebra*
   Credit: Joop Trausel and Frans Slieker (see left column, #1)

CC BY-NC-SA 4.0
9. *Nerita atramentosa*
   Credit: Des Beechey, Australian Museum
10. *Cantharidella tiberiana*
    Credit: Des Beechey, Australian Museum
12. *Pictocolumbella ocellata*
    Credit: H.Zell
    CC BY-SA 3.0
13. *Volutoconus bednalli*
    Credit: Carles Dorado, www.allspira.com
14. *Luria tessellate*
    Credit: Randy Bridges, rbridges.com
15. *Clithon diadema*
    Credit: H. Zell
    CC BY-SA 3.0
16. *Austrocochlea procata*
    Credit: Toby Hudson
    CC BY-SA 3.0

pp.126-127 Insects

1	2	3	4	5	6	7	8
9	10	11	12	13	14	15	16

1. *Harmonia axyridis*
   Credit: NYS IPM Program at Cornell Univ.
   CC BY 2.0
2. *Anoplophora amoena*
   Credit: Ben Sale
   CC BY 2.0
3. *Ceroglossus ochsenii*
   Credit: David Madison
   CC BY 3.0
4. *Psyllobora vigintiduopunctata*
   Credit: Graham Callow, Naturespot.org.uk
5. *Omophoita-lunata-Fabricius*
   Credit: Juan Enrique Barriga-Tunon, Coleoptera-neotropical.org
   CC BY-NC-SA 3.0
7. *Eupholus (schoenherri) petiti*
   Credit: Hectonicus
   CC BY-SA 3.0
8. Striped cucumber beetle (*Acalymma vittatum*)
   Credit: Jeff Hahn, University of Minnesota Extension
9. *Halmus chalybeus*
   Credit: Raewyn
   CC BY 2.0
10. Blue Banded Weevil Beetle (*Eupholus linnei*)
    Credit: TaxidermyArtistry.co.uk
12. Leaf Beetle (*Stolas decemguttata*)
    Credit: Cesar Favacho
    CC BY-NC-SA 4.0
13. *Harmonia testudinaria*
    Credit: Adam Silpinski, CSIRO Entomology
14. *Propylea quatuordecimpunctata*
    Credit: Beentree
    CC BY 3.0
15. *Amelia singulata* (Blue banded bee)
    Credit: Alan Moore
16. *Disonycha conjuncta*
    Credit: Juan Enrique Barriga-Tunon (see #5)

* World Registry of Marine Species (WoRMS), https://www.marinespecies.org/

INTRODUCTION

This book brings together decades of author's work in one place for the first time, a mix of early and new work. It combines ideas which draw inspiration from many fields – architecture, mathematics, biology, chemistry, physics, computation - with a speculative view towards the future of design arts (architecture, design, art) and some enabling technologies. It includes ongoing experiments in the search for fundamental principles of form and form-making to explore nature's principles and methods using a twin strategy. It can inform our designs so they can be in harmony with nature with the hope that our continuing explorations in the physicality of making can provide insight into how nature designs. It combines higher-level biomimicry with its opposite, a reverse biomimicry, where physical objects we build inform our understanding of how nature builds.

The three phenomena – *Coding, Shaping* and *Making* – are central to this work and relate to form. **Coding** is a symbolic representation (numbers, symbols), **Shaping** is giving spatial form to an idea (topology, geometry), **Making** converts spatial form into a physical object through real materials and physical processes (fabrication, manufacturing). These three fundamental activities have led to divergent technologies. In biology, all three phenomena, coding, shaping and making, are one. Architecture and design are heading in that direction, very slowly, a step at a time, albeit on the "front-end" since architecture is an applied field which uses fundamental knowledge from other fields. Our future technologies will integrate these three phenomena into one seamless process. Synthetic biology, nano-technology and material science lie at the "back-end" of architecture since they deal with creation of knowledge related to the materialized world at the nano- and micro-scales. These fields have been heading in that direction and will continue to lead the way in the basic sciences and technologies. The future will see a greater integration between the art and the science of physical making. Design artists and makers at macro scales on one end, working with synthetic biologists, synthetic chemists and synthetic physicists on the micro and nano-scales, knowing well, as Fuller pointed out decades ago, there are no separate departments of physics, chemistry and biology in nature.[1]

In the coming decades, matter will become encoded with shape information so it will shape itself, the way that biology does. Fundamental discoveries will continue to be made at the nano-level by scientists, engineers and technologists towards this dream. It is at the nano-scale that self-design of physical systems

---

[1] *Dymaxion World of Buckminster Fuller*, R. B. Fuller and R. Marks, (1960, republished 1973).

will be invented – physical objects will design themselves based on information encoded in matter they are made of. This knowledge will be scaled up to the micro-scale and eventually to macro-scales needed in architecture. When it happens, architecture will begin to design itself. The role of the architect, as mentioned in *Meta Architecture*[2] (Chapter 1), will be re-defined when self-architecture will begin. The experiments in form and form-making presented here continue to aspire towards this direction.

Form continuum (combinatorial continuum, binary continuum) is central to the work presented here and is defined by continuous transformations introduced in 1971 in the author's work. The readers are encouraged to think of any form changing to another through a series of transformations. These transformations are presumed to be finite (Chapter 3). This is particularly important within the higher dimensional taxonomies presented throughout the book, and especially in the *Morphoverse*[3] (Chapter 8), a universe of form first described in *Genomic Architecture*[4] (Chapter 2). The continuum extends beyond form, into processes as well as other physical attributes of form.

The first seven chapters deal with the *Milgo Experiments*,[5] my long-term physical experiments with the company Milgo-Bufkin. These experiments, five in all,[6] mark a unique collaboration between a designer-artist and a leading metal fabricator. The experiments provided a leading example in mass customization of industrial products in architecture[7] in *AlgoRhythms* (Chapter 1-3), a trade name for products in curve-folded sheet metal. Its extension to emergent designs in the *Morphing Platters* series (Chapter 4) provided another example of individuation, where the twin themes, "unity in diversity" and "diversity in unity", mimic nature which builds snowflakes, molecules[8] or roses, each different yet alike, from a fixed number of parts and processes. In addition, the platter designs provided no economy of scale while making each design different at the same cost.

While the two experiments, *AlgoRhythms* and *Morphing Platters*, had a formal, algorithmic, "genetic"[9] approach to form-generation and required 3D digital modeling, the two classes of expanded surfaces (*XURF* and *X-STRUCTURES*, Chapter 5, 6) addressed the opposite approach, that of "epigenetics", to borrow another term from biology. No 3D digital models were used to derive the resulting

---

[2] First published in 1999, republished here with permission.
[3] Originated in 1981 and in continuous development since.
[4] First published in 2003, republished here with permission.
[5] Named after the company to memorialize the place where the large body of physical experiments in these chapters were done and generously supported during an extended period (1997-2014). The term 'Milgo Experiment" first appeared in 2006 in the interview by John Lobell (Chapter 3), reproduced here with permission.
[6] *AlgoRhythms* (Experiment #1, 1997-2004), *XURF* (#2, 1998-2010), *X-STRUCTURES* (#3, 2008-2014), *Morphing Platters* (#4, 2011), *HyperSurfaces* (#5, 2006-2012), in the chronology of production in the factory. Digital work on some has continued since.
[7] Mario Carpo, in his essay *Pattern Recognition* for the Venice Biennale 2004, cites three examples of industrially produced design products: Greg Lynn (2001), Haresh Lalvani (1999), and Bernard Cache (2003), cited here in the order of appearance in the essay. For a detailed citation, see Chapter 1, Postscript 1.
[8] The chemist Roald Hoffmann describes this as "Being Same and Not the Same", the title of his book (cited in Chapter 4).
[9] This usage of the term "genetic", borrowed from genetics, was developed independently in design and computation in a formal (algorithmic, generative) sense three decades ago. Recently, with design artists increasingly working with biologists, this usage will be confusing.

form in these experiments; form emerged from the physical process of applying force in different ways. To our great surprise, it led to "self-shaping" of metal at room temperature without using heat to soften the metal. The metal self-organized (self-folded, self-undulated, self-wrinkled) and produced graded properties. This removed the designer from the shaping process. More significantly, it provided an insight into the process of self-folding which, in these instances, is related to the interplay between changes in boundary (length, curvature) and the area of the expanding surface.

The eerie resemblance of our structures, "growing" in real-time within the factory as we formed them, with living skins and shells in nature (insects, crustacea, seedpods, and various biological skins), was inescapable. It made one wonder if nature deployed a similar physical technique in its inventory of morphogenetic processes. In our experiments, we were witnessing a special type of morphing, a continuum of two binaries, as the metal sheet morphed continuously from a pliable state, enabled by laser-slitting which weakened it, to the rigid state when it could no longer be formed any further. More remarkably, the material flowed towards emergent strength and became stronger where it needed to be. The "supporting" and the "supported" differentiated in a continuum. The emergence of "eyelids" (structural beams at lintel and sill level) in the "eye" of the *X-POD 138* (p.192) is one striking example. The binary continuum in the process of flow (rheology) in shaping a physical structure adds a design strategy to complement the geometric and topologic binaries we will encounter in the *Morphoverse* (Chapter 8) and carried further in the last chapter.

The real-time video recordings of the actual forming process in the factory, cleaned-up in the studio to remove stops, starts and checks in the forming process, provided an interesting (theoretical) result of forming the entire structure in one go. *X-Tower 88.2* (pp.186-7), which took approximately 30 minutes of factory time, was reduced to 58 seconds. A stainless steel tower, 3/16inch thick, rose to 12 feet in less than a minute. This suggested instant-forming, setting the upper limit goal of forming *X-STRUCTURES* at the speed of force.[10] In gravity-forming, this limit is the speed of gravity and leads to *instant architecture*. Such rapid forming provides a promising application to emergency shelters which are needed to address the problem of climate refugees displaced by lost land due to hurricane flooding or rising sea levels. We are currently exploring versions of *X-POD 138* in bamboo, a green regenerative material with a negative carbon footprint,[11] to test the idea.

HyperSurfaces (Chapter 7) provide a transition from the earlier chapters to the last two. These are physically realized higher dimensional structures built in 3D space after projection from dimensions greater than three. Physical hyperstructures provide a counterpoint to the higher dimensional meta-structures in the *Morphoverse* (Chapter 8). Meta-structures are used as conceptual frameworks (diagrams, networks, lattices) to organize information relating to form, and extended to other disciplines in *Abiogenesis and the Origin of Architecture* (Chapter 9) and other unpublished works. In Chapter 9, higher dimensions provide

---

[10] In gravity-forming, the speed of force is the speed of gravity which is the same as the speed of light. This sets the upper limit to the speed of fabrication by gravity-forming (and also light-forming).
[11] At the Center for Experimental Structures, School of Architecture, Pratt Institute.

a way to organize the chemical elements,[12] molecules and compounds, and is extended to DNA sequences. Minimal information coding molecules, double-helical and complementary-paired like DNA but reduced to their minimum, are presented here as a designer's speculative approach to molecular morphology. These may be DNA precursors, or examples of new molecules that encode information, or may inspire new work in the search for DNA alternatives for non-biological applications. Information coding molecules are only the first step towards self-building which, in addition to being encoders, will require molecules to become builders, assemblers and replicators.

Some thoughts are presented on the equivalence between DNA and number systems, providing a bridge to number-coded forms and computation. This will lead to number-based DNA-shape scripting by designers. It will require a tie-in between complementarity in geometry and topology with complementarity in DNA.

A section in Chapter 9 deals with abiogenetic space, the chemical space for the origin of life. The search for our deep origins is of great importance for us humans. Equally important is our interest in our future. If our future technologies are tied to the early pre-biotic technologies, a link between our deep past and our future would bring a sense of closure. It will represent a deeper harmony between us and nature on a broader cosmological scale.[13] On our planet, this harmony has been deeply ruptured as our technologies have been at odds with nature. Addressing this existential planetary threat requires a philosophical re-alignment in our relationship with nature.

Design arts – art, design, architecture – have a fundamental role to play in realizing this bridge between our creations and those of nature. Science, which builds on collective knowledge, represents our collective "we", Art which is unique to each individual, represents our collective "I" with one essential difference: there is wrong science, but there is no wrong art. ArtSci, which combines the two and includes design arts, occupies the continuum between these two spaces. The integration between the two, tied by our humanism, marks the beginning of bridging the great divide, a divide between art, science and humanities we do not experience when we are children. We are at the dawn of this integrative process. During the last session of TED2004 in Monterey, the 'positive psychologist' Martin Seligman mentioned the example of Florence during the 15th century which (through its patrons) decided to invest its excess wealth in beauty. This led to the Renaissance. He called it the 'Florentine moment' and suggested that we should seize on it now so society is elevated to reach its highest aspirations of beauty and creativity. A synthesis of our knowledges provides a different type of healing among the others we need – healing between us, and our healing with nature. Creativity is an essential bridge in this healing to link beauty with the most advanced knowledge of our times. If it can solve pressing societal problems in doing so, that is a huge bonus. If not, it can create solutions for future problems that are yet to be invented.

---

[12] The concept extends to particle physics where the elementary particles define the building blocks of neutron and proton which, along with the electron, are the three sub-atomic components of the atom. Neutron and proton are built from two of the six quarks (up, down), and the electron is an elementary particle (one of the six leptons). The three generations of fermions, each comprising 2 quarks and 2 leptons can each be mapped on the vertices of three analogous 4D cubes. Adding anti-particles extends these to families organized on the vertices of three analogous 5D cubes.

[13] This is reminiscent of Carl Sagan's "we are made of star-stuff" statement in the series Cosmos, a *Personal Journey* (TV Series, 1980), Episode 1: The Shores of the Cosmic Ocean.

1
META ARCHITECTURE[1]

Architect-morphologist Haresh Lalvani has developed a technique to modulate sheet metal into a wide range of new configurations that can be easily manufactured using a patent fabrication process he developed with MiIgo-Bufkin. Here he discusses the development of his theory of Meta Architecture, his application of the term Hypersurface, and his work with Milgo-Bufkin which is currently launching his new design series.

Meta Architecture is based on manipulating morphologically structured information via algorithms and genetic codes that encipher the formal possibilities of architecture. These possibilities are determined by mapping them in a unified morphological universe [1], a higher dimensional meta space which (theoretically) encodes all past, present and future morphologies. It also maps all their transformations. The coding of structures within this universe leads to an artificial genetic code [2]. This is a universal morphological code [3] and acts as a driver for organizing shaping building and transforming architecture over short-term and long-term time scales. Coupling the code with manufacturing processes, both at the macro level of current computer-aided manufacturing and the micro level of nanotechnologies and genetic engineering, enables the direct translation of the code into the physical process of building. Coupled with biological (DNA-based or other chemical physical) building processes, the artificial genetic code enables growth, adaptation evolution and

---

[1] This chapter, with minor edits in figure captions and a Postscript added, is a re-publication of 'Meta Architecture' by Haresh Lalvani in 'HyperSurface Architecture II, Guest Ed. Stephen Perella, *AD* Academy Editions, Vol.69, 9-10/1999, Profile 141, pp.32-37. Reproduced with permission from AD-Wiley. It was re-published in the AD book *Architecture and Science* (2001) by Guiseppa Di Christina, Academy-Wiley, 2001. The Postscript describes further developments during that early period.

replication of buildings, permitting architecture to design itself and eventually liberating it from the architect. Architecture as we now know it will end when self-architecture begins [4].

Within this overall premise, several examples from my ongoing work in Meta Architecture, and the related visual product, Hyperspace Architecture (or Hyper Architecture), are presented. The work offers an alternative paradigm to 'digital architecture' which has emerged in the last decade. Increasingly sophisticated computer graphics tools have enabled architects to visualise relatively complex spatial environments in virtual space without recourse to physical models or in some instances (as in Frank Gehry's museum at Bilbao), to digitise complex built models directly. These digital visualisations, all conceived in virtual space, are admittedly visually spectacular and are conceived 'top down' both visually and spatially However they are neither informed by construction methods or the properties of physical materials, nor by any morphological principles of space and structure, which impose strong constraints on architecture. Architecture, shaped by these constraints and modelled by morphological principles – including Meta Architecture – is architecture that proceeds from the 'bottom up'.

The works presented here exemplify the bottom up approach in two different ways: one driven primarily by higher dimensional geometry and the other by combining geometry with manufacturing process in making physical form out of real material. Both examples show the unprecedented possibilities for shaping architecture opened up by recourse to basic morphological principles (geometric, topologic, structural, etc.). The images demonstrating the first approach are excerpted from my folio 'Hyperspace Architecture' which shows the various applications of higher dimensions for architecture [5], and the second approach is from an ongoing experiment currently being carried out with Milgo/Bufkin, a leading metal fabricator in New York.

Highly ordered geometry is used in the first approach (**figs.1, 2**) as a basis for generating irregular hyperstructures, in this case, hypersurfaces. The term 'hypersurface' here is used according to its original meaning defined in the strict geometrical (mathematical) sense: i.e. having spatial dimensions greater than three. This definition contrasts with the usage in this and the previous special issue of *AD*, edited by Stephen Perrella, entitled 'Hypersurface Architecture' where the term 'hyper' is used as a meta-dimension of the surface and not its spatial dimension. Interestingly in the first example shown here (**figs.1A, B**), the term has a double meaning. The two tiling designs are identical in their base geometry which comprises an assembly of identical crescent-shaped tiles [6] based on two-dimensional projections from five-dimensional Euclidean space. The crescents are thus hyper-tiles. In addition, they have a superimposed pattern of dark lines, echoing the other meaning of 'hypersurface' (as used by Perrella). While the designs appear random each tile is identically marked in both cases. The image captures the paradigm that irregular and random-looking designs can be constructed from identical modules, an idea of great significance for architecture as it visually blends order with chaos.

Another example (**figs.2A-C**) further exemplifies this juxtaposition between order and disorder in a three-dimensional structure. The regular surface, a true hypersurface projected from higher dimensional space hovers like a cloud over a space that, when extended is non-periodic. The structure can be constructed

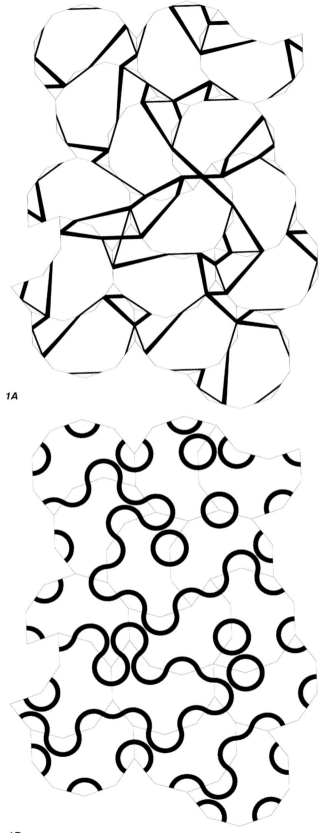

1A,B
Hypersurfaces using crescent-shaped tiles.

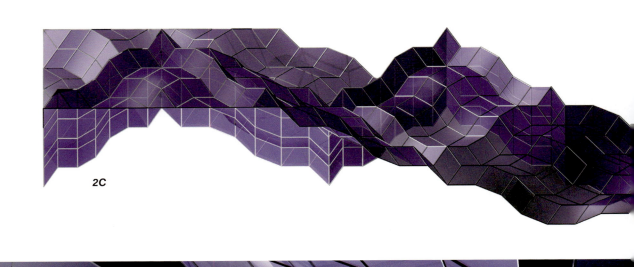

2C

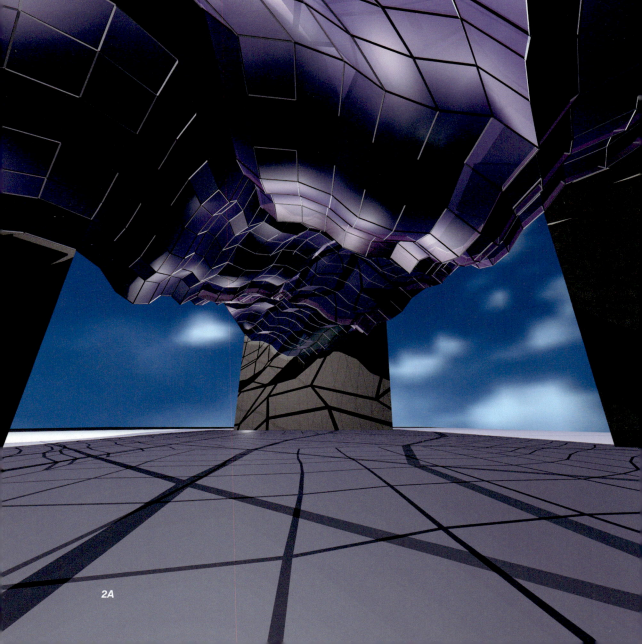

2A

Meta Architecture

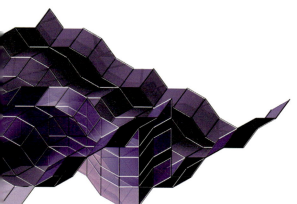

2A-C
Cloud Cover.

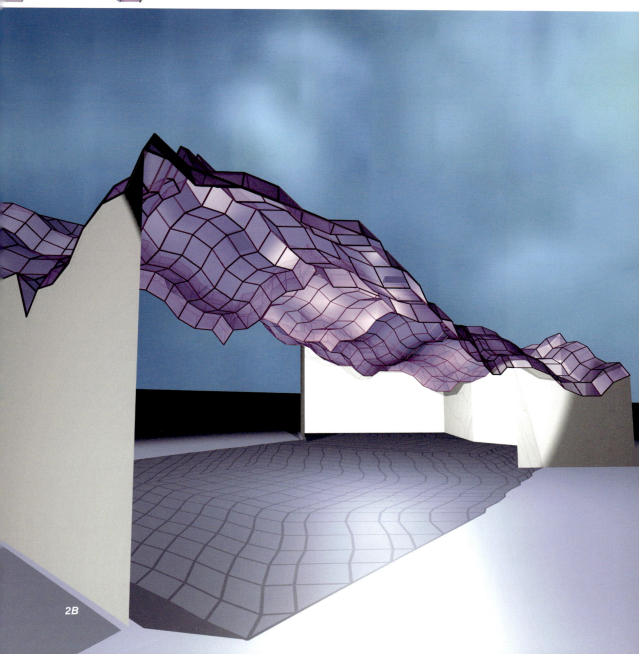

from a single-node design [7], a single-strut element and flat panels. Additional stabilizing features would most likely be needed for its structural stability. This is just one example of the unlimited and varied architectural compositions that can be constructed from this new morphological invention.

The next group of images (**figs.3A-O**) exemplifies the second approach. The project deals with software-driven fabrication of sheet metal for architectural surface structures [8] and is being carried out at Milgo/Bufkin's manufacturing facility. Though columns, capitals, wall and ceiling panels are immediate architectural applications of concern for Milgo's business interests, the project provides a unique opportunity to experiment with broader Meta Architectural concepts, especially the relationship between an artificial 'genetic code' and the manufacturing process. All sheet-metal structures shown here were generated using a morphologically encoded algorithm, which provides the possibility to generate endless 'variations on a theme' by manipulating the code. As a result, no two structures need be alike, so each individual in the world if desired could have their own unique structure. A procedure was developed whereby single continuous metal sheets could be marked by computer-driven equipment and then folded (manually for now). The resulting structures not only have a new look, but appear to be structurally advantageous at the same time. Architectural and industrial design products as well as complete environments based on these structures are currently being developed (**figs.4-8**; see Postscript **figs.A**, **B**, **D**). The algorithmic approach permits the structures to be modeled transformed and fabricated with ease. We expect that the morphologic elegance in the shaping of these structures would also translate into an economy in building.

Column Museum (**fig.4**; see Postscript **fig.C**) shows a sampling of the morphologically encoded columnar structures being prototyped and fabricated at Milgo. *Fractal High-Rise* (**fig.5**) shows a branched fractal column concept applied to a glass skyscraper. *Umbrellas* (**fig.6**) utilizes the twisted fold for a freestanding structure in an open-air environment. *Transitions* (**fig.7**) shows flat, wavy and irregularly curved walls within the same spatial layout, using the same material and fabrication technology. *WaveKnot* (**fig.8**) employs a continuous rippled surface for a ceiling or roof defined by a simple topological knot space.

The undulating look of these structures resulted from an interest in the fundamental behavior of sheet material under forces. Material 'flows' under its own weight and other forces according to predetermined morphologic laws, which pertain more to fluid motion than to static objects. Constructing architectural elements from rigid rectilinear units (such as bricks and beams) has 'frozen' this inherent flowing nature of architectural envelopes. The wrinkles on our skin, the surfaces of plants and skins of animals, waves and cloud forms, display this fluid-like quality in nature. Curvilinear architectural forms constructed using standard building methods have usually raised concerns of economy.

However, our experiments at Milgo suggest that advanced software-driven manufacturing processes, coupled with powerful morphological underpinnings, can easily and possibly economically generate a wide repertory of new curvilinear vocabulary unavailable to architects in the past. Paradoxically high technology representing the opposite pole in the man-nature dichotomy permits fluid shapes not possible earlier in a simple and elegant manner and in doing so, brings us closer to forms in nature. This is true not only in visible forms, but also in the

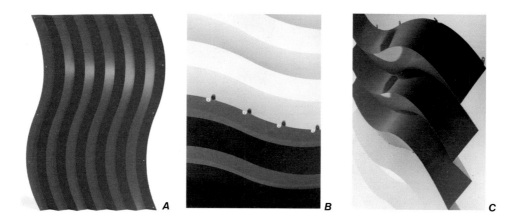

3A-C
Architectural sheet metal panels manufactured by Milgo.

concept of the genetic code, which permits each one of us to be unique yet encoded by the same basic genetic alphabets (DNA bases). These sheet metal structures are morphologically coded in a similar way.

12 Coding, Shaping, Making

3D-O (left)

3D-O
Architectural sheet metal columns and panels manufactured by Milgo.

4
Column Museum.

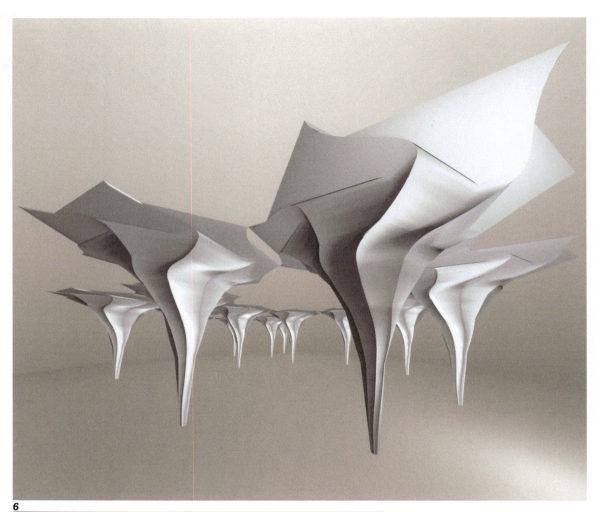

6

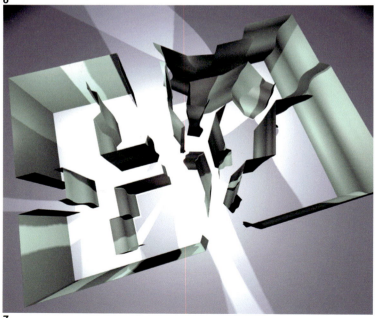

7

5

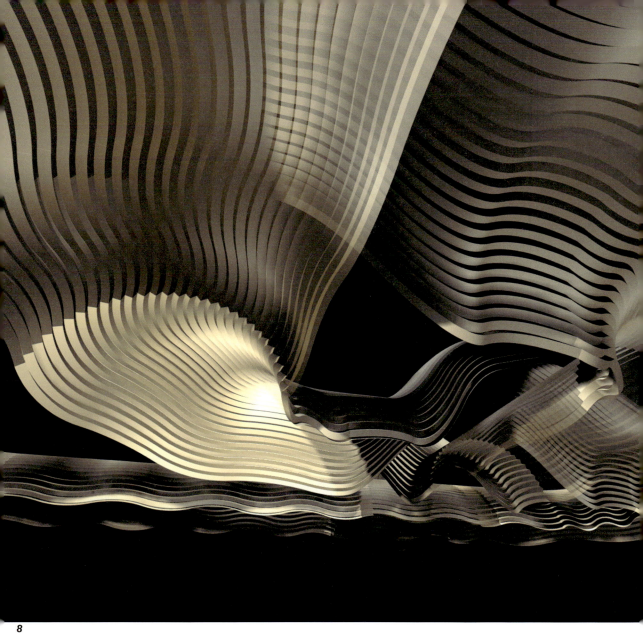

5
Fractal High-Rise.

6
Umbrellas.

7
Transition.

8
WaveKnot.

ACKNOWLEDGEMENTS

I am indebted to the following for their contribution to the project: *computer modeling and rendering*, Neil Katz and Mohamad Al-Khayer; *photography*, Robert Warren; *product development, prototyping and fabrication*, Milgo-Bufkin with Bruce Gitlin and Alex Kveton. More information on these projects with Milgo-Bufkin, Brooklyn, New York, can be found on their website <www.milgo-bufkin.com> from January 2000.

NOTES/REFERENCES

1. Haresh Lalvani, 'Morphological Universe, Expanding the Possibilities of Design and Nature' unpublished, 1998, based on a lecture presented at ACSA conference, Dalhousie University, Nova Scotia, October 1998, on the theme 'Works of Nature: The Rhetoric of Structural Invention'

2. My interest in the genetic code of architecture dates back to 1975; my first published work on morphological coding was in the context of Islamic patterns (1982). In 1993, I proposed 'architectural genetics' as an emerging science.

3. I have been developing such a code for over two decades. Early interim results appeared in *'Multi-dimensional Periodic Arrangements of Transforming Space Structures'* PhD Thesis, University of Pennsylvania (1981), self-published as *Structures on Hyper-Structures* (1982). Subsequent extensions of this work have been published in various papers, and applications to various structural morphologies have been in progress since the early 1990s.

4. For the origins of 'growing' architecture, see William Katavolos, *Organics*; Steendrukkirj de Jong & Co (Hilversum), 1961 Vittorio Giorgini, 'Early Experiments in Architecture using Nature's Building Technology' in H. Lalvani (ed), *The International Journal of Space Structures*, vol 11 nos 1 2, special issue, 1997. My work in this concept came via genetic engineering and was proposed in 'Towards Automorphogenesis, Building with Bacteria' unpublished, 1974, the source of which goes back to the question asked by my thesis of 1967· 'Why don't we build with bone and spider silk?' In recent years, John Johansen has been proposing growing architecture using 'molecular engineering'.

5. Lalvani, Haresh. 'The Architectural Promise of Curved Hyperspaces' *2nd International Seminar 'Application of Structural Morphology to Architecture'*, University of Stuttgart, 1994; 'Hyperstructures' in P. Dombernowsky and T. Wester (eds), *Engineering a New Architecture, Conference Proceedings*, Aarhus School of Architecture (Denmark), 1999.

6. Lalvani, Haresh. *US Patent 4,620,998*, 1986.

7. Lalvani, Haresh. *US Patent 5,505,035*, 1996.

8. Lalvani, Haresh. Patent pending.

POSTSCRIPT 1

This chapter is the first in the trio of related articles published in *AD*, the other two are reproduced as Chapters 2 and 3 in this book. The "morphological universe" mentioned in the opening paragraph in Chapter 1 is further described in Chapter 2 and expanded in Chapter 8. The terms "hypersurfaces", "hyperstructures", "hyperspace" in relation to architecture are addressed in Chapter 7.

This postscript, like the companion postscripts for Chapter 2 and Chapter 3, is a visual catalog of the work done during that period but not included in the published articles due to limited space. They are an important part of this body of work and are added here to give a fuller account. *AlgoRhythms* were introduced as a product line of columns, wall and ceiling panel systems by Milgo-Bufkin and included some of the photographs in Chapter 1 and this Postscript [1]. The article in *Metropolis* [2] covered the broader aspects of the work and provided the context of a unique designer-fabricator collaboration. This was followed in the next year by a TED talk and installation along with an exhibit at the Municipal Arts Society and the acquisition by MoMA which are briefly described next.

Fig.A (pp.16-17) shows the *Prague Column* (**A1**) displayed at the Prague Biennale 1 (2003) [3]. **A2** (***top***) shows four sequential (morphing) *AlgoRhythm Columns* in folded titanium which were commissioned and acquired by The Museum of Modern Art (MoMA), New York, for their permanent design collection in 2004. They were featured in the Phillip Johnson Design Galleries when the museum reopened that year. This was the first time titanium was curve-folded. Bruce Gitlin, a metallurgist by training, found the correct composition after a world-wide search that enabled their folding after several brittle attempts at folding this material. **A3** (***bottom right***) shows *AlgoRhythm Columns* (***bottom right***) at TED2004, Monterey, CA, described in their program guide as "revolutionary AlgoRhythm columns" [4], (installation and lighting by Bruce Gitlin and Uttara Asha Coorlawala.). This was a companion to the author's presentation at the same conference entitled 'Morphogenomics and The Milgo Experiment' (February 26, 2004).

Fig.B (pp.18-19) shows a collection of images of full-scale prototypes, 8ft to 12ft tall, in folded metal from single sheets. Some of these are larger versions of the ones shown in **Figs.3D-O**. They are in different painted or patina finished cold rolled steel or stainless steel with a non-directional brush finish. These required various methods of "stitching" which were invented to enable folding with the use of laser-cutter or a water-jet cutter which permitted "v-grooving". Additional methods that did not require laser- or water-jet cutting were invented and led us to think of morphing tools and, eventually, morphing machines. The latter required a collaboration between Milgo-Bufkin and a major manufacturer of metal-working machines, an idea that was abandoned for business reasons. But it pointed to a future where architects and designers could break through the fabrication grid-lock on an industrial scale since machines (and lineages thereof) designed for straight-line and flat-plane geometries were being adapted to make curved surfaces. This backwards approach required a revolution in manufacturing so machines could manufacture curved surfaces economically from Day 1. As the great Florentine architect Vittorio Giorgini, also a colleague of mine and initial co-founder of the Center for Experimental Structures at School of Architecture, Pratt Institute, stated clearly that history took a wrong-turn when it embraced straight-line and flat-plane production. At Milgo, we also discussed the possible use of robotics to enable automated curve-folding.

Fig.C (pp.20-21) shows a mini *Column Museum* for the Prague Biennale (2003) and re-displayed at the Municipal Arts Society (2004). This is a different rendition of the digital *Column Museum* in **fig.4**. See also **fig.3** and **fig.11** in Chapter 3.

Fig.D (pp.22-23) shows prototypes of a few panel systems for walls, ceilings or partition systems. Several developments are worthy of a note here. **D1** shows that panels with irregular parallel waves could be achieved from the same system, **D2** and **D3** show the introduction of wavy framing members with drop-in panels, **D4** shows a nested modular system composed of individual modules that could be varied in length and curvature, **D5** shows the limit of curvature achieved from a single metal sheet with right-angled folds requiring a forming innovation, **D6** shows the prototype of a ceiling system for a restaurant in Staten Island, **D7** shows prototypes of two door panels for an office in Texas, **D8** and **D9** show two panel systems with right-angled folds from a single sheet with more modest curvature than **D5**; **D10** and **D11** are two different ceiling panel systems in the Milgo factory, **D12** is a prototype of a single panel using a new manufacturing technique, **D13** and **D14** are images of a villa in the Caribbean showing the application of the new technique. The framing members in **D1** and **D2** were the beginnings of folded structural elements which could match panel geometries, an idea carried further in Chapter 3 and its Postscript.

This article was re-published in a collection of articles showing new work in the 1990's in the AD book *Architecture and Science*, edited by Giuseppa Di Cristina [5]. **Fig.3A-O** was re-published by Mario Carpo in his essay 'Pattern Recognition' for the Venice Biennale 2004 [6]. The author was surprised to learn from the dates published in Carpo's article that *AlgoRhythms* was the first example of industrially produced mass customization in architecture.[1]

---

[1] See footnote 7, p.2.

NOTES/REFERENCES

1. *AlgoRhythms* (a catalog of new curved folded metal products) by Milgo/Bufkin, New York, 2001.

2. Peter Hall, Bend the Rules of Structure, *Metropolis*, June 2003, p.138.

3. Andrea Di Stephano, Aion: An Eventual Architecture, In: *Prague Biennale1, Peripheries Become the Center*, 2003, p.486; the exhibition catalog of Prague Biennale 1, National Gallery, Prague, Czech Republic, June 26 – August 24, 2003.

4. Program Guide, *TED2004: Pursuit of Happiness*, February 24-28, 2004, Monterey, CA.

5. Lalvani, H., Meta Architecture, republished in *Architecture and Science*, ed. Guiseppa Di Christina, AD book, Academy-Wiley, 2001, p.178.

6. Carpo, Mario, Pattern Recognition, In: *Metamorph 9, International Architecture Exhibition, FOCUS*, La Biennale di Venezia, Venice; Rizzoli, 2004, p.44.

A1

A2

A3

A1
Prague Column.

A2
AlgoRhythm Columns at MoMA.

A3
AlgoRhythm Columns at TED2004.

22 Coding, Shaping, Making

B
AlgoRhythm columns prototypes.

Postscript 1

c
Mini *Column Museum.*

D1

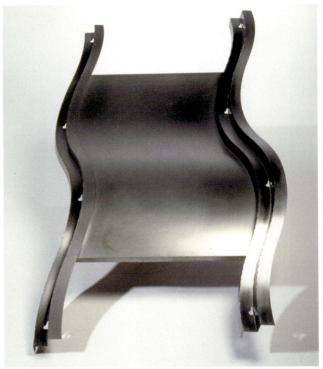

D4

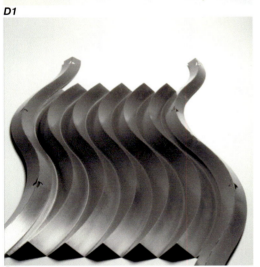

D2

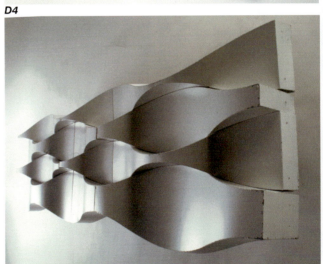

D5

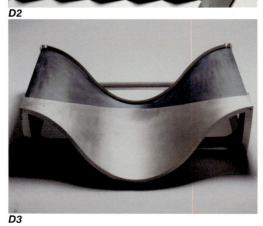

D3

D6

D7

Postscript 1

D8

D
Architectural
Panel Systems.

D9 *D10*
D11 *D12*

2

GENOMIC ARCHITECTURE[1]

Introduction

Genomic architecture is based on the manipulation of the *architectural genome*. Like its biological counterpart, this genome is universal and encompasses all architecture – past, present and future. At its root, this genome is defined by a unified *morphological genome*, a universal code for all morphologies – natural, human-made and artificial. *Morphogenomics*, a possible new science, deals with morphological informatics. It includes mapping the morphological genome as a basis for generative morphologies that underlie the shaping of architectural space and structure. Once mapped, the morphological genome will need to be layered with other genomes (also requiring mapping) to cover different aspects of architecture: physical (for example, materials construction technologies) sensorial, cognitive and behavioral. Genomic architecture, based on the layered genome, encompasses an integrated world of 'artificial architecture' (used in the same sense as 'artificial intelligence' and 'artificial life'), a world of complexity evolving in parallel with the natural world. It is a morphologically structured network of information that determines architectural taxonomies and phylogenies, permits digital manipulation of form in the design process and enables mass customization in digital manufacturing.

Limits of Organic Architecture

The meaning of the term 'organic architecture' which draws its inspiration mostly from biology keeps evolving with increasing knowledge of nature combined with foreseeable technologies. As new technologies emerge, architecture becomes *more* organic in its

[1] This chapter, with minor changes in figure captions and a Postscript added, is a re-publication of 'Genomic Architecture' by Haresh Lalvani in 'The Organic Approach to Architecture', Guest Eds. Deborah Gans and Zehra Kuz, Wiley-Academy, 2003, pp.116-126. Reproduced with permission from AD-Wiley. The Postscript describes related developments during that early period for a fuller account.

scope, intent and realisation. The upper limit to this sort of biomimicry would be biology itself. Buildings would grow [1] respond, adapt and recycle, they would self-assemble and self-organize, they would remember and be self-aware, they would evolve and they would reproduce and die. Organic architecture, were it to attain biology would design itself. It would also perpetuate itself. Architecture would then become 'life' and, paradoxically buildings would no longer need architects. Organic architecture, in this limited case scenario, would also define the end of architecture (as we define architecture now).

Extrapolating from projected technologies of the future [2], a scenario like this one is quite possible, even inevitable but it is flawed for two reasons. First, biology as a goal for organic architecture assumes that such a biology (namely, *existing* biology) is frozen in time since it is based on 'life' as we know it presently. Extrapolation of architecture from *present* biology ignores *past* and *future* biologies. Nature's ongoing experiment comprises structures that are extinct, structures that exist now and structures that have yet to appear [3]. The definition of organic' must thus encompass all biologies: past, present and the future. Second, it ignores the creation of the new; for example new materials (new chemistries) not found in nature, new technologies not found in nature and new organisms (based on known or new biologies) not existing in nature. Besides new natural biologies, the term 'organic' must thus include *artificial biology* as well. This is where the line between human designs and those made by nature becomes a continuum.

Unifying Laws

What unites the natural and the human-made (including the artificial) are fundamental laws, the *laws of nature*. Our knowledge of nature and human-made constructions evolves so that these laws become increasingly more encompassing, tending towards the natural upper limit of a single unifying law for everything (as in the current search in physics, for example). Whether this limit is attainable is an open question. The natural and the artificial are facets of organic architecture that are joined at this fundamental level. This is true of biology and buildings. The morphologic possibilities within these two worlds fall within a single *morphological universe* [4] governed by unifying laws of form that are common to both [5]. It is governed by the mathematics of space, structure and form. When physical constraints (size, material, movement, weight, stability, building method or forming process, etc.) are imposed on form, this universe shrinks through the elimination of mathematical structures that are physically unrealizable. The physics and chemistry of form delimit the morphological universe.

Hyperuniverse of Form [6]

Imagine a universe of all possible morphologies, a universe that includes all past, present and future structures. A universe that is infinite and open-ended, and one that has a fractal hierarchy composed of recursive levels within levels. A universe that one can access and navigate through in any number of ways from any level. A universe where each structure (and each type of architecture) can transform from one to another in a continuum of space and time, both within and between levels. Using simple orderly structures as a starting point for more complex structures, the results thus far suggest that this universe is highly structured [7,8]. It has an underlying structure, a meta-structure, which can be continually modelled and extended in higher dimensional space (hyperspace). This metaspace defines the HyperUniverse of Form.

In this morphological hyperuniverse, different types of morphings [9] are encountered. Simple forms can transform to complex, regular can transform to irregular, periodic to non-periodic, symmetry to asymmetry, static can become dynamic, solid can become void, tension can become compression, inorganic (geometry) can become organic (geometry) and so on, all in a continuous manner. In this universe of continuous transformations, all dialectics disappear. Within this universe, topologies are created and destroyed. elements (points, lines, planes and cells) are added and subtracted or simply appear and disappear, open lattices transform to finite closed objects, genus is created and transformed, closed becomes open, and inside becomes outside; and Euclidean space changes to non-Euclidean space. This universe is a continuum where every form can transform to another continuously. This metamorphosis follows systematic transformation pathways within the hyperspace. The structures can also undergo points of singularities in this space to enable dramatic topological transformations. Further, this universe itself evolves and grows as new morphological possibilities are discovered or invented by humans or nature.

Morphogenomics

Each structure within the hyperuniverse has a unique address. The code, defined by symbols (for example, numbers) determines this address and defines a structure uniquely. The numbers have a parametric meaning, both topologic and geometric. The morphological code (morph code) [10] defines the *genetic code of form* and serves as the basis for a unified *morphological genome*. The genetic code of form, when layered with other aspects of architecture, defines the *genetic code of architecture* as the basis for the *architectural genome*. In a broader sense, the morph code encodes all formal design possibilities in the natural, artificial and the human-made worlds [11]. In biological terms, this is an epigenetic code that exists in parallel with the biological genetic code captured in the DNA sequences [12]. The morph code leads to the possibility of *morphogenomics* mentioned earlier. It provides a formal tool in the design and manufacturing processes.

The morph code can be manipulated to generate an endless variety of forms. Manipulating this code is the same as navigating through the hyperuniverse of form. Since any location within the hyperuniverse determines a unique form, a unique structure or a unique design solution, one can run through a myriad of solutions for a design problem through this navigation process and select the best possible solution. If needed, this solution can be easily altered (transformed) through the same navigation process to suit any changes in the design. Once a design is selected, it can be linked with digital manufacturing and assembly processes, also similarly coded, to get a built structure.

The first example developed by the author 20 years ago along these lines was a pattern-code for generating Islamic and other geometric patterns [13]. Typing in a sequence of the code (alphanumeric, in this instance) yielded a new pattern on the computer screen. Though only a pen-plotter was used as the output device to make long scrolls of line patterns any device for marking, scoring, cutting or milling would have instantly linked it to pattern-making in other materials. The code provided a formal and technological tool for pattern-generation with obvious ramifications for designers, craftspersons and pattern-manufacturers in various fields (architecture, textiles, paper and various sheet material industries). It also provided the possibility of indexing all patterns within a unified digital taxonomic database as a universal (pan-cultural) pattern informatic system useful for historians, anthropologists and others interested in the evolution of pattern across cultures. Evolution of pattern would mean mapping the pathways of cultural patterns within the hyperuniverse. The same would apply to a chronological evolution of architectural form.

Windows into the Hyperuniverse

Zooming into this hyperuniverse through one window we see the world of coded squares in its many states all morphing from one state to another (**figs.1A, B**; see Postscript **fig.A**). When the individual squares are coalesced into a continuum, a pattern similar to Escher's celebrated metamorphic designs is obtained, with the difference that the result is a higher dimensional analog of Escher, a hyper-Escher pattern [14]. Looking elsewhere into another window say into the special world of spheres we see all morphologically coded geodesic surface patterns, all inter-transforming (**fig.2**). Fuller's geodesic domes are embedded in this part of

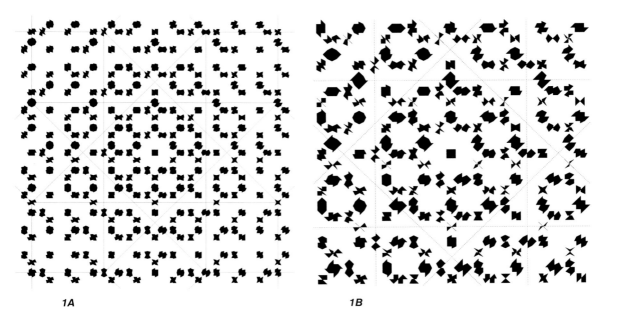

1A   *1B*

32 Coding, Shaping, Making

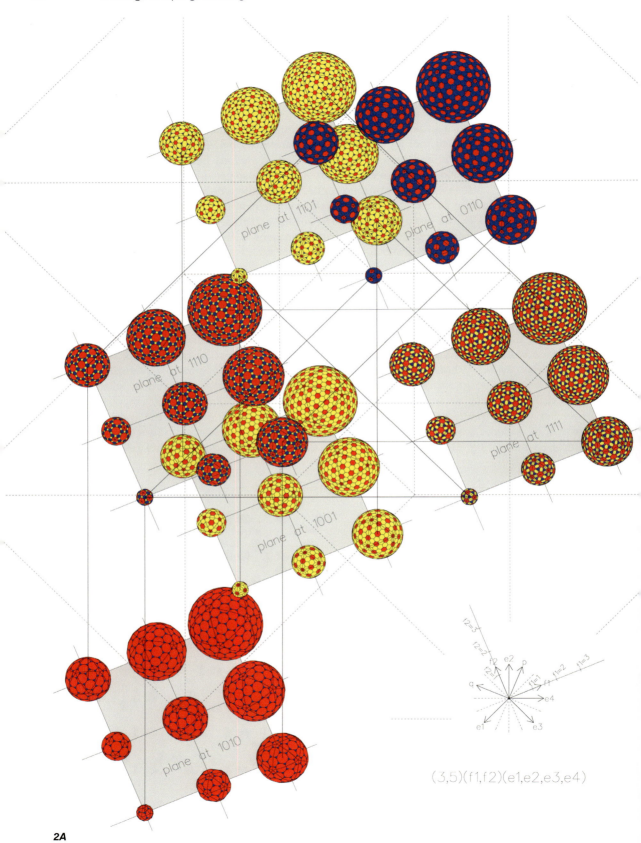

2A

Genomic Architecture   33

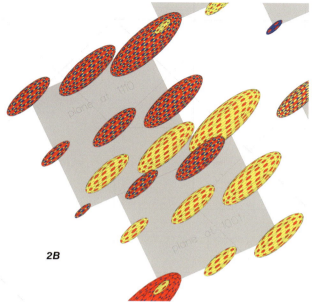

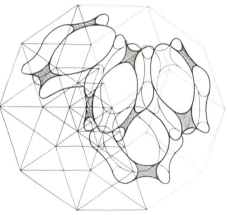

1A, B (*previous page*)
Coded Squares in Morphing States.

2A, B
Family of Geodesic Surfaces Patterns.

3
Family of Hypersurfaces.

4A, B
Spheroids.

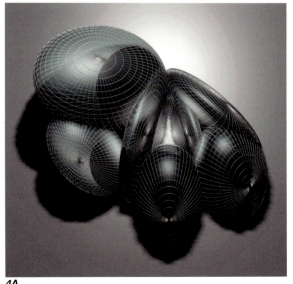

34 Coding, Shaping, Making

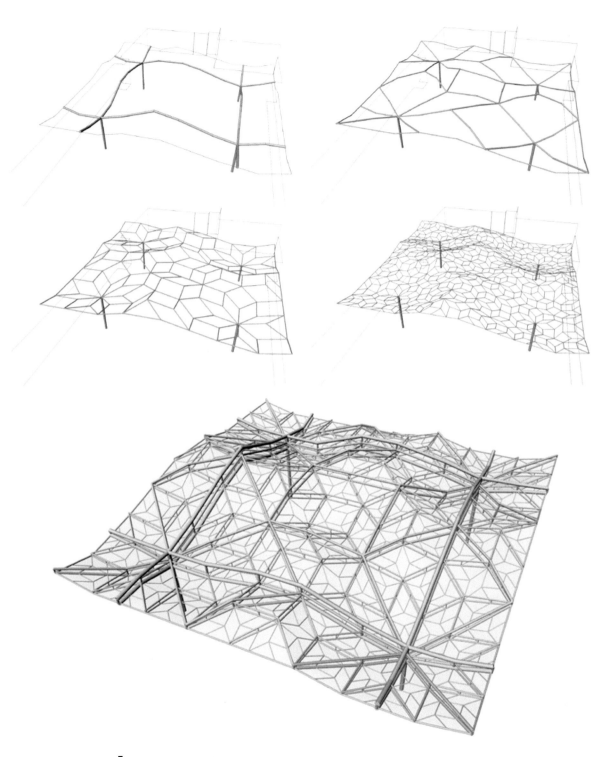

5
Fractal Fold Roof.

the hyperuniverse, so are the various spherical fullerenes [15] and the spherical viruses. A peek into higher dimensional worlds located in another part of the hyperuniverse reveals the many states of higher dimensional space structures. This example is interesting as it suggests a self-similarity between structure and meta-structure. One example is a family of cells of hyperspheres and two curved variants of the cells of a six-dimensional cube, related to the well-known quasi-crystals in nature [16] (**fig.3**). And, so on.

Navigating through the many hyperstructures in the hyperuniverse, one discovers interesting spatial possibilities for architecture. In *Spheroids* intersecting volumes with varying tilts and embedded in six-dimensional space provide architectonic possibilities for varying functions (greenhouses, convention halls, etc.) (**fig.4**). In *Fractal Fold*, a fractal geometry based on higher dimensions [17] defines an irregular roof suggested for a new assembly space within an existing building complex [18] (**fig.5**). This building system is interesting in that it permits irregularity from equal struts within a hierarchical structural system of increasingly larger and stronger elements.

Bottom up Mass Customization: The Milgo Experiment

Five years ago we began linking morphology with manufacturing. Milgo Industrial Inc, a leading metal fabrication company for art and architecture, provided the opportunity to link shaping with making. We chose a single material, in this case sheet metal, as the medium for experimentation as it provided the possibility of shaping a surface to define architectural space and form. We wanted to modulate the stiff metal surface in various ways to make rigid curved surfaces without 'deforming' the metal. This way the sheet metal would preserve its 'integrity' and become what 'it wants to be' (to borrow Louis Kahn's words). It would also become stronger in doing so. By comparison, standard methods of forming sheet metal, for example, stamping, deep drawing, etc. deform the metal sheet into desired three-dimensional shapes (as, for example, in the making of automobile bodies). We discovered a new type of forming which we have termed 'non-deformational bending'. This forming process generates its own vocabulary of curved surfaces [19], opening up a new window into a different part of the hyperuniverse.

We used this forming technique to prototype a large body of architectural elements like columns wall and ceiling systems in sheet metal. The examples (**figs.6-8**; Postscript **figs.B-E**) show related examples of continuously morphing systems developed from a morphologic algorithm and constructed from single sheets of metal using a single method of forming to obtain an endless 'variation on a theme'. Generative morphology is combined with mass customization. Just as each new design and each new variation we have been developing can be visualised by walking or flying through the digital hyperuniverse, we expect manufacturing technologies to be similarly automated to fabricate morphologically encoded one-of-a-kind designs. The *transforming columns* show iterations between groups of columnar structures, each group indicating a different direction in the hyperuniverse (**fig.6**). A *wall system* (**fig.7**) and an irregular *ceiling system* reminiscent of zebra patterns and sand dunes (**fig.8**) show panel morphologies derived and fabricated

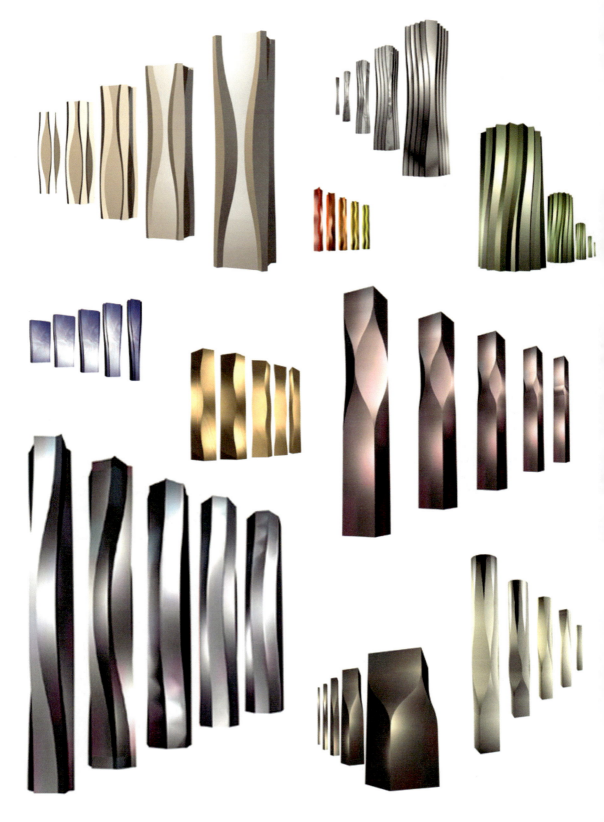

6
Transforming Columns.

7
AlgoRhythm Wall System.

from the same shape and make techniques and their details (**fig.9**). Currently we are extending our investigations to other types of design products including architectural structures, environments and spaces defined by continuous folded surfaces like *wave space labyrinth* (**fig.10**; see Postscript **fig.F** for a few examples).

Unified morphological algorithms combined with digital manufacturing are ideally suited for mass customization of design. As we relax the constraints of the material and forming methods that we have been using, related morphologies emerge and take us into another domain of the hyperuniverse. As new attributes and processes are added, new morphologies become immanent and form begins to become more fluid, to take one morphogenetic pathway as an example (**fig.11**). A digital environment based on a unified morphological genome would enable us to explore expediently and systematically the incredible diversity of forms that exist in nature, and beyond. The

8
Ceiling System.

9
Prototype details for Architectural surfaces.

Genomic Architecture

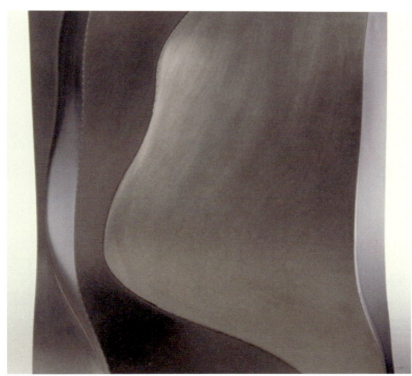

9

10
Wave Space Labyrinth.

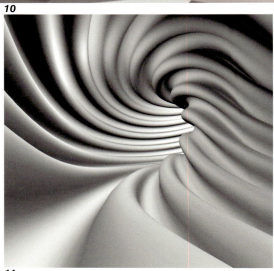

11
Study for *Milgo Gallery.*

periodic table of chemical elements and the tables of elementary particles in physics are finite. In contrast, the *periodic table of form* is infinite and open-ended, the way biological genomes are, and the way the biological 'tree of life' [20] is. This enables new forms to be continually discovered within the ever-expansive morphological hyperuniverse.

An interesting discovery in bioinformatics based on DNA sequencing is that, in many cases, different genes suggest different family histories for the same organisms suggesting that a more apt analogy for the history of life will be a net or a 'trellis' instead of a single universal tree [21]. Integrated hypernetworks of life will provide a theoretical upper limit to such trellises of life. As the new tree (or 'trellis') of life is mapped, the morphogenomics agenda proposed here provides for the beginning of artificial genomics as a parallel to biological genomics. As the morphological genome begins to be linked with physical morphogenesis (not digital, as in the case of 'artificial life') and, in time, automorphogenesis (selfreplication) [22], the bridge between the artificial and the natural will begin to disappear. In this scenario of genomic architecture, architecture and biology will become one, reaching the upper limit definition of 'organic architecture'.

ACKNOWLEDGEMENTS

The author wishes to thank the School of Architecture, Pratt Institute, for its support of the Center for Experimental Structures where the author is a co-director. Various individuals have contributed generously to this project in different ways over the years. Their tireless efforts are gratefully acknowledged. *Computer modelling*: Neil Katz at Skidmore, Owings and Merrill, New York City. *Computer rendering*: Mohamad Al-Khayer and Ajmal Aqtash, *Product development* at Milgo: Bruce Gitlin and Alex Kveton, *Photography* at Milgo: Robert Wrazen. The author also wishes to thank the biophysicist Loren Day for his constructive remarks on the paper.

The project AlgoRhythm Structures (with Milgo/Bufkin) is currently being funded by NYSTAR (New York State Technology and Academic Research, 2002-3). Some of the ideas in this article were published in the author's 'Meta Architecture' *Architectural Design* (Sept/Oct 2000), and republished in Giuseppa di Cristina (ed), *Architecture and Science*, Wiley-Academy, 2001.

NOTES/REFERENCES

1   For origins of growing architecture, see William Katavolos, *Organics*, Steendrukkirj de Jong & Co (Hilversum) 1961 Vittorio Giorgini, 'Early Experiments in Architecture Using Nature's Building Technology' in H. Lalvani (ed), *The International Journal of Space Structures*, 11, 1 and 2 (1997).

2   See, for example, K Eric Drexler, *Engines of Creation78: The coming era of nanotechnology* Anchor Books (New York), 1986, describing the future of building ultra-small and ultra-large objects atom by atom, Marvin Minsky, 'Will Robots Inherit the Earth?' *Scientific American* (October 1994), pp.109-113, where the author argues that they will, but as our 'mind-children'

(using H. Moravec's term); Ray Kurzweil, *The Age of Spiritual Machines*, Penguin Books (London), 1999, where the author predicts that by 2099 there will no longer be any clear distinction between humans and computers. On the software end, there is sufficient literature on artificial life describing attempts at re-creating life 'in silico'.

3   The paleontologist Stephen Jay Gould describes three possibilities in the landscape of imaginable lifeforms: 'can't work well', 'can't work at all' and 'just haven't been there yet' in 'Stretching to Fit', *The Sciences* (July/August 1998), pp.12-14

4   Haresh Lalvani, 'Morphological Universe, Expanding the Possibilities of Design and Nature', unpublished manuscript, 1998, presented at the ACSA conference, 'Works of Nature: The Rhetoric of Structural Invention' Dalhousie University, Nova Scotia, October 1998.

5   This universe encompasses structures ranging from the microscopic to the macroscopic, from living to nonliving, chemical to biological, and those realised by humans and (eventually) by machines. See D'Arcy Thompson's classic *On Growth and Form*, Cambridge University Press (Cambridge), 1942.

6   The author has been working on this hyperuniverse since the late 1970s (first published in 1981 see note 7). The term 'morphological hyperuniverse' in this context was first used in print in the author's contribution to *Cyberspace: The world of digital architecture*, Images Publications (Australia), 2001, p.38.

7   Haresh Lalvani, *Structures on Hyper-Structures*, Lalvani (New York),1982, based on the author's doctoral dissertation Multi-Dimensional Periodic Arrangements of Transforming Space Structures, University of Pennsylvania (1981), University Microfilms (Ann Arbor), 1982.

8   Beginning with orderly structures like tilings, polyhedra, geodesic spheres, space-fillings, etc., this universe is being continually mapped and extended by the author to be more encompassing. This extension includes a variety of curved structures as well as architectural and structural morphologies. The formal languages of selected architects (e.g. Gaudi, Wright, Calatrava, to name a few we have been studying) provide fertile material for such mapping.

9   In 1971 the author began experimenting with various types of morphings, some related to D'Arcy Thompson's celebrated method of transformations, and others inspired by different types of growth in nature (e.g., crystal growth) and movement. During the mid-1980s, while Tom Brigham was experimenting with 'morphing' at the Computer Graphics Laboratory, New York Institute of Technology, the author, in collaboration with Robert McDermott and Patrick Hanrahan (also at CGL) was involved with a different type of continuous morphing, a structured morphing based on the reference in note 7.

10 The term 'morph code' (or shape-code) raises the interesting question for biological form, whether such a code is embedded in the DNA sequences or whether biological form is purely the result of non-genetic factors (e.g., physical or chemical forces). If yes, then mathematics in nature must be a by-product of the physics and chemistry of life, raising the intriguing question of how the biological genetic code triggers the formation of highly mathematical structures like the logarithmic spirals of seashells, to give one example.

11 A universal morphological coding would work at different scales of magnitude from the microscopic to the macroscopic. For example, the generative classification of transformational polyhedra embedded within the hyperuniverse would apply to the formal classification of crystal morphology. The corresponding sphere-packed configurations would apply to atomic arrangements in physics and chemistry. In principle, the morphological system of continually transforming space structures lends itself to modelling the constructions of kinetic nanoscale deployable structures (see also note 22).

12 It is possible that genetic codes other than DNA, RNA exist though none have been found so far. These (theoretically possible) codes would define the upper limit for all forms of life in the universe. The study of all possible biologies will fall within the domain of hyperbiology.

13 H. Lalvani, *Coding and Generating Islamic Patterns*, National Institute of Design, (Ahmedabad), 1982. Other articles by the author on this include: 'Pattern Regeneration in S. Doshi (ed), *An*

Impulse to Adorn, Marg Publications (Bombay) 1982, and 'Coding and Generating Complex Periodic Patterns', *Visual Computer 5*, Springer-Verlag (Munich), 1989, pp.180-202.

14   H. Lalvani, 'Structures and Meta-Structures', *Symmetry of Structure*, International Society for the Inter-Disciplinary Study of Symmetry (Budapest), 1989.

15   In 1993, the author suggested the existence of skewed fullerenes and hyperfullerenes (higher dimensional analogs of fullerenes) as part of the 'periodic table of fullerenes' in his presentation at the 3rd International Conference on Space Structures, University of Surrey, 1993. Parts of this table were later published in the proceedings of the conference *Katachi U Symmetry*, Tsukuba University, Japan, 1994. Relating to note 11, it is interesting that the skewed fullerenes have the same underlying morphology as that of some of the spherical viruses with skewed icosahedral symmetry. The morphological hyperuniverse transcends scale at its root level, though each sub-universe will bring its own specificity.

16   In 1984, when Schectman et al at NIST (National Institute for Standards and Technology) reported the existence of a rapidly cooled alloy of manganese and aluminum based on icosahedral symmetry, till then denied in crystallography, a new class of natural non-periodic structures (termed 'quasi-crystals') opened up. Interestingly, non-periodic structures were independently discovered in different fields: mathematics, physics, crystallography and architecture. The author's independent work during the period 1981-5, and subsequently, was amongst those within the field of architecture along with the works of Baer and Miyazaki.

17   This particular example, related to the Penrose tiling, is a projection from four dimensions. The dimensionality of the structure increases with the extent of subdivision of the surface and results in increasing irregularity.

18   This roof concept was developed in 2001 in collaboration with Maria Sevely and Archronica architects on their project for a pharmaceutical company.

19   US Patent No. 6,341,460.

20   A major effort is currently being planned for assembling the tree of life to construct a phylogeny of the 1.7 million described species of life.

21   Francis S. Collins and Karin G. Jegalian, 'Deciphering the Code of Life' in Editors of Scientific American, *Understanding the Genome*, Warner Books (New York), 2002, p.37.

22   The idea of using morph coded designs of Drexler's nanotech 'assemblers' and 'replicators' (cited in note 2) is a more direct application of morphogenomics. Morph coded shapes (and materials) that change from one to another using transformation pathways within the hyperuniverse can be constructed using Drexler's proposal for building atom by atom. On a related note, the author, in the early 1970s in an unpublished paper, had suggested achieving 'aulomorphogenesis' using genetically engineered bacteria as an alternative method of self-replication that relies on a biological process of building.

POSTSCRIPT 2

This postscript includes additional related material from that period and adds the animation component to the Transforming Columns in **fig.6** and additional related structures like transforming rippled surfaces related to **fig.8** (also **fig.3A-C** in Chapter 1), morphing umbrellas related to **fig.10** in this chapter (and **fig.6** in Chapter 1), and so on, to show the continuum model for mass customization where any "still" can be chosen for fabrication. Our intention was to have this available to architects at the time so they could interactively select a product of their choice. This was the intention behind the invention of *AlgoRhythms* as a mass customized product line with continuous transformations as a way to enter the universe of columns (or walls, or ceiling panels) and choose a preferred one. These stills from animations are snapshots from the morphological universe described in this chapter as the *Hyperuniverse of Form* and extended further in Chapter 8.

The 3D models used in Chapters 1 and 2 were developed during the period 1997-2004; these were used to make the animations in 2005 and are shown here **figs.B, D**, and **E**.

Fig.A (pp.42-47) shows various details of the 625 morphing squares corresponding to **figs.1A,B** (p.27) and organized within the full 4D cube (**A1**), each in 5 states along the edge of the hypercube. Four close-ups of **A1** are shown in **A2-5**. **Figs.A7-9** show a subset of **A1** with the individual states in various stages of increasing size till the adjacent ones meet in a contiguous tiling pattern (**A6**) described by the author as a 'hyper-Escher' pattern earlier, a higher dimensional analog of Escher's metamorphic designs [1,2] (**video 1**). A more detailed zoom-in of **A1** is shown in **A10** and individual stills from an animation [3] of continuously morphing squares are shown in **A11**.

Fig.B (pp.48-49) shows three sequences: a beam (**video 6**), a panel system, and a morphing canopy (**video 7**). The morphing *AlgoRhythm Beam* (**B1**) is one example of the many structural elements we developed during that period. This particular sequence shows a "fixed" beam folded from a single sheet morphing to an articulated one which is able to receive panel elements on top and bottom, and on left or right, or both as in a double-wall system. Though the last image in the sequence was developed in the context of a curtain wall system with double glass, it applies equally well to double-walled systems in other materials needed for insulation (sound or heat). The morphing *AlgoRhythm Vaulted Wave Ceiling Panel System* (**B2**) shows the development of a flat-folded system to an increasingly curved one. The morphing *AlgoRhythm Canopy/Roof System* (**B3**) starts with the even rippled system we saw in Chapter 1 (**fig. 3A-C**) and gradually extends to make a canopy system. Simple extensions in the widths between folds can be approximated by a system of parallel telescoping elements. However, the changes in the fold angle to achieve global curvature will require a new invention which

Note: The product names in the figure captions are taken from the Milgo website: http://www.milgo-bufkin.com/algorhythms/index.html

permits shifting fold locations. This image is a companion to *Transitions* (Chapter 1, **fig.7**, this Postscript, **fig.F9**), the irregular wall system (Chapter 2, **fig.9**; Chapter 3, **figs.5** and **7**).

Fig.C (pp.50-51) shows the morphing *InterRipples* (see **fig.8** in this chapter for a prototype of a small portion of the starting image on *top left* in this sequence) shows 18 stills from an animation that increases ripple widths and heights (**video 8**). Inspired by water and sand ripples, *InterRipples* provides a solution for a modular system that permits varying ripple configurations from very few modules.

Fig.D (pp.52-53) shows morphing sequences from three different animations, rolling and unrolling an *AlgoRhythm* column (**D1**), a simple *AlgoRhythm* column morphing to an *Umbrella* (**D2**; also a companion to **fig.6** in Chapter 1, **video 4**) and an *AlgoRhythm* column morphing to columnar unit of the *Wave Space Labyrinth* in **fig.10** of this chapter (**D3**, **video 5**). In the evolution of form, this shows the continuum between a column and a ceiling system, and with a continuation of this sequence the column can split into stalactites and stalagmites which can further become reverse-morph to a ceiling and a floor. The three principal elements of column-slab architecture, a column separating the floor from the ceiling, are a morphologic continuum.

Fig.E (pp.54-55) shows four morphing sequences of *AlgoRhythm* columns, some corresponding to **fig.6**; see also **fig.3D-O** and **fig.4** in Chapter 1 (**video 2, 3**).

Fig.F (pp.56-57) shows three versions of *Milgo Gallery* as a set-up for display of a showroom for Milgo products with the simplest folding in *Gallery 1* (**F1**) and a more complex folding in *Gallery 2* (**F2**) having an irregular section for the vaulted space. *Gallery 3* (**F3**), like the *Waveknot* in Chapter 1, was a reminder of where we would like to go by including doubly-curved versions of our folded systems, a problem that was unresolved. **F5** and **F6** are preliminary studies related to **fig.11**, **F7** is related to *Transitions* (Chapter 1, **fig.7**), **F8** is another rendition of *Wavespace Labyrinth* in **fig.10**.

NOTES/REFERENCES

1   Lalvani, H. (1989a) Structures and Meta-Structures, In: *Abstracts, Symmetry of Structure*, Budapest, Hungary, August.

2   Lalvani, H. (1993) Metamorphic Tiling Patterns Based on Zonohedra, *U.S. Patent 5,211,692*, May 18.

3   Lalvani, H. (2003) *Meta-Morphing*, powerpoint presentation, Pratt Institute, February 6.

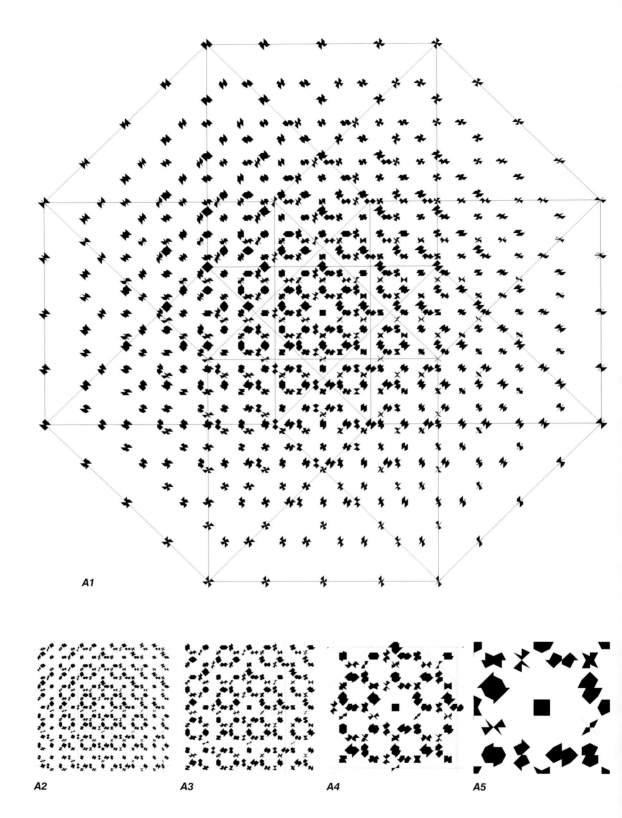

A1-9
625 morphing squares within a 4D framework, with each square in 5 states. A6 shows a Hyper-Escher pattern.

Postscript 2   47

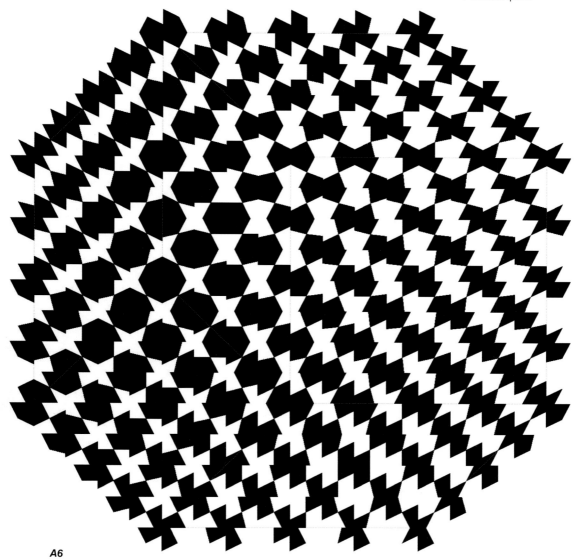

A6

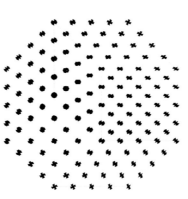

A7   **A8**

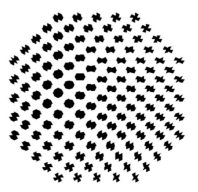

A9

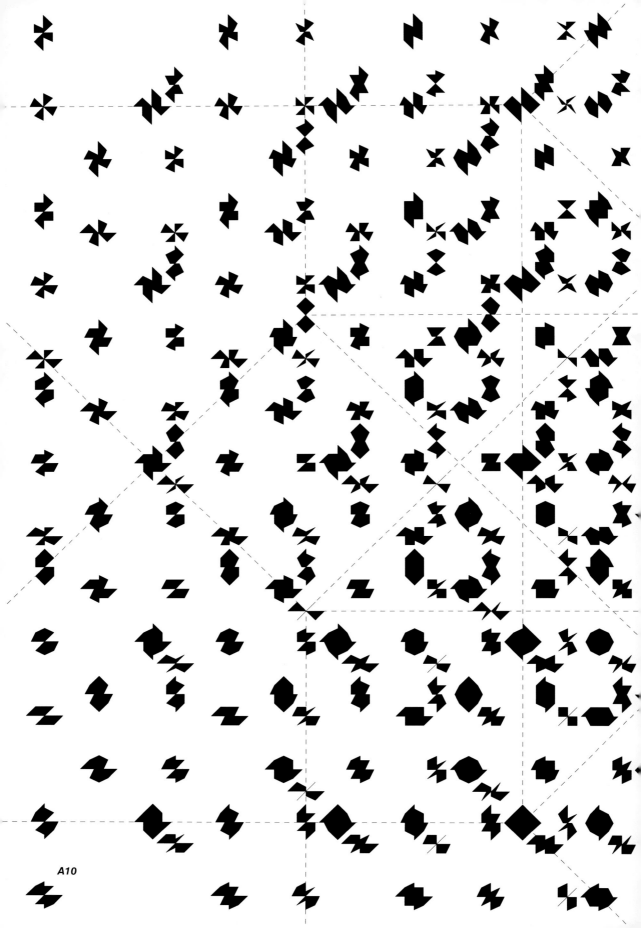

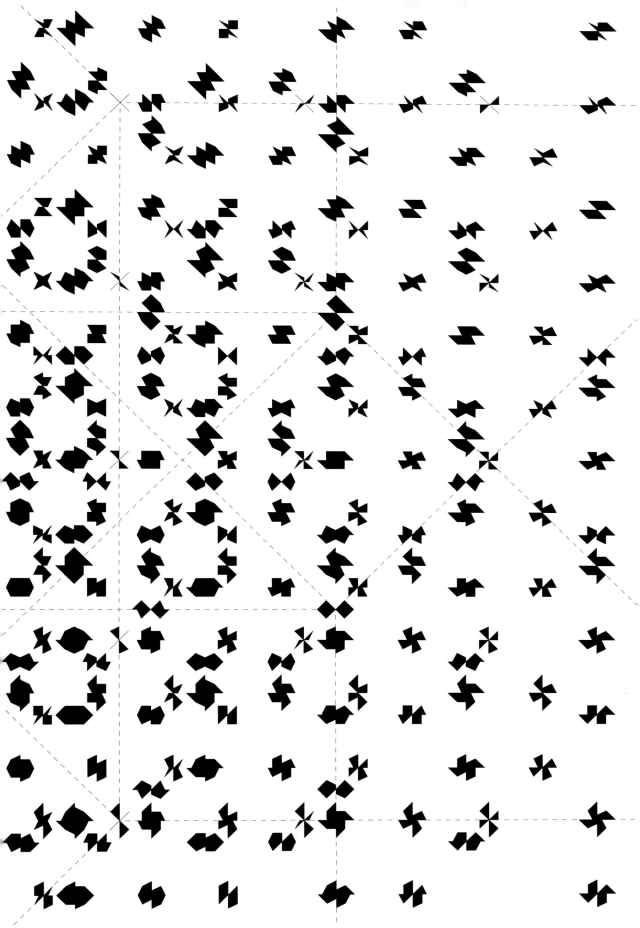

video 1

A10 (*previous page*)
A close-up of A1.

A11
A sequence of stills from an animation of a continuously morphing square (**video 1**).

B1
AlgoRhythm Beam (**video 6**).

B2
AlgoRhythm Vaulted Wave Ceiling Panel System.

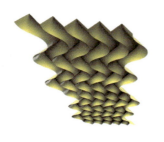 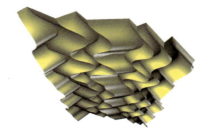 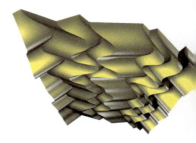

 video 6

B3
AlgoRhythm Canopy/Roof System (**video 7**).

Postscript 2

 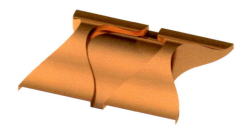

video 7

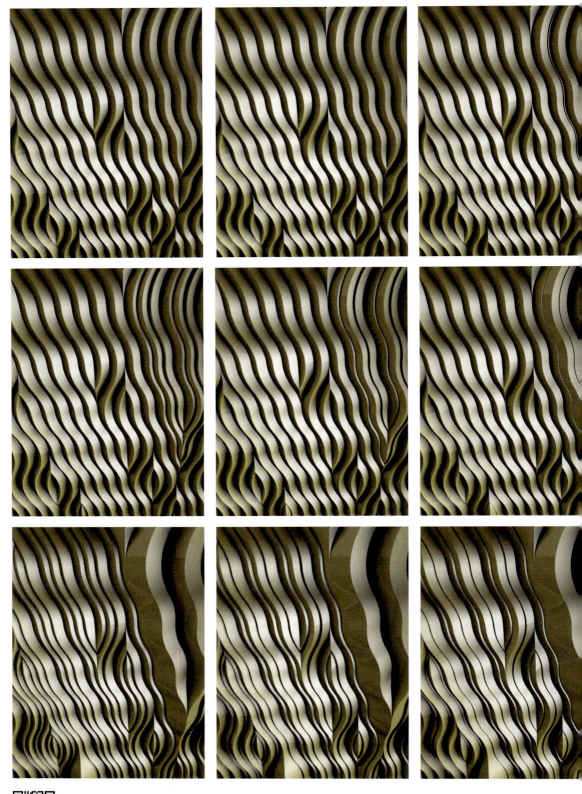

video 8

C
Morphing *InterRipples* (**video 8**).

Postscript 2

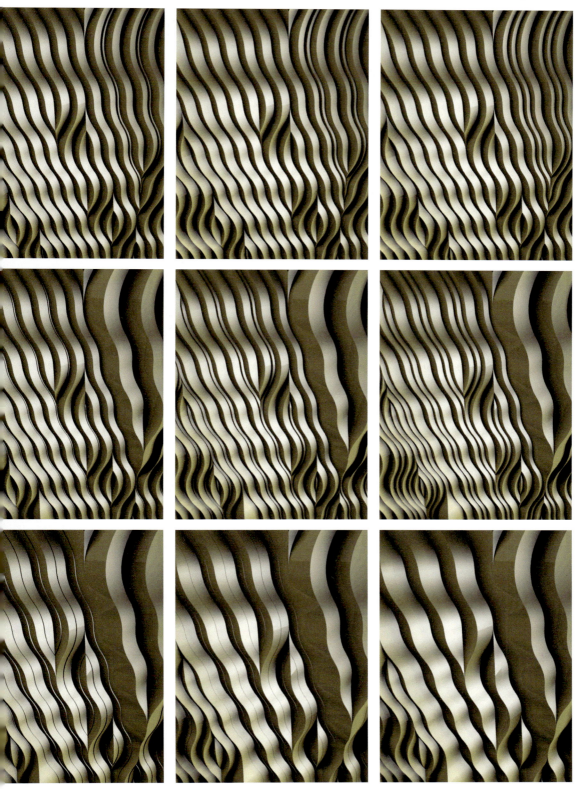

D1
Rolling an *AlgoRhythm* column.

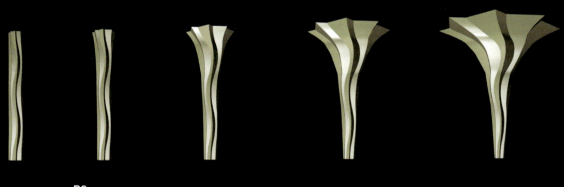

D2
Column morphing to an umbrella (**video 4**).

video 4

D3
Column morphing to a unit of *Space Labyrinth* in fig.10 (**video 5**).

Postscript 2

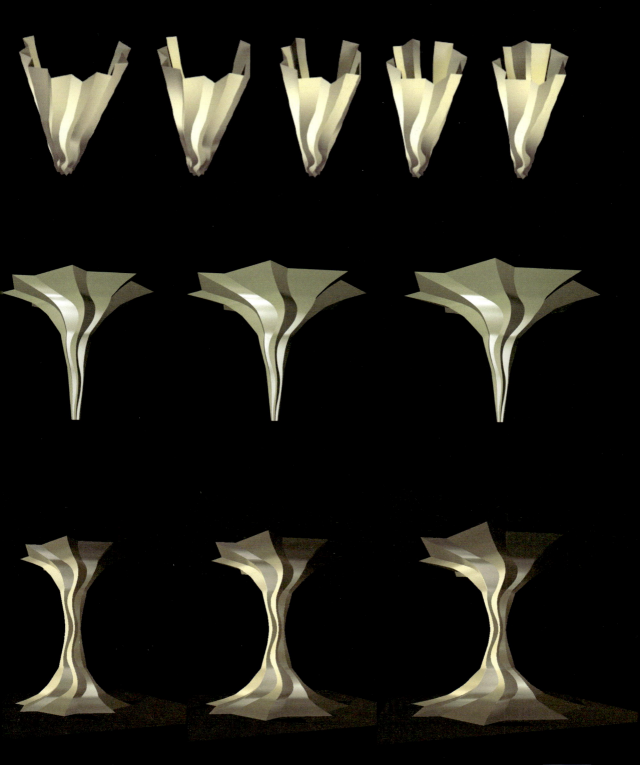

video 5

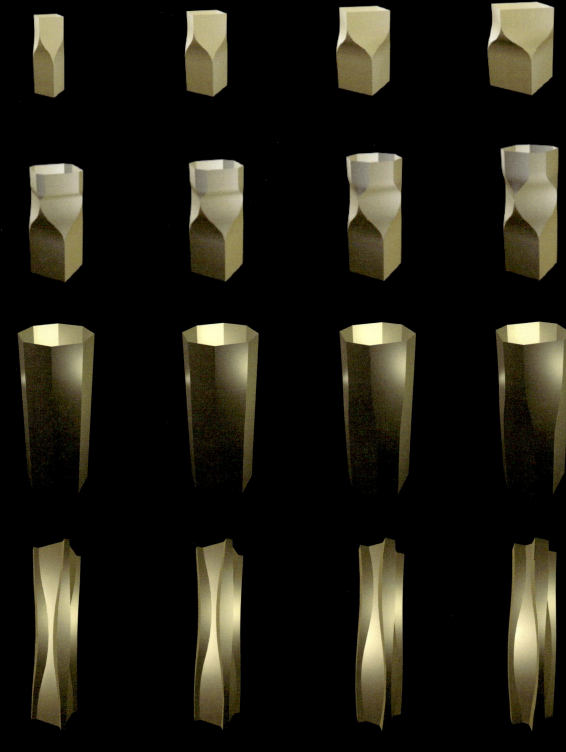

video 2

E
Four sequences of morphing
AlgoRhythm columns (**videos 2, 3**).

Postscript 2

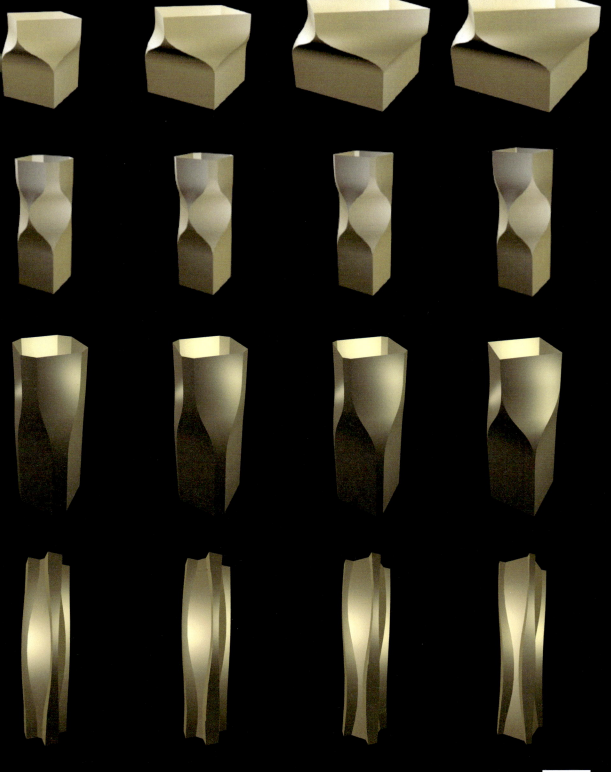

video 3

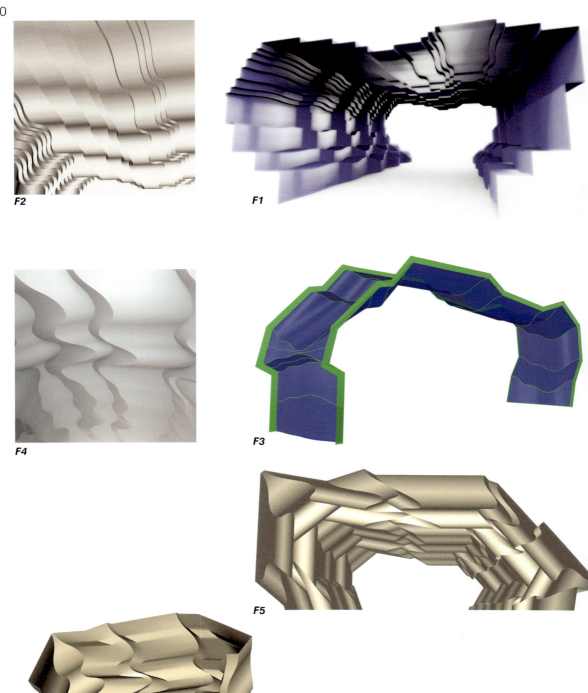
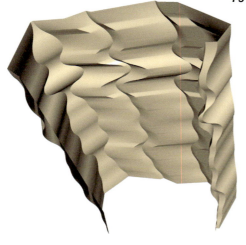

F
Milgo Gallery 1 (F1) and study (F2)
Single slice of *Milgo Gallery 2* (F3)
Milgo Gallery 2 (F5) and study (F4)
Milgo Gallery 3 (F6) and studies (F7, F8)
Transitions (F9)
WaveSpace Labyrinth (F10)

F6

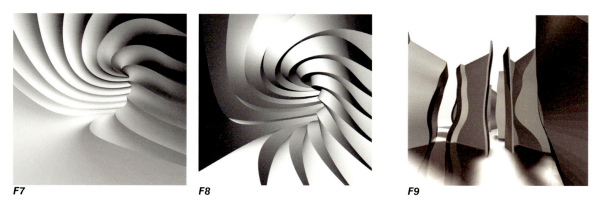
F7 *F8* *F9*

F10

THE MILGO EXPERIMENT:
An Interview with Haresh Lalvani[1]

For just less than a decade, Haresh Lalvani has been working on a series of projects with Milgo/Bufkin, a leading architectural metal-fabrication company, seeking to economically integrate the shaping and making of metal surfaces into elements that create a seamless whole. Here, John Lobell interviews Lalvani about his pioneering work with computer-aided design and manufacturing and how it has been informed by his development of the Morphological Genome, a universal code for mapping and manipulating form, whether natural or man-made.

> The dilemma posed to all scientific explanation is this: magic or geometry.
> *Rene Thom, 1972* [1]

> A more important set of instruction books will never be found by human beings.
> *James D Watson (on the human genome), 2003* [2]

In recent decades there has been an increase in the use of curved surfaces in architecture, although to date more often in computer simulations than in realised projects. The most common use of curved surfaces is cylindrical column covers that are created from sheet metal on press brakes, with a skilled worker continually moving the sheet of metal under the machine's blade by hand. This process curves the metal without warping or deforming it.

Compound curves, along which one cannot place a straight edge in any

---

[1] This chapter is a re-publication of 'The Milgo Experiment, An Interview with Haresh Lalvani' by John Lobell in 'Programming Cultures: Art and Architecture in the age of Software', Guest Ed. Michael Silver, AD, Wiley-Academy, 2006, Vol. 76, No.4, July/August, pp.52-61. Reproduced with permission from AD-Wiley. A Postscript has been added to include related developments during that period to provide a fuller account of the experiments.

direction (for example, a bowl), are far more difficult to fabricate for architecture. They are commonly used in the automobile industry, where they are created by pressing the metal between curved male and female dies, which stretches and deforms the metal. However, such dies are expensive - a cost that can be justified only by making large quantities of identical pieces. Compound curves in sheet metal can also be created by hand, for example with an English wheel, though this technique is also expensive and it would thus be impractical to make an entire building using this method.

In 1997 Haresh Lalvani began working with Milgo/Bufkin, a leading metal-fabrication company, to find a way of creating developable curved sheet-metal surfaces using an economical process that recognises both the characteristics of sheet metal and the digital capabilities of metal fabrication, such as laser cutting (**fig.1A**), water-jet cutting, press braking, digital punching. The fabrication of sheet metal (**fig.1B**) at a scale economical for architecture inhibits compound curves and requires developable surfaces - curved surfaces that can be folded or bent from a flat sheet without deforming the material. This work launched the AlgoRhythms project that Lalvani terms the 'Milgo experiment'. According to Lalvani:

'There is more than one Milgo experiment in progress, *AlgoRhythms* being the first. All relate to genomic architecture [3]. The objective of these experiments is to integrate shaping (morphology) and making (fabrication) into a seamless whole. In nature, the two aren't separate. In these projects, I am interested in architecture as surface, not mass. Mass focuses on material performance (strength of material) while surface depends on its geometry (strength of form). This approach has several motivations. First, it is deeply connected with sustainability, as limited resources require us to maximise performance. Second, the bulk of the morphological universe (morphoverse) is inhabited by curved forms that need to be discovered, mapped and accessed. Straight lines and flat planes are in a minority within this universe. Third, new building technologies, closer to those found in and beyond nature, need to be invented to deal with these new vocabularies.

We focused on metal because Milgo is a metal-fabrication company though our experiments extend to other materials and forms. The question was as follows: Using sheet metal as it is used in architectural metal fabrication, could we form surfaces to define architectural space using techniques available at Milgo? We wanted to modulate the stiff metal surface to make rigid curved surfaces without deforming the metal. This way the sheet metal would preserve its integrity and become what "it wants to be" to borrow Kahn's words. It would also become stronger. Strength "emerges" from curvature when 2-D becomes 3-D during the forming process. We have called this new kind of forming "non-deformational bending". By comparison, standard methods of forming sheet metal, for example stamping, deep drawing and so on, yield stronger 3-D surfaces by deforming the metal sheet as, for example, in the making of automobile bodies.

As we proceeded with these projects, three things happened. First, on a more practical level, we found that we could bend sheet material in a new way. We produced AlgoRhythms, a line of architectural products (**figs.2A-C, 3**), column covers, wall panels and ceilings, and proposed other products.

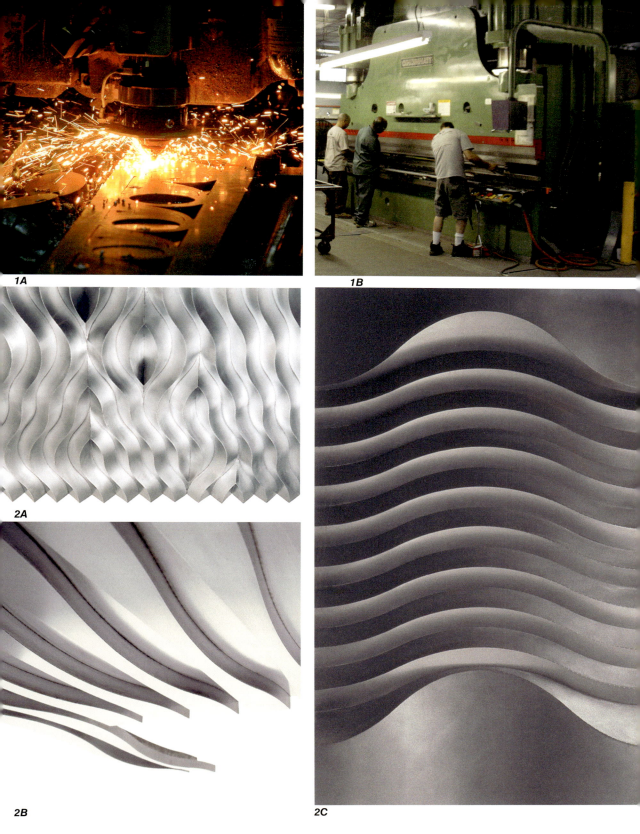

1A-B
Milgo's laser-cutter, one of the several digital-fabrication devices used for different Milgo experiments.

2A-C
Close up of the first line of *AlgoRhythm* products.

Second, it was satisfying to demonstrate that theoretical ideas could be translated into practice. The Milgo experiment is an application of the "Structures on Hyper-Structures" work I did in 1981 [4] and establishes that meta-structural thinking can lead to real products. My work at NASA-Langley from 1989 to 1990 had touched on this in the context of structures for outer space. Third, this experiment has opened a window in the morphoverse where process is added to form. This is the universe of form I have been mapping. Thus the project within-project became mapping the family of AlgoRhythmic forms as an example of forms generated by a physical process and not a computational one. The Milgo experiments attempt to marry software (morph genome) with hardware (fabrication genome!) in the making of physical objects.

It is important that the products we produce have an intellectual integrity throughout the process, from the mathematics of surfaces to fabrication techniques, to the detailing we use in installation. Working with Neil Katz, we derived AlgoRhythms through a computational algorithm based on my generalized morphological model that defined new families of surfaces from "solid" forms. The result allowed infinite variations of form, as one iteration could continually morph into another.'

> There is currently much 'computer architecture' around - exotic curved forms on the screen that may not be buildable. Lalvani's work with AlgoRhythms, although based on higher mathematics, is more rooted in the realities of construction than many other approaches:

'The interesting question is whether there is a fundamental link between form and form-making. I assume there is. The form of a seashell or a bone, though based on different growth principles, is intimately connected with its process of growth. Form is process [5]. In *AlgoRhythms* we linked the approaches to form and fabrication from the very beginning. Our approach leads to a new kind of mass customization that should eventually make it possible for us to make many units, each one different, as efficiently as we can make them all identical.

We have installed *AlgoRhythms* in various buildings and showrooms. Most prominently, we have an iteration of *AlgoRhythm Columns* in folded titanium in the permanent design collection of the Museum of Modern Art (MoMA). And in March and April 2004, at an exhibition at the Municipal Art Society in New York, we showed the *HyperWall* system (**fig.5**), an *AlgoRhythms* beam and an *AlgoRhythms* structural truss (**fig.6**) to extend the concept into forming structural shapes.

As we scale up *AlgoRhythms* and apply our theories to more building components — for example, to curtain-wall systems and glass envelopes, to beams and trusses and, eventually to entire buildings — we encounter real structural problems. We have designed the entire exterior and lobby of a 12-storey apartment building in Chelsea, New York. The important next step is to go beyond surfaces and begin investigating entire building systems, especially structures. We are now working on a residential building we are calling *Project X* that is *AlgoRhythmic* from the structural system out to the façade [6]. Since the late 1990s I had been thinking of AlgoRhythmic skins (such as in glass) for tall buildings [7] so when Stanley Perelman asked me to design a 25-storey apartment building in Manhattan

3
The first line of *AlgoRhythm* products (columns, walls and ceilings) using nondeformational bending of sheet metal.

4A, B
Project X, an apartment building in Manhattan, designed for Stanley Perelman. The morphology of the building skin and structure are derived from *AlgoRhythm columns*.

5 *(right page)*
HyperWall system, an irregular modular wall system using AlgoRhythmic principles. Exhibited at the Municipal Art Society, New York, from March 23rd to April 30th 2004.

4A

4B

(**fig.4A, B**), here was an opportunity to mine the universe of columnar forms (**fig.11**) we had been developing at Milgo for a completely different application.

We are interested in developing *Soft AlgoRhythms* (**fig.8, 9**) where the hard edges of the folded surfaces disappear, but here the fabrication technologies are in the way We are looking into *Kinetic AlgoRhythms* (**fig.7**) as well, but here we need some bright ideas, as we don't know how to proceed without making the continuous into discrete.'

> *The approach behind AlgoRhythms began with Lalvani's work on the Morphological Genome, a universal code for mapping and manipulating all form: natural. man-made and artificial. By conceiving the morphological universe as a hyperuniverse, and by using higher dimensional geometry, we gain the power tounder stand and manipulate it. Lalvani says:*

'The morphological universe is infinite and open-ended, with a fractal hierarchy composed of recursive levels within levels through which one can access and navigate in a number of ways from any level. This is a universe where each structure (and each type of architecture) can transform from one to another in a continuum of space and time, within and between levels. This hyper-universe map provides the basis for the Morphological Genome.

6 *(right page)*
An *AlgoRhythm* truss, a NYSTAR project, inspired by the flow of milk and derived from AlgoRhythm columns.
Photo taken at Milgo factory.

7
The computer sequence shows an example of *Kinetic AlgoRhythms*, a kinetic version of the *HyperWall* system folding dynamically (**video 9**). This is another example where the forming technology is yet to be invented.

video 9

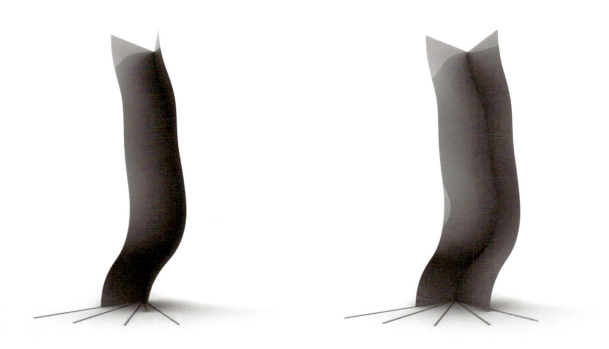

The Morphological Genome is a direct inspiration from biology, something I began wondering about in the mid-1970s. In biological terms, it is an epigenetic code [8] that lies outside the biological genome. I have always wondered if there is a "shape gene" in nature and, as far as I understand, the answer today is still "no". If the answer were to become "yes" it would be easy to reverse-engineer biology and construct every shape in nature through genetic engineering. The absence of a shape code in nature provides the need to invent a shape genome. This is the agenda of the Morphological Genome [9] project.

The Morphological Genome is a model experiment for mapping other genomes. It is composed of morphological genes (morph genes), where each gene specifies a family of related parameters, and each parameter is controlled by a single variable of form equivalent to a "base" in DNA. We are working on this universal structure of form. similar to Crick and Watson's structure of DNA, and mapping and manipulating it at this genetic level.'

In deriving AlgoRhythms from the Morphological Genome, Lalvani and 11 is associates took a series of algorithmic steps. At the first level is the genomic concept, a meta-algorithm, which defines a family of inter-related, inter-transforming shapes tied to a fabrication process. At the second level is a computational algorithm of developable surfaces, surfaces that can be formed from flat sheets by bending without deforming.

'I want to point out that the issue of mapping all form is like counting to infinity, many times over, since there are infinite infinities within the morphological universe. So as a practical matter, the morphoverse is not mappable. However, what is possible is to provide a framework and show the workings within it with some examples. The expedient way is to identify the morph genes. The key idea here is that the infinity of possible forms can be specified by a finite number of morph genes. I am expecting this to be a small number [10]. Besides the difficulty of identifying these as universal morph genes. you run into unsolved mathematical problems. Only a few years ago I discovered that there exist topological surfaces that violate Euler's famous equation that relates the number of topological elements [11]. The fact that new morphologies (which lead to new mathematics, and new architecture) are possible, is a feature of the morphoverse. This makes mapping the morphoverse more challenging to accomplish.'

Interestingly. several architects now refer to their work as 'genetic', Lalvani speculates:

'All these genetic and algorithmic approaches to architecture, including our genome project, mark an evolving path in our relationship with the computer – from a laboursaving device to an intelligent problem-solving machine, to an autonomous entity capable of independent creativity. We hold the latter as sacredly human, but this may change [12]. The point I want to make is that as we pass on our

The Milgo Experiment

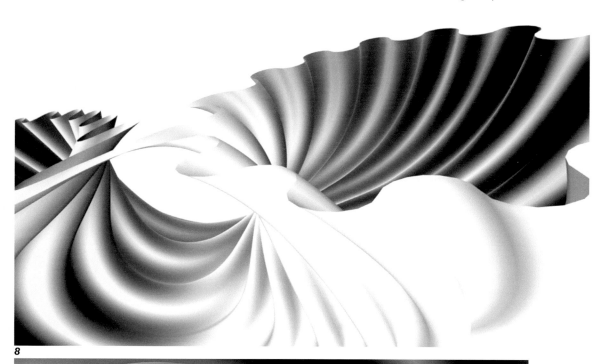

8

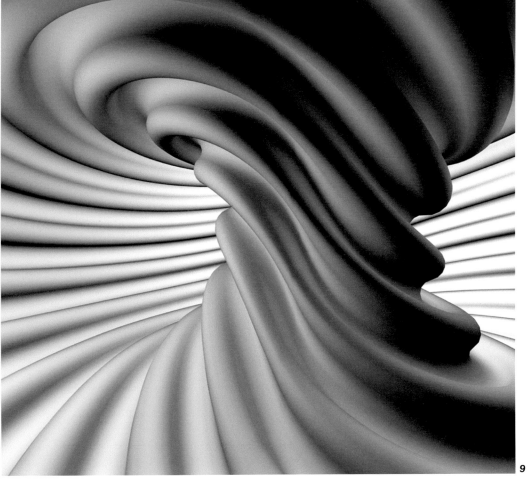

9

skills, knowledge and intelligence to the computer we must be prepared to pass on creativity as well. This trajectory inevitably leads to architecture without architects, in a different sense from Rudofsky [13], of course. In "Meta Architecture" [14] I had referred to this as the end of architecture (as we define it presently), an architecture liberated from the architect.'

> *Lalvani's morphogenonmic model seems to imply continuity, but at the moment there appears to be a bias cowards seeing the universe in terms of binary discreteness. On whether reality is continuous or discrete, Lalvani comments:*

'The binary view of the universe, exemplified by discrete computational sequences of 0s and 1s, does not appear to capture the continuum we see in life. In my morphological model, there is a continuum, infinite states between 0 and 1. Yet I am unable to carry this through to all aspects of form. I would like to believe that all topology, the bottom-most level of form, is a continuum and it is also emergent. But the quantum nature of integers that describe many topological features makes it discrete. For example, the number of sides of a polygon is, say either three or four, both discrete states. Yet a few years ago I discovered that the space between three and four is filled with fractional regular polygons with sides in between these two numbers. So what appears discrete may be hiding a continuum.

However, there are clear instances, like cellular automata, that resist the continuum [15]. Additionally, when you apply the bottom up cellular automata rules, a one-gene phenomenon in the morph genome, to smooth topological surfaces we need to impose a top down topology from a completely different gene [16]. The emergence of top down from bottom up is an intriguing idea. Nonetheless, the coexistence of discrete and continuous systems leads me to think that the universe may have a variable dial that can be set to binary or continuous states, or any state in between. After all, movies, which have 24 discrete frames per second, appear continuous when we run them. Physical reality may have a discrete basis at the root level, though at the experiential level it appears perfectly continuous [17].'

> *Lalvani's work seems to have broader implications for architecture:*

'At the very least it would be satisfying to show that the "alphabet" of form is not

8,9
Soft AlgoRhythms, surfaces with softer, rounded bends as morphological extensions of bent-surface *AlgoRhythms*. Their forming technology is yet to be determined.

10A-D (*next page*)
Wolfram's cellular automata rule 30 applied to a sponge-like surface (10A), a sphere (10B) and torus (10C, D). Such surfaces require the activation of at least two morphological genes.

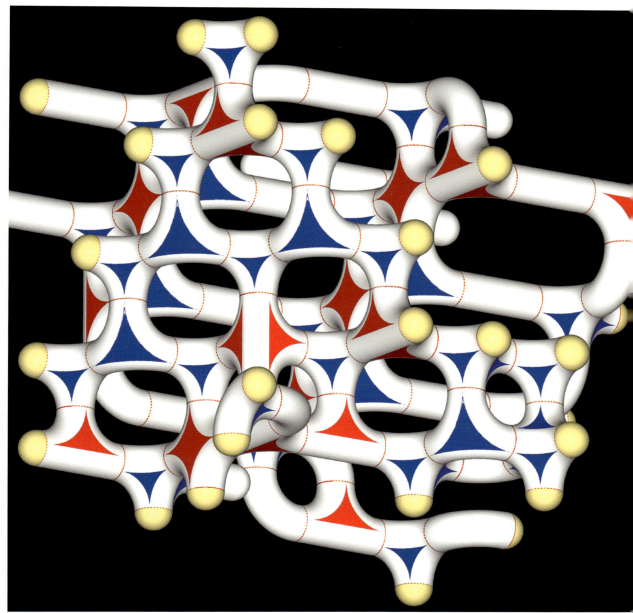

10A

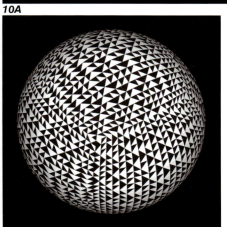

10B

10C

10D

restricted to simple shapes like the cube, cylinder. cone and sphere, but that it is an extensive family of inter-transforming alphabets embedded in an integrated framework. This framework, an open-ended, expansive, continually extendable higher dimensional periodic table of all forms, is an interesting idea. You are no longer limited to existing architectural vocabularies; you can generate your own vocabularies, morph any vocabulary to any other, or existing vocabularies (architectural contexts) to new ones. Any language becomes an instance of the other. Here is a universe where the particular and the universal coexist, as do complexity and diversity in a dual relationship to one another. This understanding can be liberating, and empowering.

The known issues of architecture. programmatic, structural, poetic, visual and spiritual, will not go away. The challenge is to infuse new formal vocabularies with meaning. The morphology of meaning (and hence of language and thought) is one of the most interesting unsolved problems. A coherent mapping of form and meaning is our challenge and architecture is the best experimental medium for that.'

And his Morphological Genome appears to have implications that go far beyond architecture. He has been applying this approach to dance, chemistry, metrics in physics, and so on:

'The Morphological Genome could impact how we approach different areas of knowledge. At this time, we are pressing forward in several areas outside architecture. The challenges are enormous. Each new area has its own knowledge base and to overlay an extrinsic model would be naive. The challenge is to find the intrinsic structure of native knowledge.

However, the Morphological Genome does have significant ramifications in all areas of shaping and making. It can be embedded in a morph chip and linked with new fabrication techniques including nanotechnology (for example, in Drexler's 'assembler' at the nano level) [18], or in some universal forming machines (that do not exist) at the architectural level. This forming ability must include the integration of bottom up and top down mentioned earlier. Think of this in terms of the stem cell, a universal cell that differentiates into different specialised cells that perform completely different functions in a co-coordinated manner. At the purely formal level, the Morphological Genome has the same intent, a universal code that can produce the infinite variety of forms we find in nature, technology and beyond. At this point we are far from realising this. The genome is in progress,

11 (*next page*)
A portion of the morphoverse suggesting a continuum of *AlgoRhythm columns*. These columns, fabricated from single metal sheets, are genetically coded and are based on one algorithm, one material and one method of forming. Adding process to form opens a new window in the morphoverse and requires the expression of 'fabrication genes' in addition to selected 'form genes'.

The Milgo Experiment

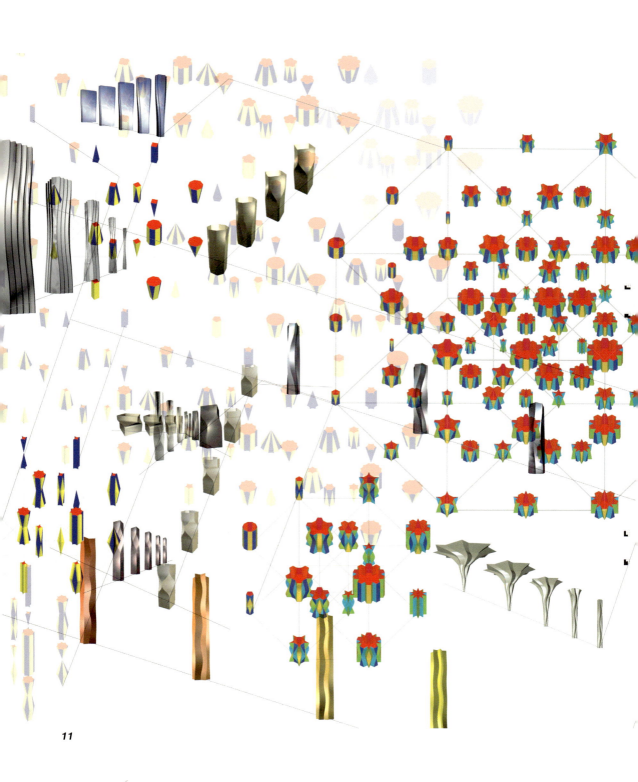

11

and the Milgo experiment is just one case study. The goal is automorphogenesis, human products (including architecture) that design themselves physically not just virtually [19] out of encoded forms, processes, materials — something the biologists are likely to achieve first as a neat counterpoint to artificial life [20]'.

ACKNOWLEDGEMENTS

School of Architecture, Pratt Institute, for its support of the Center for Experimental Structures; Bruce Gitlin, Alex Kveton and Robert Wrazen of Milgo/Bufkin; Neil Katz, Ajmal Aqtash, Che-Wei Wang and John Gulliford for computer-modelling/rendering as members of Lalvani's studio team; NYSTAR grant (2002-04) to experiment with *AlgoRhythms Structures* involving Pratt students at Milgo [21]. Some ideas in this paper were first published in [3] and [7].

NOTES/REFERENCES

1   From Rene Thom, *Morphogenesis and Structural Stability*, WA Benjamin (New York), 1972.

2   From James D. Watson, *DNA: The Secret of Life*, Alfred Knopf (New York), 2003.

3   Haresh Lalvani, 'Genomic Architecture', in Deborah Gans and Zehra Kuz (eds), *The Organic Approach to Architecture*, Wiley-Academy (Chichester), 2003, pp.115-26.

4   Haresh Lalvani, Structures on *Hyper-Structures*, Haresh Lalvani (New York), 1982; based on author's doctoral dissertation entitled *'Multi-Dimensional Periodic Arrangements of Transforming Space Structures'*, University of Pennsylvania, 1981.

5   In nature, process results from force. D'Arcy Thompson's dictum 'Form is a diagram of forces' is another way of saying the same thing.

6   This project began in September 2004. Vince DeSimone has analyzed the structure and proposed a method of construction, and Israel Berger has been advising Milgo/Bufkin on the exterior skin construction. Che-Wei Wang, working in Lalvani's studio, has been assisting with the modelling and visualisations shown here.

7   The first example, called 'Fractal High-Rise', was published in Haresh Lalvani, 'Meta Architecture', in Giuseppa di Christina (ed), *Architecture and Science*, Wiley-Academy (Chichester), 2001; article reprinted from Architectural Design, HyperSurface Architecture 2, Guest Ed. Stephen Perella, Sept/Oct 1999.

8   D'Arcy Thompson, in his classic *On Growth and Form*, has already pointed out physical forces and mathematics as the guiding principles of form lying outside genetics. See D'Arcy Wentworth Thompson, *On Growth and Form*, Cambridge University Press (Cambridge, UK), 1942. Dr. Loren Day, a structural biophysicist in New York working on the structure of viruses, who reviewed this text, sent Lalvani the following remark on this speculation of his: 'I think "mathematics and physical forces" do, in fact, guide genetics. It's more than a possibility'.

9   The Morphological Genome defines just one chromosome of architecture and other chromosomes of architecture could be added through genomes of materials, processes and so on, and the most complex of all, the genome of meaning to complete the entire architectural genome.

10 At present, Lalvani is focusing on a dozen or so genes. Several of these genes are infinite-dimensional, yet rounding that off to a small manageable number gives an exact count of the number of dimensions active in the morph genome.

11 The simplest of these surfaces is one with a singularity (i.e., with a single vertex at the centre) where all edges and faces meet. This is an Infinite class of surfaces obtained by taking all Eulerian structures (like a cube, tetrahedron, dihedron, and so on, that satisfy the Euler relation Vertices - Edges + Faces = 2) and collapsing the vertex to its neighbours or to a common centre. The resulting structures violate Euler's equation. Adding holes to these generates an additional infinite class of new topologies not defined by the more general Euler relation where the 2 on the right-hand side of the equation is replaced by 2 -2g, where g is the number of holes.

12 For the exponential change in the power of the machine and its ramifications for society see, for example, Ray Kurtzweil, *The Singularity is Near*, Viking (New York), 2005. Kurtzweil argues that

'We won't experience the 100 years of progress in the 21st century — it will be more like 20,000 years of progress [at today's rate] ... Within a few decades, machine intelligence will surpass human intelligence, leading to The Singularity ... [that will include] the merger of biological and non-biological intelligence'.

13   Rudofsky used this term for indigenous architecture: Bernard Rudofsky, *Architecture without Architects*, University of New Mexico Press (Albuquerque, New Mexico), 1987 (reprint).

14   Haresh Lalvani, 'Meta Architecture', see [7].

15   Lalvani asked Neil Katz (who modelled the CA surfaces for him) to take Wolfram's one-dimensional cellular automata (see Stephen Wolfram, *A New Kind of Science*, Wolfram Media (Champaign, IL), 2002) and index his 256 rules in an 8-dimensional cube as part of the morphoverse so that each vertex is one rule, and check for a continuum between any two adjacent rules. Surprisingly, the result showed no such continuum.

16   This top down component comes from the topology of the surface, for example the surface of the cube has six squares with three squares meeting at every corner, a rule that is independent of cellular automata rules and is specified by a different morph gene. This gene is the same gene that establishes the square grid of the cellular automata, an a priori grid that has to be 'imported' to demonstrate the bottom up nature of a cellular automaton. A fully bottom up phenomenon requires the emergence of (primary) topologies as well. This means a minimum of two genes is required for a cellular automaton.

17   Loop Quantum Gravity theory suggests something similar for physics. See, for example, Lee Smolin's article 'Atoms of space and time', A Matter of Time, special edition of *Scientific American*, Vol 16, No 1, 2006.

18   Eric Drexler, *Engines of Creation*, Anchor Books (New York), 1986.

19   This is related to the idea of growing architecture. For the origins of this idea see William Katavolos, *Organics*, Steendrukkirj de Jong & Co (Hilversum), 1961; Vittorio Giorgini, 'Early experiments in architecture using nature's building technology', in H Lalvani (guest-editor), The International Journal of Space Structures, 11 :1 & 2, Multi-Science, 1997. Lalvani's own ideas were proposed in an unpublished article 'Towards Automorphogenesis: Building with Bacteria', 1974. For more recent work, see John Johansen's *Nanoarchitecture: A New Species of Architecture*, Princeton Architectural Press (Princeton, NJ), 2002.

20   The genomics pioneer Craig Venter foresees creating life in the laboratory from gene sequences in 10 years. On a related note, earlier, in the 1990s, the idea of originating 'from nothing' had led the physicist Alan Guth to suggest the possibility of creating a 'universe' in the laboratory.

21   NYSTAR, New York State Office of Science, Technology and Academic Research, gave Milgo/Bufkin a two-year grant to experiment with two different AlgoRhythm structures: AlgoRhythm glass-panel systems and the AlgoRhythm truss system. The following participated in the various prototyping experiments at Milgo: Henry Harrison and Jenny Lee (production managers), Neil Katz (consultant) and Pratt students Ori Adiri, Ezra Ardolino, Ajmal Aqtash, Daniel Barone, Christopher Devine, Thorsten Foerster, Brandon Gill, Gershon Gottlieb, Seo Kiwon, Jeff Mitcheltree, Matthew Peterson, Sabrina Schollmeyer, Reza Schricker, Jarrett Shamlain, Che-wei Wang, Hiroshi Yamamoto and Albert Zuger.

POSTSCRIPT 3

This Postscript builds on the experiments in Chapters 1, 2 and 3, and addresses the question: Could ONE fabrication technology, ONE material, ONE morphological algorithm that generates INFINITE possibilities be used at ALL scales of design? The question is in the spirit of looking for universal scale-free design principles. The scale-free component of *AlgoRhythms* was tested by designing, developing and prototyping miscellaneous products in varying scales from small scale product designs to larger structures like curtain wall and structural systems, and experimental artworks. Structural systems offer the greatest challenge since they require curve-folding of thicker materials which puts an upper limit to bending thick sheets, especially in metal, and soon reach their limiting size. This puts an upper bound on the size of curve-folded surfaces. So far, we have successfully curve-folded ¼ inch thick stainless steel sheets, a challenge in itself. The opportunity to try greater thicknesses did not present itself, but we were ready to experiment with self-supporting free-standing wall-type elements as in the *HyperWall* (**fig.5**) which was designed as a double-walled module, or experiment with thicker structural columns and beams. New beams, framing systems and trusses were invented, some already shown earlier (**fig.6**). The insightful engineer Vincent DeSimone on his visit to Milgo to see the work had this to say about an application of *AlgoRhythms* technology (excerpted from *Metropolis,*[1] p.192):

> *"With Lalvani's technique, in theory an entire building could be made of load-bearing folded metal. The difficulty with using metal for structure is that it has a tendency to perform badly in the intense heat of a fire. DeSimone proposes filling the folded metal forms with concrete. "The panel would be an external form of reinforcing," he says. "You could come up with a designer's dream, which would be an exposed structural-steel metal building." He adds, "I can see walking into a building where the lowest structural columns are magnificently folded pieces of titanium and they're real: the titanium is not just an appliqué but integral to the strength of the concrete." (Vincent De Simone, 2003)*

In 2002, New York State Office of Science, Technology and Academic Research (NYSTAR) gave us their first ever grant to architecture in NY State for a collaboration between Pratt Institute and Milgo-Bufkin for an academia-industry collaboration to develop research applications that impact the local economy. This was an opportunity to extend our experimental *AlgoRhythm* technologies to larger architectural structures over a 2-year period (2002-04). It led to two projects: *AlgoRhythm Glass Panel Systems (Curtain Walls)* and *AlgoRhythm Truss System*. Some results are illustrated in **figs. A** and **B**. The designs were developed by Lalvani Studio and all resulting prototypes were fabricated in metal by Pratt students at the Milgo factory using the *AlgoRhythm* curve-folding techniques. (For student credits, see under Notes, #21, Chapter 3).

---

[1] Peter Hall, Bend the Rules of Structure, *Metropolis*, June 2003.

Fig.A (pp.80-81)
AlgoRhythm Curtain Walls (2002-2003)
During the first year (2002-2003) we developed curtain wall and glass panel systems and prototyped several of them on a full-scale using folded sheet metal frames. The curve-folded frames enabled two classes of curtain wall systems based on the shapes of framing members, namely, curved beams with straight or curved axes. **A1-7** show curved frames with straight axes configured as rectilinear systems. These, like the folded columns, were single sheets folded into closed sections and used flat glass panels with overall orthogonal geometries as the simplest starting point. The single window frame (**A2,3**) leads to a simple curtain wall system (**A4**) which was extended to the exterior of a 13 storey building (**A5**) in Chelsea, Manhattan with architect Garrett Gourlay and the developer Stanley Perelman (2004-05). Curved variants with angled frames (related to Penrose tilings in the examples shown) provided additional possibilities (**A6,7**). Examples (**A8-14**) show curved frames with wavy axes. These undulating frames were constructed from curve-folded and required glass panels curved in one direction.

Fig.B (pp.82-83)
AlgoRhythm Truss (2003-2004)
In the second year of the NYSTAR project (2003-04), we prototyped a truss system of curved wavy struts with each strut having a rectangular cross-section and fabricated from two curve-folded parts joined as in the curtain wall frames. The concept was an extension of close-packings of curved polyhedra that the author had developed in the early 1970's. This is related to Fuller's octet truss composed of tetrahedra and octahedra. A single truss module (**B1,2**) and double module (**B5**) were constructed to make the truss in **fig.6** of this chapter. An inventory of small scale laser-cut assemblies from flat wavy elements was also built to explore alternative curved truss morphologies emanating from *AlgoRhythm* geometries. This led to the digital renderings (**B6-12**) which represent a sample of possible morphologies shown with wavy tubular elements.

Fig.C (p.84)
AlgoRhythm Beams, Wavy Standard Sections
The wavy framing members had closed sections and were constructed from two or more curve-folded wavy angles (similar to the hangars we saw in Postscript 1, *Chapter 1*, **fig.B2, 3**) joined to create a variety of cross-sections which are as curved variants of "standard sections" in **C3**. The individual sides of the square frames in **figs.A1,2** are also the shapes of new beams – simply-supported (**B3, B4**), fixed, cantilevered or continuous. Here, **C2** and **C3** show a fractal beam comprising several stacked beams having the shape of fixed and continuous beams defined by the varying number of waves and reminiscent of Le Ricolais' automorphic (self-similar) queen-post structure. It can be used in the orientation shown as an upturned external beam or a framed structure or in its upside-down state as an internal beam. A simple cubic frame system (**C1,C2**), a 3D version of the curtain wall frame in **fig.A4**, uses identical elements in orthogonal relationships and can be adapted to angled relationships or applied to a curved surface as in **A6-7**.

Fig.D (p.85)
Fractal AlgoRhythm Ripples (2002?)
The *AlgoRhythm Ripple* system we saw earlier (p.7, 22, 34, 50-51) can be converted into a fractal system by changing the widths of the waves in the binary series 1, 2, 4, 8, 16, ... shown in digital images (**D1,D2**) and a prototype (**D3**). The prototype, 8ft tall, requires half-slitting the sheet as the wave doubles its width leading to successive openings also in a binary series. The openings can be used for light, sound, air movement for temperature control.

Fig.E,I (pp.86-91)
AlgoRhythm Products, 1999-2004, 2006, 2008
A few examples of smaller scale designs using curve-folding are shown. The seating systems (**E3-7**; 1999-2003) permitted single or multiple units and a proposal for a customized bench for the student lobby at Rutgers University (**E1,2**;2008) applied the principles of *HyperWall* to a continuous bench.

Several door systems were developed, some using *AlgoRhythm* panel systems. From these, two examples are shown and include a single-panel door with a matching frame (**F1**), both from folded metal sheets. A corner door (**F2**) to connect spaces corner-to-corner (as opposed to the current face-to-face connection between spaces we are familiar with) was an interesting surprise. Additional doors were designed including double-panel doors and an irregular curved swivel door with top and bottom pivots. Free-standing shelving systems (**F3,4**), with the outer frame fabricated from a single metal sheet folded into 3D, was another surprise. The standard rectilinear shelf panels, made separately and inserted, gave it structural integrity. Additional products like suspended lights (**F5**), a simple table lamp or a dustbin, all from single folded sheets, and various other containers were among the many products designed at the time. A spin-off system of using the wavy elements of *AlgoRhythms* as stiffeners for containers (e.g. cans) were also considered as examples of self-stiffened tubular structures (**H1-5**).

The drop-boxes (2003) for a courier company (*top right* image, **G2**), provided an opportunity to design a high visibility product for their drop-off locations. The three designs we developed with curve folding are shown (**G1, G3-5**).

The *Algo Signage* project (2006) was to develop 3D font designs that could be made from folded sheets in small and large billboard sizes or bigger as street signs (**fig.I1**). An example is shown with lowercase 'a' (**I4**); all 26 alphabets in lower and upper cases could be fabricated with similar variations from curve-folding flat sheets. The author's interest was in developing a genomic alphabet which captures font (script) shapes within a periodic table of alphabets, similar in spirit to the space of columns in **fig.11** of this chapter. The matrix (**I3**) shows a few familiar English letters in lowercase in a stylized form. The letter p (*top middle*), q (*left middle*), b (*bottom middle*), d (*right middle*), o (*center*), a (*center left*), stylized n (*center top-left*) and stylized u (center bottom-right) are visible in the 25 shapes shown. Others are possible designs for letter shapes that may have occurred in some past languages or are awaiting their adoption in future languages. **I2** shows the 4D cube mapping some lower case English alphabets in one stylized version. The method applies to any language.

84　Coding, Shaping, Making

A1-7
Curved frames with straight axes (2003).

A8-14
Curved frames with wavy axes (2003).

Postscript 3

A12

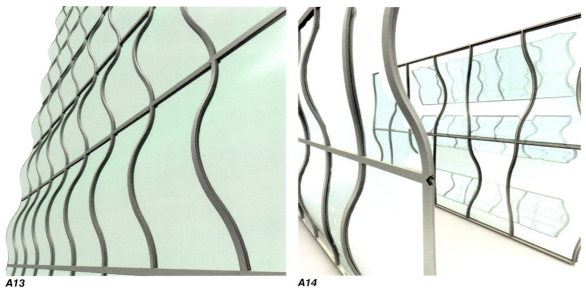

A13

A14

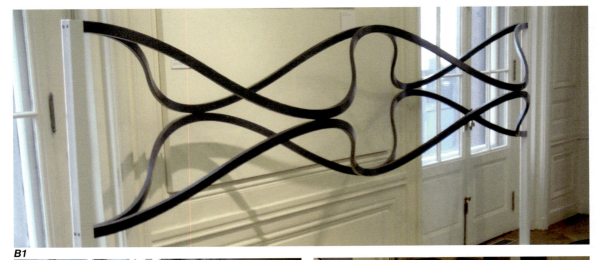

B1

B3

B4

B2

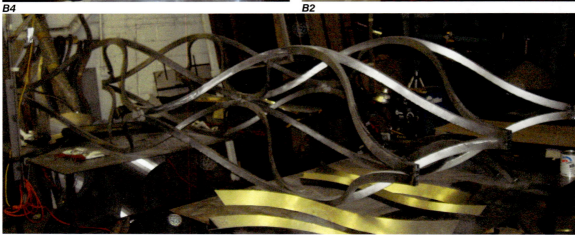

B5

Postscript 3

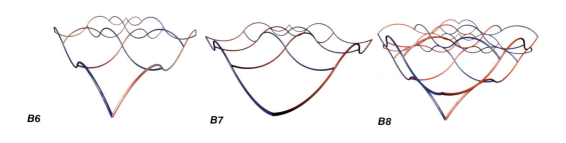

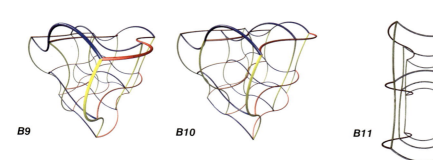

B1, 2
Single *AlgoRhythm Truss* module (2003).

B3, 4
AlgoRhythm Beam (2004).

B5
Several modules of the *AlgoRhythm Truss* in the factory.

B6-12
Cubic frame systems with curved and wavy elements (2003-04).

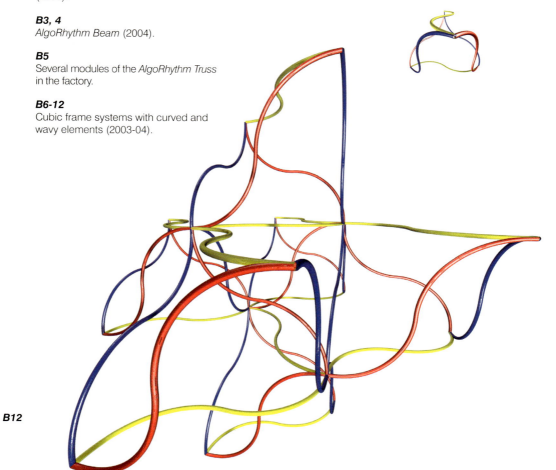

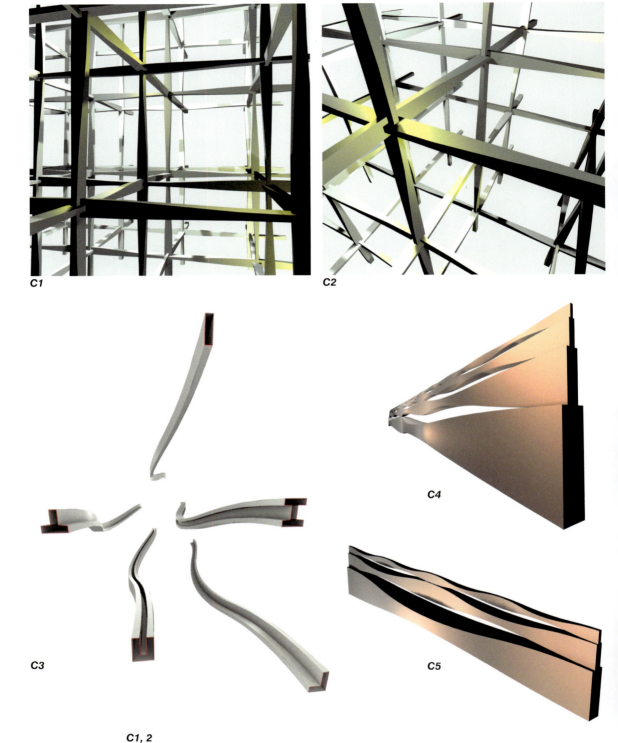

C1, 2
A cubic frame with *AlgoRhythm beams*.

C3
New "standard" sections from folded parts that are joined (2002).

C4, 5
Fractal *AlgoRhythm Beam* (2002).

D1, 2
Fractal *AlgoRhythm Ripple system*.

D3
A full-scale prototype from a single metal sheet.

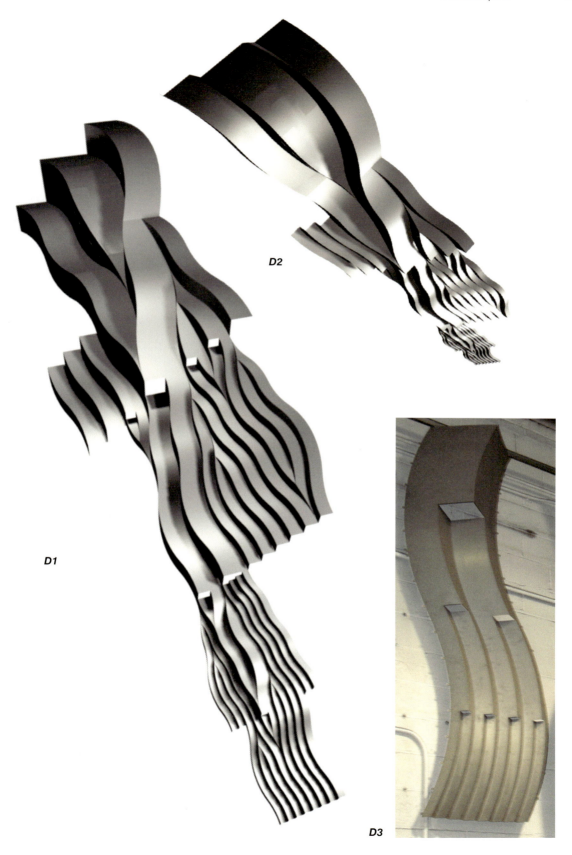

D1

D2

D3

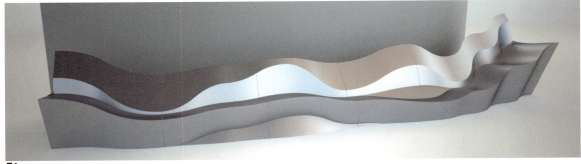

E1

E2

E3 *E4* *E5*

E6 *E7*

E1, 2
Bench for Rutgers University's student lobby (2008).

E3-7
Seating systems made of units folded from single sheets.

F1
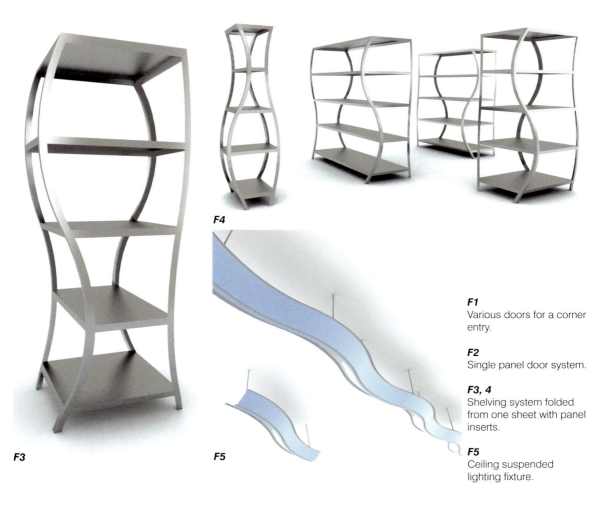

F2

F3

F4

F5

F1
Various doors for a corner entry.

F2
Single panel door system.

F3, 4
Shelving system folded from one sheet with panel inserts.

F5
Ceiling suspended lighting fixture.

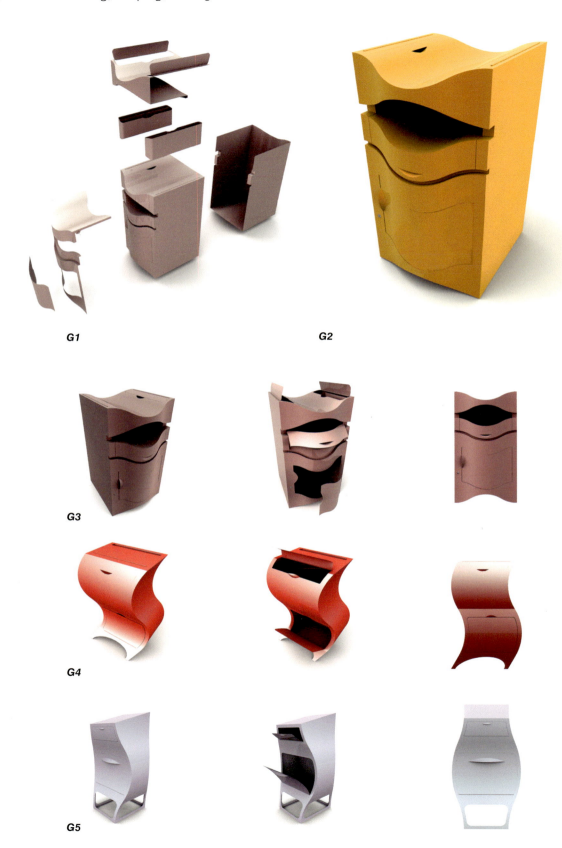

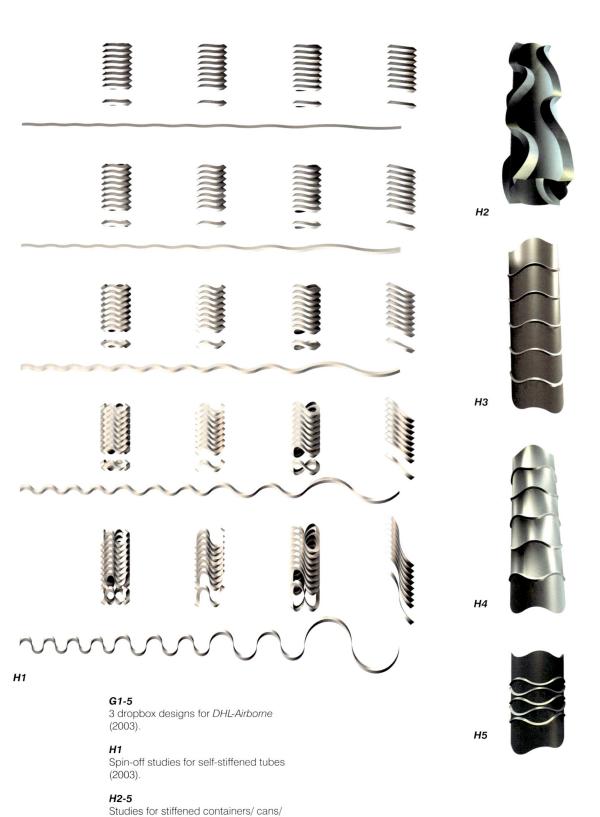

G1-5
3 dropbox designs for *DHL-Airborne* (2003).

H1
Spin-off studies for self-stiffened tubes (2003).

H2-5
Studies for stiffened containers/ cans/ tubes (2003).

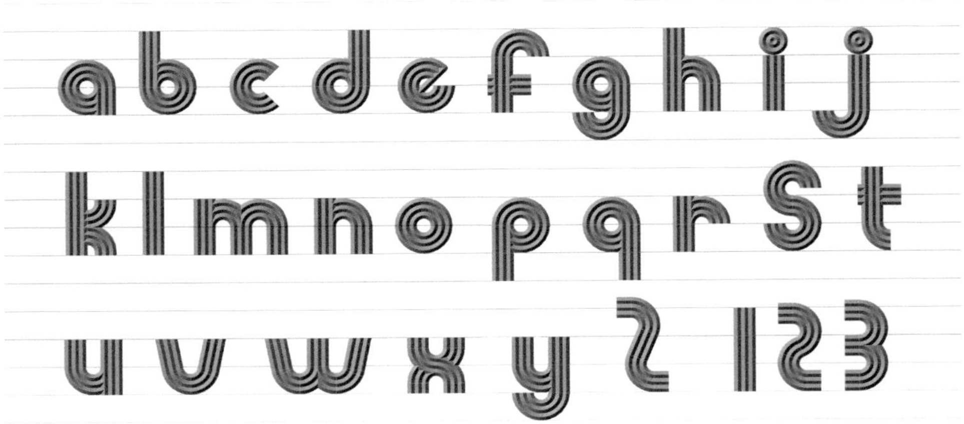
a b c d e f g h i j
k l m n o p q r S t
u v w x y z 1 2 3

I1

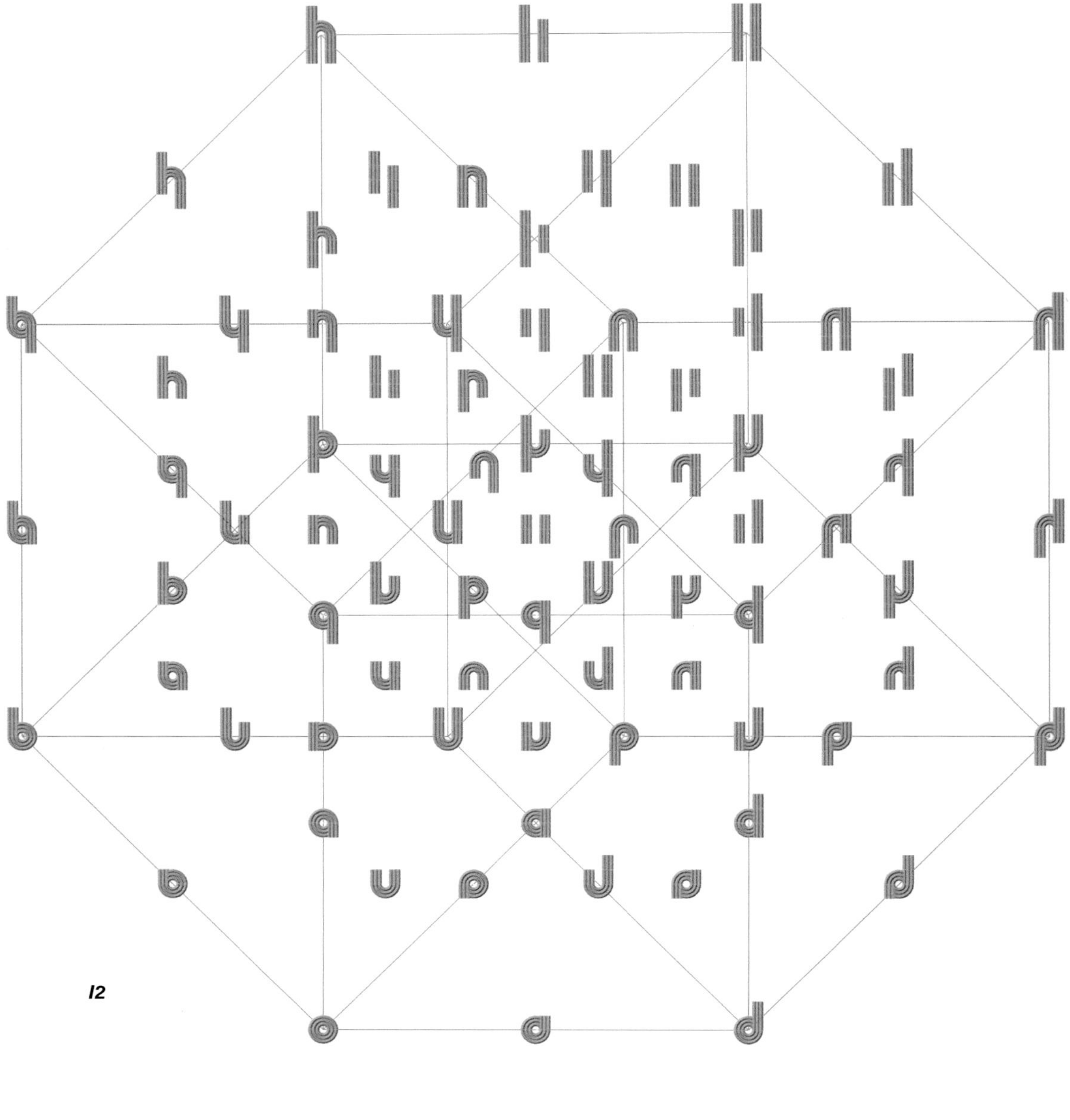

I2

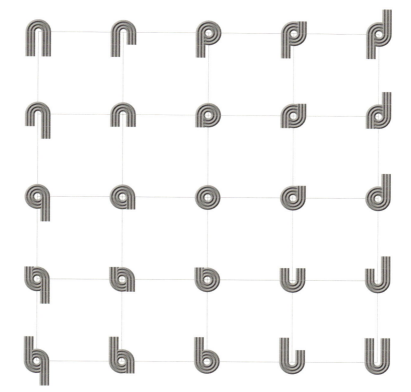

I3

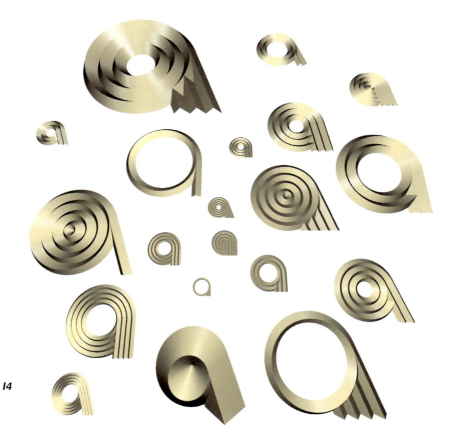

I1, 4
Signage design in folded metal.

I2, 3
Portion of the morphogenetic space for origin of fonts and scripts.

I4

THE PATTERN CONTINUUM:
MASS CUSTOMIZATION OF EMERGENT DESIGNS

"Order for free"
Stuart Kauffman[a] (1995)

Murray Moss, on a casual visit to see our work at the Milgo factory in 2011, picked up an innocuous laser-cut piece, unfinished and rusting. It represented our determined attempt to develop mass customized products, in this case a fruit platter, from our ongoing experiments over 15 years but with a new requirement that the designs were emergent. The designer did not know what the next design in the series will be. Not a word was exchanged on the underlying idea, yet it was "spotted" amongst the several hundred pieces we had, all scattered over any available space at Milgo. It is remarkable that this was the only "functional" object Murray selected for showing at his Soho gallery/showroom, the others being the entire *HyperSurface* series (see Chapter 7). It also happened to be one of my favorites. Over the following months, the underlying idea morphed into a solo exhibition of over a hundred pieces, all "same" yet each "different", like all of us. The *Morphing Fruit Platters, 1D Series 300*, was launched by Moss at *Design Miami 2011* on November 29th with the startling claim of introducing mass customized industrially produced objects with emergent designs and capturing two fundamental principles: *conservation of cost and conservation of mass.*

[a] Kauffman, Stuart, *At Home in the Universe: The Search For Laws of Self-Organization and Complexity*, Oxford University Press, 1995

We showed 100 different platters, each in laser-cut steel and painted, each unique, from a folio of 1,000 platter designs. Our algorithm enabled us to test for 100 million platters before our computer crashed, falling short of our goal of more than 7 billion platters, one for each individual on the planet, all from the same algorithm yet each different from the other. This was an extension of our first example of mass customization with *AlgoRhythms* described in Chapters 1-3 which had a similar goal. It retained the key feature of continuous transformations as a design generator but replaced it with a much simpler form, a circular dish-shaped platter, and a simpler fabrication process, laser-cutting in this instance. It was a model test case, a more compact one, and added several important design strategies for mass customization. This included emergence as a key feature of bottom up design complexity. This feature demonstrated that design, controlled by the algorithm, showed regular and irregular, symmetric and asymmetric, as part of a process of continuous transformation. A highly ordered process showed designs with recognizable regularity, for example, with visible 3-armed spirals or others with more arms, and some others with cells that appeared randomly packed. All were instances in a pattern continuum. Visible order emerged from a lack of order from time to time. Irregularity was the norm and regularity was captured in fleeting moments. The unpredictable smooth but gradual switching between emergence of order and disorder was a big surprise in this continuum.

Infinite Designs from One Variable

In *AlgoRhythm* columns, changes in form were controlled by a combination of topological and geometric parameters. Here, the many variables in our earlier example of mass customization were reduced to one. In addition, *AlgoRhythms* required a designer to select a particular design from an interactive palette of design options, a top down process. In *Morphing Platters*, the designs were emergent, a bottom up phenomenon. It answered the question: could changing one number produce infinite emergent designs? This question relates to the origin of form, especially complex form. To a mathematician, the answer would deal with two phenomena, space and time. A physicist would add physical phenomena like mass, temperature and current.[1] These five metrics provide the starting point for a quantitative basis for all physical phenomena and structures. Our platter project dealt with only three, space, time and mass. The space component was kept fixed by an underlying invariant structure and, as it turned out, mass was fixed as well. The latter was unknown to us when we began and was another surprise. The only variable in design was time which was captured digitally in an animation. In each design, a moment in time was expressed by a distance between two points in space which changed continuously from one design to another. The combination of discrete numbers (integers) and continuous numbers (real numbers having

---

[1] These are the five of the seven S.I. units measured by meter (length, representing a measure of space), second (a measure of time), kilogram (mass), kelvin (temperature), ampere (electric current), mole (amount of substance, a measure in chemistry) and candela (luminous intensity, a measure of the amount of light). https://en.wikipedia.org/wiki/International_System_of_Units

decimal places)[2] in *AlgoRhythms* were replaced by continuous numbers in the *Morphing Platters* series. Each successive number marked the passage of time. A new automaton, a length-automaton, was developed. It controlled length, the distance between two points on a continuous curve.

Length-automaton (Lautomaton)

The length-automaton, *Lautomaton*, controlled the distance between two successive points on *any* continuous curve, a 1-dimensional line. An Archimedean spiral (a whorl, as in a rope coiling on itself) was chosen as the simplest plane-filling curve to serve as our model experiment (**video 10**). Changing the distance between equally spaced points on this curve produced emergent designs of points (and associated cells, e.g. the continuously morphing Voronoi cells) distributed in a spiral space. This could be extended to any 1D line in 2D or 3D space and some 3D examples and animations were developed (**video 11**). All fractals constructed with a continuously winding self-similar line are a possibility. This will allow the emergent property of the length-automaton to be combined with the emergent property of fractals leading to doubly-emergent designs having an increased complexity of design.

The resulting morphology originating from the whorl was transcribed into a laser-cut product, an important control since laser-cutting produces a linear cut, a line, and the cost of laser-cutting is determined by the amount of laser time which is directly related to cutting length.[3] Keeping all other costs of manufacturing constant in the production of each platter (e.g. material, finishing, set-up time, any post-cutting forming method, etc.), this experiment tied production cost directly to laser-time as the sole determining factor for the cost. The idea of using a linear curve for the base design and a linear tool as the principal production component, a one-to-one connection between geometry and the production method defined the project. There were more surprises to come which led to two important results in industrial production, economy of scale and conservation of material.

Twenty-five stills from an animation of the *Lautomaton* applied to 100 points on a continuous whorl are shown in **fig.1**.[4] The whorl is evenly divided with equal distances between successive points on the curve. These distances increase

---

[2] Real numbers are "continuous" as opposed to integers which are "discrete" and jump by one digit at a time. Yet they have a level of resolution determined by the number of decimal places after the dot. Most of our computational designs go up to a maximum resolution of 16 places after the decimal point. This is acceptable for what we build at the "gross" visible level (as in art, design, architecture), yet at the nano-level such differences have an impact. In physics, this departure from the mathematical ideal is acceptable to a very fine level – up to 35 decimal places in length measures, 43 decimal places in time measure, and so on, at the Planck level. This is far too refined for what we build and make in art, design and architecture, at least at the moment. The morphing integers in the *Morphing Fruit Platters* computer animation are a "representation" of a continuous change in length. The limit of computational resolution, and the limited ability of the eye to perceive global changes in integers caused by minute increments in real numbers, informed this decision. So, even though integers are changing in our animation, they "represent" continuous changes in length.

[3] The other cost factor in calculating digital production time is the number of stops and starts in laser-cutting.

[4] Based on some initial development of the model in Rhino by the author, a robust Grasshopper script was written by Peter van Hage at Lalvani Studio and applied to a platter design. This procedure was extended to the production of the *Morphing Platters* by Patrick Donbeck, also at Lalvani Studio.

The Pattern Continuum

continuously from one design to another and are indicated by the numbers next to each still. The variation in designs in this sample was startling. Each design was scaled down to the same area forcing the spiral to become tighter as the number of turns in the spiral increased.

A selection of groups of 24 designs derived from **fig.1** is shown in **fig.2A,B**. In the platters, each point in **fig.1** has been converted into a hole that can be laser-cut. Though the numbers are continuous and have decimal places, only whole

1
Length-automaton applied to equally spaced 100 points on an Archimedean spiral (whorl)(**video 10**).

100 Coding, Shaping, Making

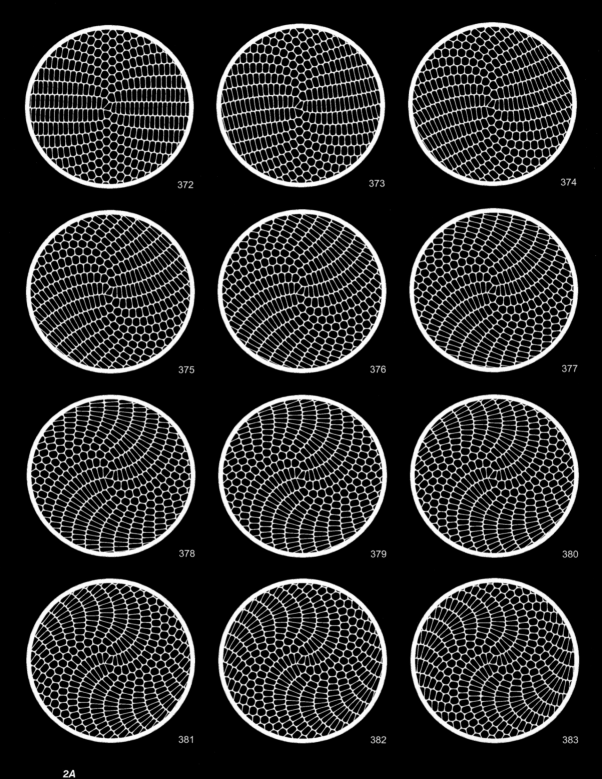

372
373
374
375
376
377
378
379
380
381
382
383

2A

 video 12

2A, 2B
Two sequences of designs, 093-104 (2A), 372-383 (2B) from 1000 designs showing continuous morphing from mirror symmetry to rotational symmetry (**video 12**).

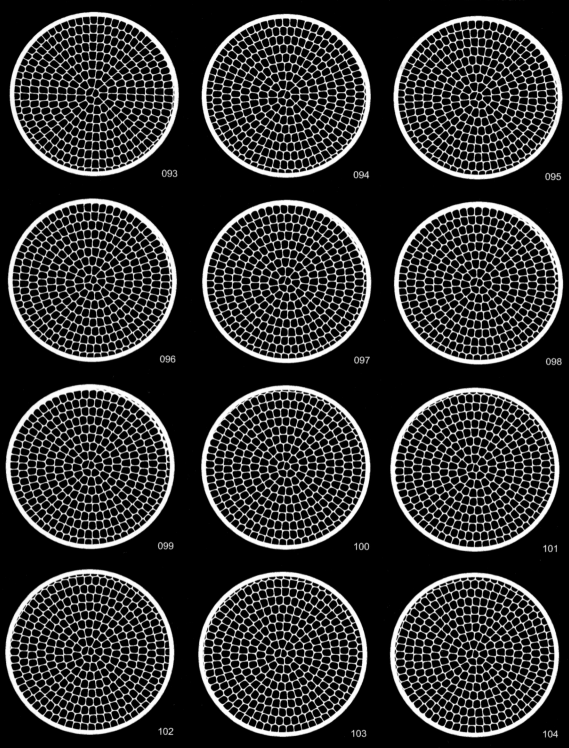

3

4

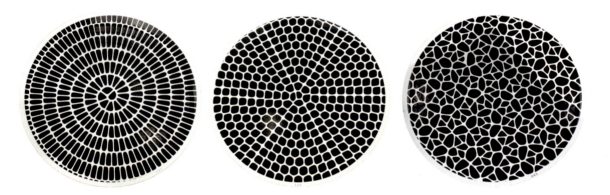

3
Morphing Platters displayed at Moss gallery, Design Miami (2011).

4
Morphing Platters, Moss Gallery, New York; gallery shot (*above*) and three platter designs (*below*).

numbers (integers) in the continuum are shown. This facilitates the selection of a particular design by a potential client interested in acquiring a platter of their choice. An animation was produced showing a continuum of 1000 platter designs that changed from one to another, with a successive integer appearing with its associated design (**video 12**).

Physical platters, displayed at the Moss Gallery booth, *Design Miami 2011*, are shown in **fig.3** and at the main Moss Gallery in Manhattan, New York in **fig.4** (***top***). The three individual platter designs (**fig.4**, ***bottom***) are photos of real platters showing a sample from the range of variability generated by the Lautomaton.

Conservation Laws in Morphing Platters

Design objects, industrially produced with zero-footprint on the planet, must function in accord with the fundamental laws of nature. This goes beyond digital biomimicry of nature or bio-mimetics, and must connect with nature at a deeper level. Nature is governed by fundamental laws, the "laws of nature", that are scale-free. The *Morphing Platters*, a limited industrial application to a trivial object, introduced two laws in the mass customization of industrially produced objects, one applicable to economy of production, and the other to conservation of resources. The first deals with conservation of cost, the second with conservation of mass. The project provided a test-bed to establish a model for more sophisticated applications to art, architecture and design. The first, *conservation of cost*,[5] is essential to break the barrier of 'economy of scale' which prohibits production of industrial objects having differing designs at the same cost. Our *Platters*, enabled by their performative morphology, provided an example. The cost of each laser-cut steel platter was the same (within permissible margin set by the fabricator)[6] regardless of the design difference. The total perimeter in each design is (nearly) conserved through the entire series (**fig.5**). *Perimeter-preserving* morphing ties directly to cost of fabrication, here tied to linear cutting length which translates to laser-time and hence cost, with other production costs remaining the same. This is neither trivial, nor obvious, yet is enabled by the *Lautomaton*. Such a direct tie-in of geometry to

---

[5] 'Conservation of cost' is a term being suggested here for what Murray Moss described to me as 'economy of scale' used in relation to production of industrialized objects. In broader terms, this would translate roughly into the ecological footprint (or carbon footprint as its subset). A more precise definition could be in terms of the number of carbon atoms (and other atoms) that are used in human production of objects, processes, and technology. This will lead to how the periodic table of chemical elements is distributed over our planet, and beyond, and the conservation of this distribution. The true cost of production will be the number of carbon (and other) atoms that are exchanged in any process of transformations of matter for human use. The underlying principle of conservation of cost in industrial production will remain the same, but our economy will be tied to nature's economy which is constrained by the constants in nature (fundamental constants) in its many transformations within matter and between matter and energy. The S.I. units have moved to nature as a measure of our fundamental metrics (length, mass, time etc.). This is the beginning of a paradigmatic inversion from man as a measure of nature to nature as a measure of man. The key question in linking our economy to nature's economy is to answer the question (for example): what is the cost of a carbon atom?

[6] The perimeter-preserving (and area-preserving cited later) are within allowable variation permissible by the fabricator (Milgo) for conserving the cost. Our aim is to fine-tune our algorithm so all designs have *exactly* the same perimeter and same area. At the time of this writing, this remains to be done.

47 [651 in]   317 [600 in]

727 [660 in]   811 [637 in]

5
Perimeter-preserving morphing in our Platters series; the numbers within brackets is the total perimeter of all the holes. The average over 1000 platters is 611". Our aim is to make the perimeters exactly the same in all platters.

cost of fabrication in a one-to-one relationship guarantees that the manufacturing cost will remain constant for *all* designs however different they might be. It also embeds the cost in the formal features of the design where the designer has some "control" over the cost and not the manufacturer.

The second, *conservation of mass*,[7] is essential as a strategy to conserve the resources of the planet whereby designs produced industrially, or otherwise, are done such that *total mass* of all designed objects on the planet remains the same or if the total mass increases, it does so within prescribed limits that sustain the balance between human production and nature.[8] This may not be possible, yet certain targets may be reachable. Recycling mass is a part of this equation. Within this strategy, mass customized designs could target an industry at a time, a material at a time, or a production process at a time, with the eventual goal that the combined mass (and energy) of human designs is conserved in relation to nature. As the population and human needs grow, this may not be possible, yet sustainability paradigm would require us to manage this ratio of humanly consumed mass-energy to the total planetary mass-energy available for human use of materials and energy. This would most likely require an algorithmic approach to conservation. Here, the case of *Morphing Platters* provides a model where the *Lautomaton* enables conservation of mass since each different platter in our infinite series has the same amount of material used (**fig.6**). Each design has (nearly) the same "solid" or "void" area. *Area-preserving* morphing, enabled by our algorithm, ties directly to the amount of material used in each platter and guarantees each platter to have equivalent mass. Hence, same weight! Our aim is to fine-tune our algorithm so all designs have *exactly* the same perimeter and same area. At the time of this writing, this improvement remains to be done.

The 3-dimensional analogs of these two conserved quantities are surface area-preserving and volume-preserving morphings. These need to be combined with the use of least material[9] for greatest performance (size, strength, surface-to-volume ratio, etc.), and do so with the least imprint on the environment. The surface-to-volume preserving nature of a generalized *Lautomaton* potentially suggests a way to remove the limit on size of structures that can be built. As we learnt from D'Arcy Thompson's *principle of similitude*, it was this ratio that put an upper limit to the size of dinosaurs [4]. Their weight (related to length-cubed) grew faster than the bone sizes (related to length), converging rapidly to the breaking strength of bone (related to length-squared). As size grows, the weight grows faster than the strength of the physical material it is made of. Beyond a certain size, the bones of

---

[7] 'Conservation of mass' is a law of nature, whereby total mass/energy within a closed system is conserved. Mass/energy are neither created nor destroyed, just transformed from one state to another. For design, this means thinking beyond local production into planetary terms. Precedents include Bill McDonough's *Cradle to Cradle* strategy for design [2], and global sustainability efforts (see, for example, the Global Footprint Network), though not exactly in the same manner. In comparison, our *Platters* are a restricted experiment since the ecological footprint of production has not been addressed. However, our model does demonstrate that algorithms, guided by morphology, can address one fundamental issue of production, that of conservation of material in producing different designs. Same mass is re-distributed over the same area, as in the expanded structures in Chapters 5 and 6.

[8] This recalls Edward Wilson's 50-50 solution as the nature-human proportion in response to the Anthropocene era [3].

[9] One solid steel sheet of the same area as a platter has approximately the same mass as three platters. The platters are off by approximately 10% from the (theoretical) minimum material needed for the same area. Our aim is to reduce this difference to zero.

The Pattern Continuum 107

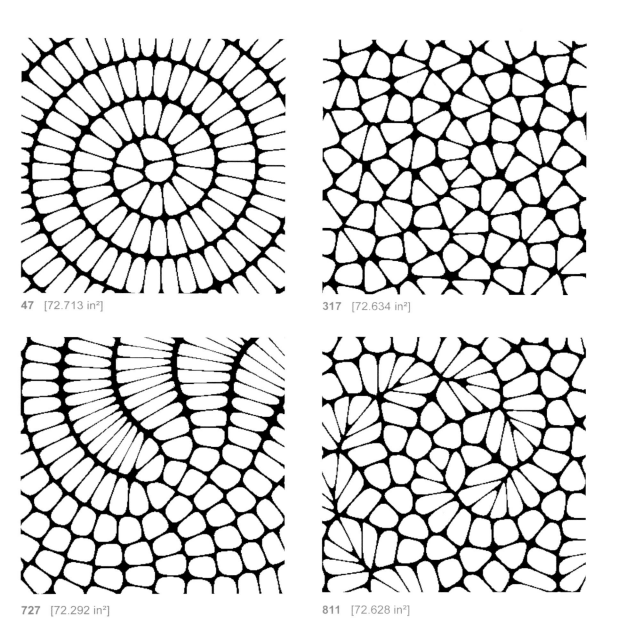

47 [72.713 in²]

317 [72.634 in²]

727 [72.292 in²]

811 [72.628 in²]

6
Area-preserving (or conversely, void-preserving) morphing in our Platters series; the numbers within the brackets is the total void area in each platter; these are practically the same, and our aim is to make them exactly identical.

dinosaurs would crush under their own weight. This is the same principle that limits the use of available tensile materials for the space elevator where promise is provided by the high strength allotropes of carbon, initially buckytubes [5,6], and now its newer relatives like carbon nano-threads [7]. However, if the surface and volume were to grow at a fixed rate with increase in size (mass) in a constructive system, our structures could be as large as we'd like.[10] Our limitation would be material, not size.

The Pattern Continuum

The pattern continuum covers a range of emergent designs that transition from regular to irregular or random-looking patterns. The latter are not random since the automated procedure and the underlying spiral structure are both highly ordered. Even the regular patterns are not regular since the center region seems to behave differently from surrounding regions. Leaving aside definitional precision, achieving regularity and irregularity from a single algorithm is no longer surprising due to the broader acceptance of chaos and complexity theories. However, the range of patterns generated by the *Lautomaton* is a surprise, as is their continuum.

In a pattern continuum of (theoretically) infinite patterns — one pattern for each number — it may be futile to list how many different types of patterns emerge. However, the emergent morphological features are important for their physical implications in built structures. These features are also visual markers that enable pattern-recognition which is fundamental for communication in organisms and humans. The interest in the fundamentals of pattern has led to taxonomies, especially generative taxonomies (addressed in Chapter 8), that are needed to organize and simultaneously generate the vast inventory of possible patterns. A quick survey of platter designs shows at least 8 classes of patterns (**fig.7**), some with global symmetry, others with distinctive morphological features.

The continuous spirals (**fig.7**, 061 as an example), in the early stages of the animation, have a more uniform distribution of cells in contrast with right-handed and left-handed multi-arm spirals which emerge a bit later (184 and 188, respectively). These chiral cases appeared for all numbers 1 thru 8 representing the number of arms along which some cells are aligned. The mirror-symmetric multi-arm patterns (186) also appeared, fleetingly, as a transition between left- and right-handedness for all these numbers. Mirror-symmetry, representing stasis, in dynamically changing spiral patterns was an instant in a pattern continuum where rotational symmetry or asymmetry was the norm. Its appearance, along with the continuous change in handedness from left to right, were surprises even though they occurred rarely. The spirals that appear to be polygonal or angled (207), and show near-alignments of straight edges, was another rare occurrence of a

---

[10] An interesting question is *how* to build infinitely large structures. A recursive (fractal) morphology provides an answer. The Sierpinski tetrahedron is an interesting candidate for infinitely large structures since it is also non-redundant at each stage of recursion, it has no extra elements needed for stability at any fractal scale. If built "from rigid" joints, a fractal diamond lattice (topological dual of Sierpinski tetrahedron) is a candidate minimal solution. Other variants can follow. Once the morphology is established, our problem will shift to finding the right material and inventing a method of construction.

The Pattern Continuum 109

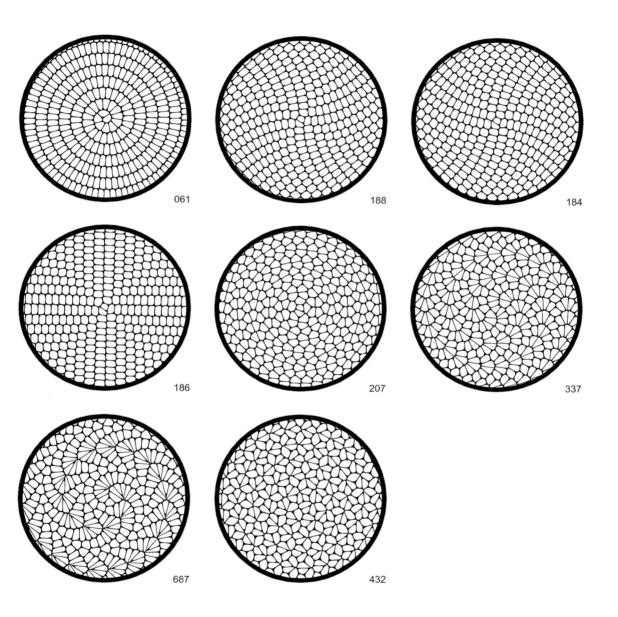

7
Eight classes of patterns in the *Morphing Platters Series*.

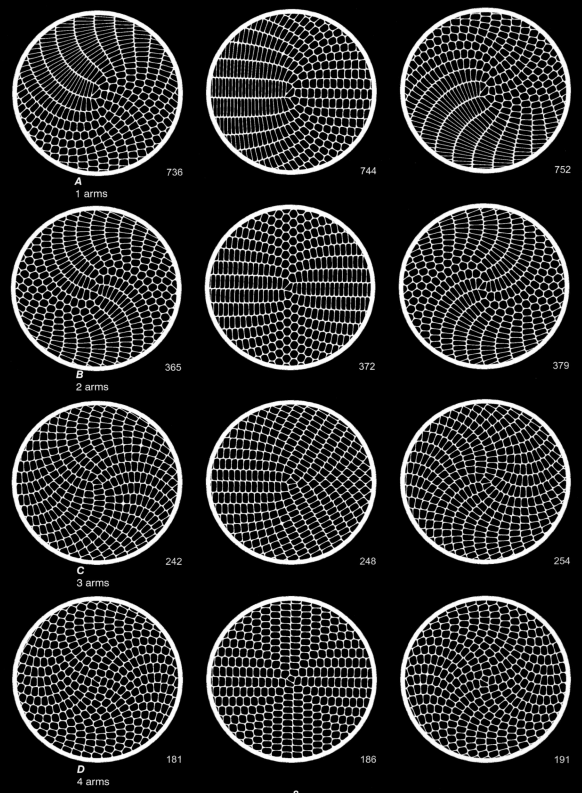

8
4 examples of left-handed and right-handed spirals with 1 through 4 arms, each shown with an intermediate having approximate mirror symmetry.

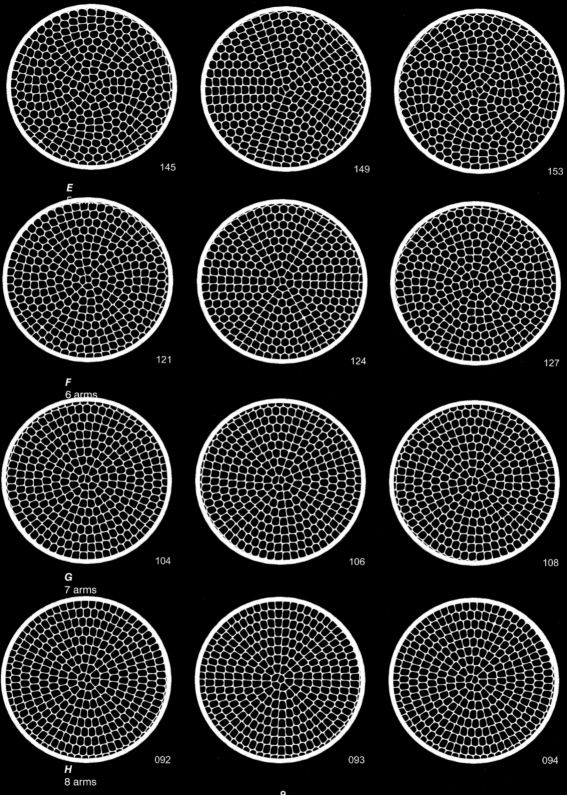

9
4 examples of left- and right-handed spirals with 5 through 8 arms with an intermediate having approximate mirror symmetry.

small distinct group of patterns. The visually spectacular designs are the scalloped spirals (337) that occur in several groups through the animations. The rarer case of zig-zag spirals (687), another dramatic group of patterns, exhibit right-rotating arms emanating from the center region joined to left-rotating arms which are joined to right-rotating arms in a zig-zag manner. Then there are groups of irregular patterns (432) that appear several times within the 1,000 designs. **Figs.8,9** show sets of left-handed, mirrored and right-handed spirals with different number of arms from 1 thru 8.

GPS as Pattern-Generator

The Metropolitan Transportation Authority (MTA) Arts and Design commission for permanent art in the New York City transit system offered an opportunity to further experiment with automated art-design production within a new context. The design of 24 art panels (**figs.10,11**) for the 88th Street train station (A line, Queens, New York), each panel 48in x 30in in 3/16in laser-cut stainless steel, extended the *Lautomaton* within a site-specific context. In the *Morphing Platters* series (2011), infinite designs were generated by the *Lautomaton* to control a single

10

10 (*previous page*)
Morphing88, a permanent public artwork comprising 24 panels for MTA train station on 88th Street (Queens, NYC), each with a unique pattern based on the GPS coordinates of 82nd St through 93rd St representing the streets serviced by the station.

11 (*previous page*)
4 panels from the set of 24 panels of *Morphing88*.

number based on *length* (distance in space). This, in principle, enabled a unique design for each person on the planet. Here, the same automaton was extended, in principle, to generate a unique design for any geographic *location* (point in space) on the planet. The continuum of patterns embedded in the automaton captures *time*, a fundamental attribute of travel as well as the continuous linear path of transport. The resulting artwork, *Morphing88*, captured the unique global coordinates (latitude and longitude from Google Maps) of 82nd Street through 93rd Street, the streets serviced by the train station. Each pattern in each panel was determined by the distance between streets and their angle to the equator, the east-west line. The 365 holes within each panel captured the perennial nature of year-around travel. The system could be adopted to a group of stations or an entire transit line, or to any mode of land transportation anywhere in the world.

The broader concept underlying *Morphing88* applies to any location-specific art, design and architecture. The idea works for all locations in space, anywhere. Immediate examples include buildings with mass customized patterns for curtain walls, frit designs, grills, screens and spatial layouts. products, objects and materials with patterns, textile designs, tiles, and so on; and mass customized art for each, by each. With the DNA shape-scripting suggested in Chapter 9, this will lead to DIY genetic art and design.

Patterns in Nature, Endless Variations

The similarity between some of these dot patterns in **fig.1** and pigmentation designs on some volute sea-shells, particularly the *Volutoconus Bednalli* (**fig.12**), is striking. It raises an interesting possibility in the physical production of these patterns by one-dimensional length control on the pattern-making apparatus during morphogenesis. This, in the case of sea-shells, is overlaid on a second physical method, a self-similar spiral growth well-known from D'Arcy Thompson's description of "terminal growth" [4]. In the case of spiral sea-shells, this is the underlying physical method whereby a curving cone "grows" from a point in a manner that preserves its shape as it becomes bigger. This, as Thompson showed, is achieved by adding a gnomon, a shape added to an existing shape to retain self-similarity. Physical pigmentation process is an overlay on this different physical process. The current known models for pigmentation patterns include the very successful diffusion-reaction model that dates back to Turing who provided a chemical basis for morphogenesis [8]. This has been successful in modeling a large number of sea-shell patterns [9] and other patterns in nature. A completely different model of shell pigmentation, based on cellular automata, has been suggested by

12
The similarity between the *Lautomaton* dot pattern (*left*) and the pigmentation pattern on the volute shells like *Volutoconus bednalli* (*right*) is striking.

Wolfram [10]. This was driven by neighborhood conditions, and hence topology-driven. The *Lautomaton* provides a different candidate model for generating emergent patterns, one that is geometry-driven.

The *Lautomaton* we have been using for the fruit platters is just one of the infinitely many such *Lautomata*.[11] Each distinct *Lautomaton* would produce its own infinite family of emergent designs, yet each family is just one small window into the infinite *morphoverse*, the integrated universe of form. It is also one application of the hypothetical morph genome. Though we are far from establishing the unification of morphology enabled by the morph genome, with each new example we continue to extend and map a tiny portion of the *morphoverse*. Pigmentation patterns is just one example of our interest in design in nature. Our current work is continuing to extend this space with a focus on the morphology of sea-shells, seedpods/fruits, crystals and carapaces (insects, crustaceans, etc.). Some examples will be presented in Chapter 8.

The following sections show a simpler class of pigmentation patterns and presents a new model for design of non-emergent patterns. We (jointly with Peter van Hage) have been exploring ways to tie this model with emergent designs enabled by the Lautomata and other generative systems.

---

[11] Our current *Lautomaton* is based on a 1-dimensional operator. We have used the Archimedean Spiral as the base structure for the potentially infinite fruit platter designs, yet we know there are other spirals as well. We also know that the spiral is just one example of the infinite family of space-filling curves in 2d and 3d and includes various fractals (curves first defined by Koch, Peano, Cesaro, Sierpinski and others). These define future extensions from 2d to 3d space-filling designs based on the *Lautomaton*.

Combinatorial Continuum

In the context of pigmentation patterns, a simple class of pigmented sea-shells from four different species of the genus *Phalium* is shown in **figs.13-16** to show underlying principles of pattern-generation for this class. This example is part of a larger body of work on design in nature and extends to continuous cellular automata (p.332).

The model for pigmentation in nature is assumed to be continuous colors, similar to the idea in color systems, both pigment and light. A full color is designated 1, no color is 0, and the space between 0 and 1 is represented by decimal places as a continuum between the two binary states. For convenience, 9 intermediate stages of a color continuum with a single-digit decimal space are shown in **fig.13A** for one color (brown) with varying lightness selected from the continuous 3D color space of all colors. For patterns composed of single unit cells, e.g. 4-sided quad regions of a square grid, each cell can be in discrete state 0 or 1, or any continuous stage in between. In a unit cell that is subdivided into 4 smaller sub-cells (2x2 array), each sub-cell can exist in an on or off state, 0 or 1 in its discrete mode. These binary states lead to 16 discrete combinations which can be mapped on the vertices of a 4D cube (**fig.13B**) as in author's previous work [11,12]. For these 16 combinations in their continuous states as in 13A, each sub-cell can exist in 11 states (9 intermediates and the two end states, 0 and 1). The total combinations of these states equal 2^{11} (=2048). These can be mapped on the vertices of a subdivided 4D-cube, a finite portion of a 4D-cubic lattice (11x11x11x11 4D array), which defines the space of continuous combinations. A close-up view of these continuous combinations lying on the vertices of outer faces of the 4D-cubic lattice is shown in **fig.14**. This mapping illustrates a combinatorial continuum with 9 intermediates between the 2 discrete states.

The combinatorial continuum of colored cells in 4D space are applied to a spiral sea-shell in **fig.15**, shown here with only one intermediate for visual clarity, and in a close-up in **fig.16**. An application to seashell pigmentation is shown in **fig.17** with 10 examples from 4 species in the genus *Phalium*, namely *Phalium areola*, *Phalium bisculatum*, *Phalium bandatum* and *Phalium decussatem*. For each example, two diagrams are shown on its right, the upper one showing an extracted portion of the seashell and the lower one with its code of color combinations. Nature's colors are more nuanced than our choice of one brown color with varying lightness. The subtleties in variations can be obtained by varying the saturation and lightness for any hue in the RGB space by extending the model to 7D space. The continuous color scheme presented here will apply to all 3D spiral shells independent of the species and will also apply to all colors in the color space.

Cell Variations

Pattern variations can be produced by altering the cell shape by various means, for example, by any combination of transformations in lengths, angles or curvature. The examples shown here are restricted to changes in length only and are thus very simple morphologic changes. Yet, they are sufficient to demonstrate

The Pattern Continuum

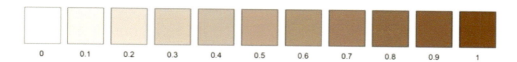

13A

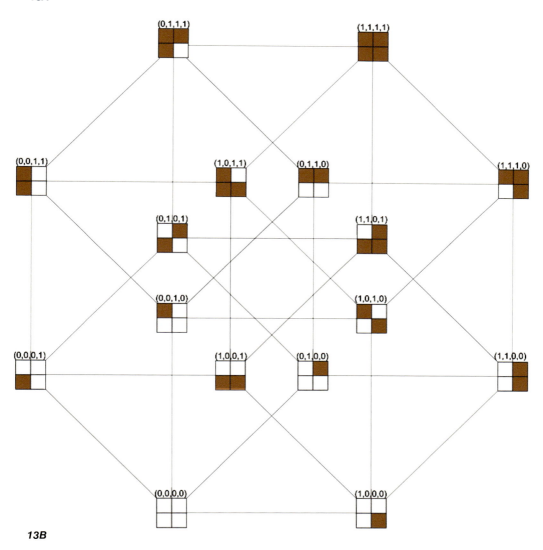

13B

13A, B
Color continuum in a single cell indicated by decimal places between the discrete states 0 and 1 (A).
16 combinations of 4 cells with each in discrete states of 0 and 1 mapped on the vertices of a 4D cube (B).

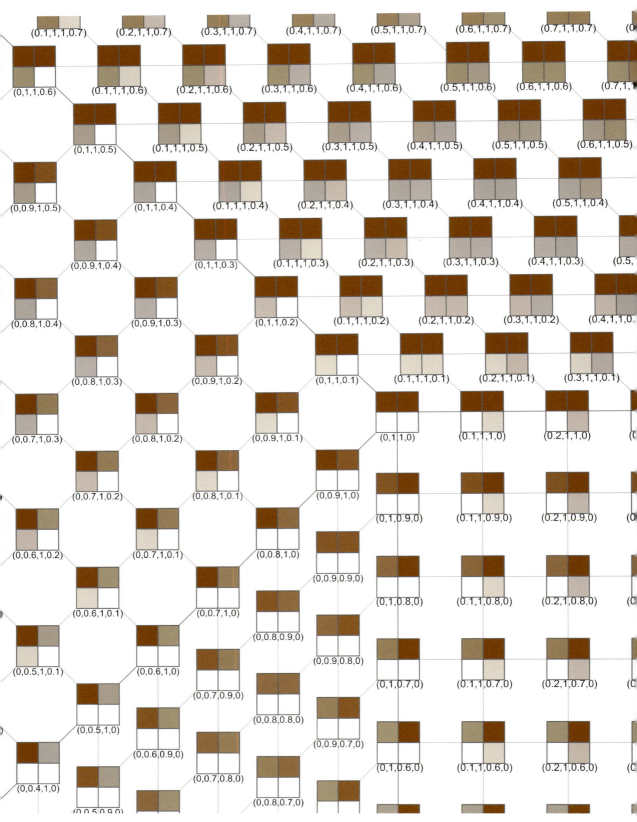

The Pattern Continuum   119

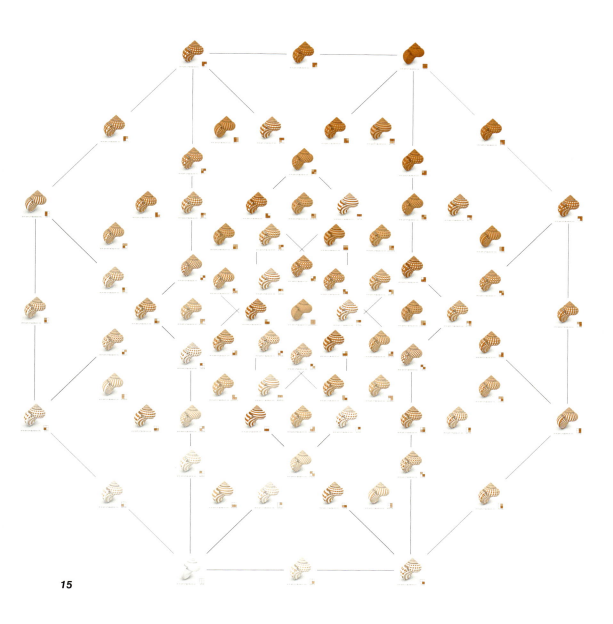

15

14
A close-up of a portion of a 4D-lattice showing a color continuum of 4 cells with each cell in 9 intermediate stages between 0 and 1 as in fig.13A.

15
A 4D cube of pigmentation patterns of a sea-shell with one intermediate along each edge of the hypercube.

120 Coding, Shaping, Making

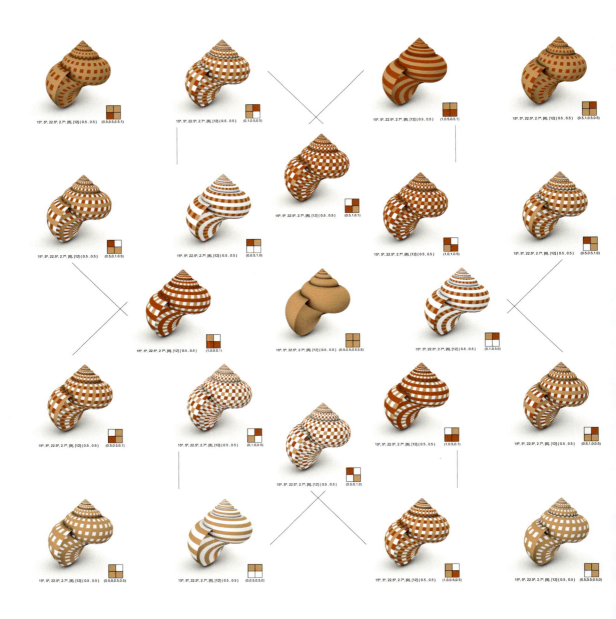

16
A close-up of the 4D cube in fig.15 showing each sea-shell and its associated color code and coordinates on its *bottom right*.

The Pattern Continuum   121

Phalium aeriola (0.4,0.1,0.1,0.1)

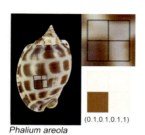
Phalium areola (0.1,0.1,0.1,1)

Phalium bandatum (0.4,0.5,0.4,0.3)

Phalium bandatum (0.1,0.1,0.2,0.4)

Phalium bandatum (0.1,0.2,0.3,0.1)

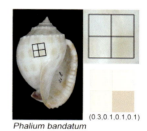
Phalium bandatum (0.3,0.1,0.1,0.1)

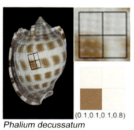

Phalium decussatum (0.1,0.1,0.1,0.8)

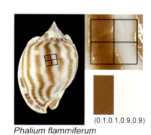
Phalium flammiferum (0.1,0.1,0.9,0.9)

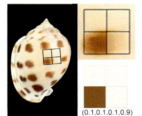
Semicassis bisculata booleyi (0.1,0.1,0.1,0.9)

Semicassis bisculata (0.7,0.1,0.1,0.1)

Phalium muangmani (0.1,0.1,0.7,0.8)

17
The combinatorial color continuum model applied to 11 examples from four species of the genus *Phallium*. For each, a portion of the seashell is extracted in the *top right* diagram with the coded diagram *below*.

122 Coding, Shaping, Making

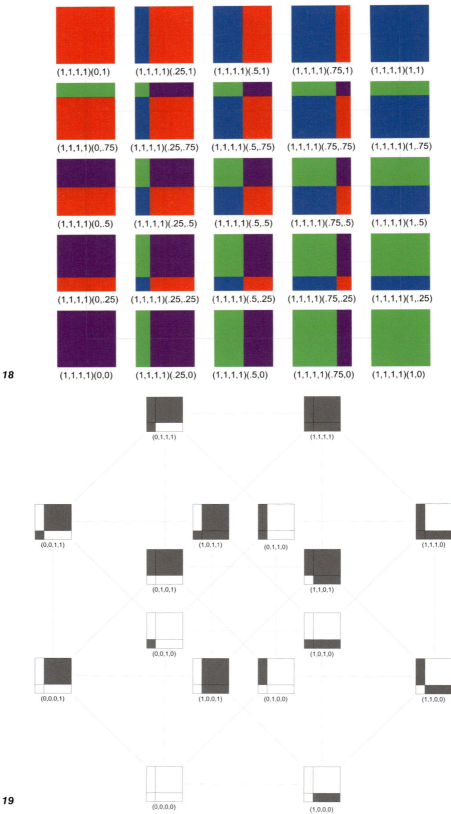

18

19

The Pattern Continuum

123

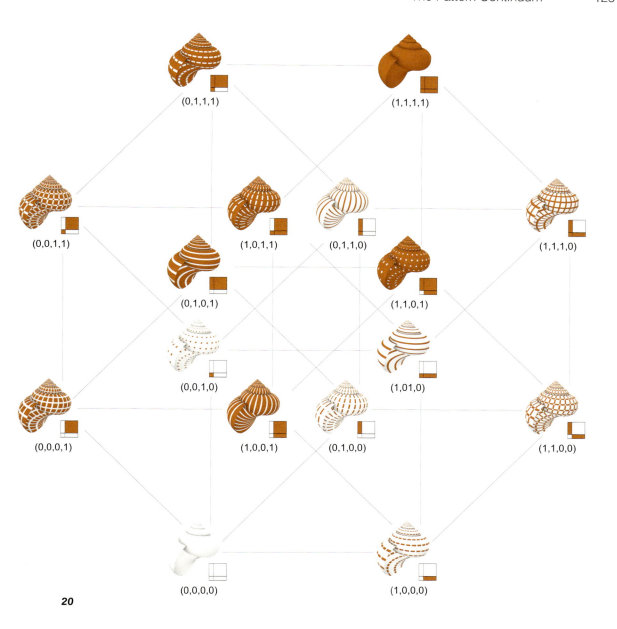

20

18
A 2D continuum of the 4 sub-cells by moving a vertex to any point location within the unit cell. Adding these changes to the cell combinations in fig.20 extends the morphological space to 6D.

19
The 16 discrete states of a 2x2 cell subdivision of the unit cell arranged on the vertices of a 4D cube as in fig.15B, but with the difference that the point (vertex) where the 4 sub-cells meet is off-centered.

20
A family of 16 classes of pigmentation patterns by repeating the unit cells in Fig.19 on the surface of a shell. Each pattern has a complementary (anti-symmetric) pattern located on the opposite vertex of the 4D cube leading to 8 complementary pairs of pattern classes.

underlying principles and underpin a design strategy for applying other types of geometric or topologic transformations to a single or multiple cells or to more complex morphologies. They are also sufficient to generate a vast class of patterns from simple combinatorial and morphing rules. The 4D model of combinatorial color continuum can thus be extended by adding additional transformations. One example is shown in **figs.18-20** with related patterns in nature in **fig.21**.

The 2x2 cell subdivision of the unit cell into 4 equal sub-cells shown in **fig.13B** is one geometric state, a centrally-symmetric one, from a continuum of states shown in **fig.18**. In this diagram, it lies at the center. The 21 states shown here are obtained by moving the central vertex (where the 4 sub-cells meet) to any location within the space of the unit cell. This adds two dimensions (x, y) to extend the 4D morphological space to 6D. It leads to patterns with squares of different sizes combined with rectangles of different proportions. One state from this set, with an off-centered vertex location within the unit cell. is selected as an example and is applied to all 16 classes of subdivided unit cells in **fig.13B** to generate corresponding unit cells in **fig.19**. When arrayed on a spiral shell surface, the unit cells generate 16 classes of pigmentation patterns (**fig.20**). Now the underlying pattern (meta-pattern) in these patterns begins to reveal itself along with the inter-relationships which are briefly described next with some examples.

Anti-Symmetry in Pigmentation

An important feature of the 4D and higher dimensional representations we have been using is that all patterns, forms and structures, are spatially organized so their complements lie diametrically across the center in these centrally symmetric projections of higher dimensions. This is one of the advantages of these particular orthographic views of the hypercubes. In **fig.20**, each pattern has a complementary pattern located on the opposite vertex of the 4D cube. In symmetry terms, each pattern located on the opposite vertex of the 4D cube. In symmetry terms, each pattern has a pattern with anti-symmetry, a black-and-white pattern on one vertex has a white-and-black pattern located on the opposite vertex. This leads to 8 pairs of complementary (anti-symmetric) classes of pigmentation. It turns out that all patterns in **fig.19** and **20** have a one-to-one correspondence with the 16 Boolean operators which can be organized in an analogous 4D-cube. This makes it a candidate part of the *morph tool kit* (Chapter 8), here applied to pigmentation patterns.

An assortment of examples of seashells and insect carapaces are shown in **fig.21** for each class. The 8 examples in the top rows are complements of the corresponding 8 in the bottom row. Examples for (1,0,1,1) and (0,0,1,1) were not found during this limited search.

The color can be replaced with corresponding examples of 3D surface modulations to produce shells with structural ribs and bumps which confer advantages of survival on the organism (strength, defense mechanism, camouflage, mating, and other biological functions). This structural aspect of the work is in progress. Additional transformations that change the cell shape in various ways, some similar to the ways described by D'Arcy Thompson and newer methods, lead to more ways to generate non-emergent and emergent patterns. As well as symmetry-breaking.

Unifying Form Generation

The explorations of pigmentation patterns presented here need to be developed further and are one of the four experiments currently in progress to look at unifying generative principles of form in nature ranging from the living to the non-living. Models for morphogenesis in crystals, fruits and seedpods, insect shells, and seashells are being developed and provide an example each from the non-living world, plant world, the animal world, a hybrid between living and non-living.[12] Some of these will be shown in Chapter 8 on the *morphoverse*. These are hardly enough to capture the extreme diversity and complexity in nature, and even within each of these examples the formal subtleties are beyond what limited explorations like this one can capture. However, the hope is that some common principles of form generation can be defined. It would be interesting to see if the combinatorial continuum model, its mapping in hyperspace, principles of complementarity, and the conservation principles in continuous form transformations in [11,12] will continue to hold as we explore and mine the infinite space of new design possibilities through the explorations of design in nature. Whether these generative principles are used in nature is a separate question and requires collaboration with scientists working in these areas.

These models are attempts by those in the design arts to understand nature's design strategies so we (as designers, architects, artists) can come up with our own designs in harmony with nature as we continue to navigate the complex intertwined landscape of human-nature interactions. Such morphological experiments indicate a human sensitivity to nature and continuing wonder about nature's design marvels. As species continue to disappear at a rapid rate, design references in nature are also being lost rapidly, forever. Nature's inspiration has continued since the birth of art, design and architecture, and has always been an open-ended endless source. But now, for the first time, the open-endedness of this inspiration is being curtailed by our own actions. In philosophical terms, it calls for us to evaluate our deep relationship with nature. It is also a reminder we are a part of it and are contributing to nature's future paths while we shape ours.

Mass customization in Nature and Economy of Scale

No two snowflakes are alike. Is this true? The consensus among scientists is that *"the likelihood of two large snow crystals being identical is zero"*. And *"Each winter there are about 1 septillion (1, 000, 000, 000, 000, 000, 000, 000, 000 or a trillion trillion) snow crystals that drop from the sky!"* [13] An example of infinite design variation on a theme.

The noted chemist Roald Hoffmann asks: *"Are there two identical molecules?"* [14] His answer is *"No, for a really large molecule, probably there are no two identical molecules in that Burmese cat."* Another example of infinite design variation in nature.

Humans are individuals, but so are snowflakes and molecules. And roses and leaves, and fruits and trees, and dogs and fish, and so on. Individuation is

---

[12] This project began as a potential collaboration with a major museum of natural history but was interrupted by the pandemic.

126 Coding, Shaping, Making

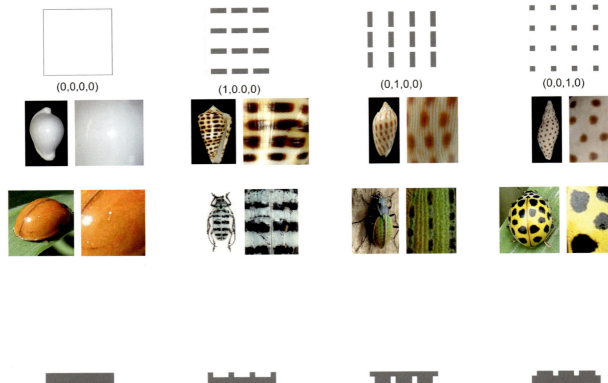
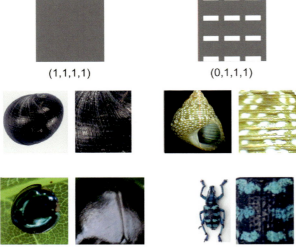

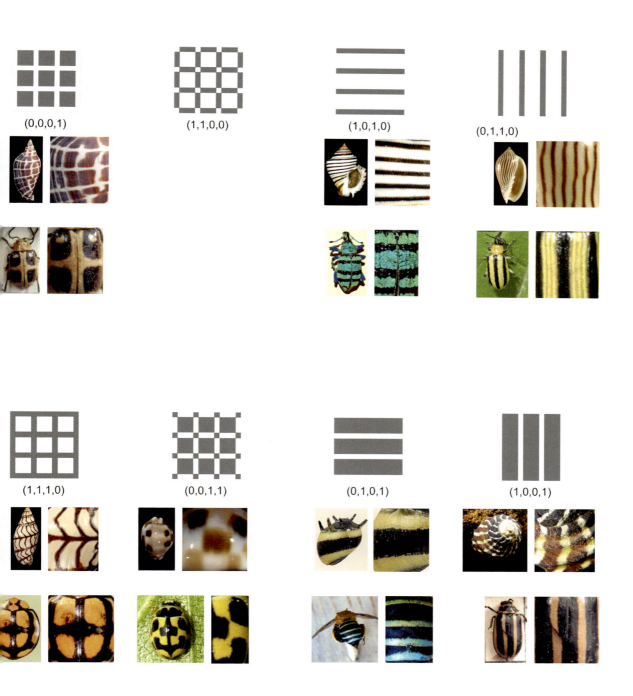

21
The 8 pairs of complementary classes of pigmentation patterns with examples from seashells and insect carapaces. The 8 example in the top rows are complements of the corresponding 8 in the bottom row. Examples for (1,0,1,1) and (0,0,1,1) were not found during this limited search.

how nature defines diversity, and hence identity, which lies on one end of the design spectrum. On the other end lies unity which is underpinned by the laws of nature and the invariant patterns we keep discovering across different scales. How nature achieves infinite diversity within unity is one of the greatest marvels of design. In living organisms, genetics is a unifier and evolution a diversifier. In the recent decades, these two overarching concepts of genetics and evolution have come together dramatically in Evo Devo, the new field of evolutionary developmental biology. The startling discovery is that *"all complex animals – flies and flycatchers, dinosaurs and trilobites, butterflies and zebras and humans – share a common "tool kit" of "master genes" that govern the formation and patterning of their bodies and body parts"* [15]. Infinite design variation on a theme in biology is governed by a universal genetic tool kit. Shouldn't there be one for human-made designs?

The term mass customization, strictly speaking, is not applicable to nature since the term originated in the context of industrial production of human-made designs, yet nature is a supreme mass-customizer. Industrial mass customization is a response to mass-production, a staple of the Industrial Revolution that has brought us endless copies of the same object at lower costs. The guiding principle, and the reason for its success, is economy of scale. The larger the number of identical copies made, the cheaper it is to make them. It's corollary is that different designs are more expensive to make. This rules out the economical production of variations on a theme. Yet this is the very idea that nature uses freely and ubiquitously. The *Morphing Platters* project demonstrated that variations on a theme could be achieved at the same cost by making conservation of cost a design-production principle for infinitely variable designs. Murray Moss, who presented the Morphing Platters series to the public, had this to say:

"Each platter bore a pattern that was different from the one that came before and from the one produced after; the process did not produce any "economy of scale" benefit, a cornerstone of the Industrial Revolution – meaning that there would have been no savings in cost had the process been programmed to produce identical platters."

(Murray Moss, 2015 [16])

The upper limit of mass customization is to have each design unique independent of the number of designs needed. This was our goal with the *AlgoRhythms* columns, *Morphing Platters* series, the art panels for MTA, and also underlies our continued exploration of design in nature. This, in a way, is a mimicry of nature with the difference that design is tied to an industrial production, a different process. The divide between design and physical manufacturing will continue. With nano-technology, especially, synthetic biology, a seamlessness between design and production seems possible but this needs to reach larger physical sizes needed for human-made objects including buildings. In biology, design (software) is the same as production (hardware) since DNA carries design information and instructions for fabricating the designs. We are, very slowly, heading in this direction. This will be further addressed in Chapter 9 on abiogenesis and the future of architecture.

REFERENCES

1. See, for example, *The World of Buckminster Fuller*, film by Robert Snyder, 1974.

2. McDonough, W., Braungart, M. *Cradle to Cradle, Remaking the Way We Make Things*, Northpoint Press, 2002.

3. Edward Wilson, 50-50 Wilson, E.O. (2016) *Half-Earth, Our Planet's Fight for Life*, Liveright Publishing.

4. D'Arcy W. Thompson, *On Growth and Form*, Cambridge University Press, 1945.

5. Lisa Zyga, Long, Stretchy Carbon Nanotubes Could Make Space Elevators Possible, *Phys.org*, January 23, 2009.

6. Aron, J. Carbon Nanotubes too Weak to get a Space Elevator off the Ground, *New Scientist*, 13 June 2016.

7. Anthony, S. New Diamond Nanothreads Could be the Key Material for Building a Space Elevator, *Extreme Tech*, September 23 2014.

8. Turing, A. The Chemical Basis of Morphogenesis, *Philosophical Transactions of the Royal Society of London*. Series B, Biological Sciences, Vol. 237, No. 641. Aug. 14, 1952, pp. 37-72.

9. Meinhardt, H. *The Algorithmic Beauty of Seashells*, 3rd ed, Springer-Verlag 2003.

10. Wolfram, S., *A New Kind of Science*, Wolfram Media Inc. 2002.

11. Lalvani Haresh Lalvani, *Multi-Dimensional Periodic Arrangements of Transforming Space Structures*, Ph.D. Thesis, University of Pennsylvania, 1981; published by University Microfilms, Ann Arbor, Michigan, 1982.

12. Lalvani Lalvani, H. *Structures on Hyper-Structures*, (self-published) Lalvani, New York,1 982; a slightly revised version of [11].

13. Is it True that No Two Snow Crystals are Alike? Science Reference Section, *Library of Congress*, 11/19/2019. https://www.loc.gov/everyday-mysteries/meteorology-climatology/item/is-it-true-that-no-two-snow-crystals-are-alike/

14. Roald Hoffman, *The Same and Not the Same*, Columbia University Press, Chapter 8, pp. 32-35. 1995.

15. Sean B. Carroll, *Endless Forms, Most Beautiful, The New Science of Evo Devo*, W.W. Norton, 2005.

16. Murray Moss, PS (monthly column), To Be or Not to Be, PS, *Interior Design*, March 2015, p.279.

FORM FOLLOWS FORCE:
THE EPIGENETIC CONTINUUM

> *"... the form of an object is a "diagram of forces"*
> D'Arcy Wentworth Thompson[a] (1917)

The essay *Meta Architecture* (Chapter 1) dealt with morphologically coded form-making tied to *genetics*[1,2] of form based on geometry and topology. These formal issues are tied to "continuous transformations" of form referred to in *Genomic Architecture*[3] (Chapter 2) and are part of a *form continuum*, i.e. all forms can morph to others in a continuum. *Form continuum* is based on representing form in interconnected higher dimensional networks that unfold systematically as new features are added to an open-ended morphological universe [11, 12, 13]. This representation suggests another important concept, a *form-process continuum*. *Form* occupies nodes of these networks and *process* links any two nodes to permit one form to transform to another through that link. These networks, in their visual representation, capture a continuum between form and process and will be presented in *Morphoverse* (Chapter 8).

---

[a] Thompson, D'Arcy W., *On Growth and Form*, Cambridge University Press, 1942.

[1] It would be instructive to document the history of the term 'genetics' in architecture. William Katavolos [1] and Vittorio Giorgini [2], early pioneers in organic architecture from late 1950s and early 1960s, mentioned the connection between their works and genetics (personal communication, 1970s). Rudolph Doernach's drawings of "edible architecture" from the 1960s and "biotecture" from the 1970s [3] and Wolf Hilbertz's structures "grown" by bio-mineralization in sea water [4], Lalvani's "automorphogenesis" using bacteria to build architecture [5], John Johansen's "nanoarchitecture" [6] are other examples.

[2] The term "genetics" in architecture, no longer in vogue now, is used very loosely, sometimes metaphorically, the way "DNA" is used in cultural (non-scientific) contexts. Parametric modeling comes a step closer since it starts to codify, in a *descriptive* way, features of form. These are seeds of morphological characters and are usually symbolized by alphabets. When these characters are grouped together and tied to a procedure or algorithm, they are *generative* and underlie the use of the term genetics in architecture. In the context of history (footnote 1), it will be also be instructive to understand the influence of Aristid Lindenmeyer [7] and John Holland [8], the originators of digital genetics, on architecture.

[3] See also Karl Chu's work on 'Genetic Architecture' [9] which parallels author's work. The author distinguishes the terms 'genetics' (derived from 'gene') and 'genomics' (derived from 'genome') and uses it in the same sense as biology where genes are subsets of the genome which includes all genes [10, 11].

This chapter presents a *process continuum* and focuses on *force* as a model physical process to shape form in any material. Form follows force, a reminder of D'Arcy Thompson's quote in the opening of this chapter, is experimentally realized here in one unlikely and unforgiving material, metal. We used cold rolled steel and stainless steel sheets, and all experiments were at room temperature with no metal-softening processes like heat were used. A variety of structures are shown where force was used in different ways as a forming process for fabrication. These experiments, *Milgo Experiment 2*, were carried out during the period 1998-2008 [14, 15] in the industrial setting, of the factory Milgo-Bufkin. Later digital models and diagrams have been added [16].

Force is an epigenetic phenomenon and offers a counterpoint to the formal genetics of form in earlier chapters. Epigenetics[4] deals with phenomena lying outside of genetics of form and includes external physical factors (environmental factors such as gravity, light, temperature, humidity, pressure, etc.) and other aspects of design (social, cultural, contextual, meaning, site, and so on). Force is one of the most important factors in shaping and making form in living and non-living systems and is used in these experiments. It is a model phenomenon and provides lessons which can be applied to other epigenetic phenomena.

Form Follows Force

This series of experiments began in 1998 with a new invention [17], a spin-off from unpublished experiments in kirigami between 1972-76 described briefly in Postcript 5 at the end of this chapter with one example published in [18]. Those early hand-cut paper models had now given way to laser-cut metal sheets and these new experiments were taking place within an industrial production environment with a view towards building large-scale structures at an architectural scale. This opened up a new direction in the work driven by epigenetic form-making. It started becoming clear that these experiments could also provide insight into the repertory of nature's design strategies.

The best known example of epigenetic shaping is Antonio Gaudi's hanging chain method [19], a pioneering architectural invention where nature (gravity, in this case) shapes the form. No human hands are involved, no prior models are needed, gravity defines the final form. Gaudi's is the first example of self-shaping in architecture. Individual 1D (one-dimensional) flexible elements, strings representing tension cables, were suspended in arrays or rotated around an axis to generate gravity-shaped surfaces which, when turned upside down and converted into a hardened state, defined architectural space. These hanging chain models led to the vaulting in the attic of Casa Mila, the asymmetric roof grid and tilted columns in Colonia Guell, and the merger of vaulting with branched columns in Sagrada Familia. Remarkably, all were built in brick and stone. This marked the beginning of the interplay between two opposing phenomena, flexible and rigid, in building technology. Flexible forms of hanging chains were transcribed into rigid architectural structures.

---

[4] In biological terms, it relates to phenotype changes (for example, changes in visual appearance) by "gene expression" rather than changes in the genetic code. These are caused by external factors (e.g. environment, gravity, etc.) that lie outside the genetic code.

The second invention in this lineage is Frei Otto's grid shell [20] where Gaudi's 1D elements (cables, strings) were replaced by a flexible 2D grid (mesh) which was suspended, its gravity-shaped form was captured by various techniques including photogrammetry, and an upside-down version was constructed in a rigid or rigidified material. The roof for Mannheim Multihalle (1970-75), the first large-scale grid shell, was constructed from linear wood elements which were inter-laced to make a pliable grid, the grid was hoisted from the ground and rigidified into its final form by tightening the two-way sliding joints which allowed the wooden grid to expand. Otto's grid shell used 1D elements as in Gaudi, but the elements were inter-connected into a pliable 2D surface. The rigidification of form was the last step in the process and was more integral to the building process than Gaudi's.

Flexible-Rigid Continuum

The third invention in this lineage is presented here wherein flexible and rigid are a continuum and the resulting form is self-rigidized. No joints are tightened; they self-tighten as the structure changes from flat to curved. Since we are dealing with two opposite states, this is an oxymoronic continuum which is different from other continua. Otto's pliable grids comprising 1D elements (edges) in his grid shells are here replaced by pliable tilings comprising 2D rigid elements (polygons) which appear to be joined "point-to-point"[5] to permit a rotation and the surface self-hardens at a desired peak state (**fig.1**). It is a dimensional increase in the structural elements[6] from linear to planar and- where the physical pivot connections between elements are self-controlled. These apparent "point-to-point" connections, reminiscent of Fuller's "jitterbug" [22], permit a continuous flexible surface to be manipulated into regular and irregular building envelopes, skins and structures. Though these surfaces can be constructed from pliable materials, it is surprising that a rigid material like a metal sheet has produced the most promising results for larger-scale structures. Experiments in other materials revealed that the "point-to-point" connections are mirco-folds (Postscript 5).

The process starts with a flat rigid metal sheet which is laser-cut with a slit pattern to weaken it, force is applied to expand the slits, the expansion changes the geometry of the surface while making it stronger and more rigid. Space, curvature, porosity, lightness and strength are produced at no additional cost during the process of forming [23]. Something from nothing? Nearly so. Openings, like the other features mentioned, are emergent and are an integral part of the structural surface. Remarkably, since no new material is added, the same amount of material is being re-arranged on a larger surface. This makes the structure less dense (lighter per

[5] This description is misleading since the "point" has a size which we can vary and it has rigidity of the metal as well which gives it structural integrity. So, in a strict mathematical sense, it is not "point-to-point" which implies a zero-size connection, a freely moving hinge which will not have the structural integrity these pieces have. The micro-geometry of the connections reveals that the point connections are micro-line connections which changes its topology.

[6] The natural next dimensional increase will involve 3D elements, e.g. a cube or another polyhedron, that are connected point-to-point or edge-to-edge. Besides enabling a variable spatial density, this is also a model for expansion of physical space where the units of space (3D cells) are quantized. The latter ties to concepts in quantum gravity where space is not continuous but quantized into discrete units at the Planck scale; see, for example, Lee Smolin [21].

1
Early experiments (1998-2003) in shaping flat surfaces into domical forms in different sizes ranging from 1 foot to 4 feet and different mesh sizes from 1/4 square inches to 1.

unit area) since the same amount of material is being re-distributed over a larger area. This provides a way to make structures that are lighter per unit of space as they become bigger. Applied recursively, as in a fractal, the density of structures will continually become less, reaching the limit condition of zero density for infinitely large structures. This is reminiscent of Le Ricolais's "infinite span, zero cross-section" [24], and also Fuller's "ephemeralization" [25], both in the context of building lighter structures to conserve materials and thus planetary resources.[7]

Built Experiments

A selection of examples in this type of form-making are shown. Numerous methods of physical shaping were explored, all requiring external physical force to enable a change from flat to compound curves in three dimensions. The earliest examples were in small size ranging from a few inches to a maximum of 16 feet, all carried out in various increments, in overall size, size of mesh unit and thickness of metal (**fig.1**). The goal was to build larger structures and explore different geometries. Since the width of available metal sheets was constrained to a fixed size, between 4 feet and 6 feet wide, it required inventing ways to seam the sheets together so when force was applied, the expansion was smooth and produced a seamless surface without wrinkles (**figs.2-5**). It was remarkable that a continuously seamless surface was achieved in our first attempts. This was enabled by a detail connection invented by Milgo. Fabricating large-scale rigid surfaces from metal sheets seemed possible and had become a tailoring problem which was easily solved, something that was not obvious when we started. The thickest we attempted was 1/4 inch stainless steel, an unforgiving thickness at the small sizes we were dealing with, but surprisingly it too expanded very smoothly and continuously. This was also unknown. In all these experiments we found ourselves working on the boundary between the known and unknown. Each new experiment was a surprise, no 3D digital models were used or needed, and with each success we were advancing the state of the art.

These few experiments provided proof-of-concept that relatively large structures on architectural scale could be built economically, and they would be light, porous, and will hold their shape without external elements except for stiff boundaries.[8] Self-rigidized, structural, metal surfaces could be built on a large scale using the counter-intuitive principle of weakening a rigid material and strengthening by expanding it into curved morphologies to make architectural and sculptural surfaces (**figs.6-8**). **Fig.6** is an example of a 7-directional expanded surface based on a non-periodic tiling. **Figs.7** and **8** are close-ups of 4ft diameter

---

[7] The two examples of Le Ricolais and Fuller, along with Le Corbusier's *Modulor* [26] and Constantinos Doxiadis's "logarithmic scale" [27], are the first examples of fractals in 20th century architecture introduced independently of mathematics and computation and were proposed before fractals captured popular imagination through Mandelbrot's book [28].

[8] Our goal of making free boundaries (without adding boundary frames) was not fully achieved but promising beginnings were made in successively building modules having curved edges that expanded with the surface (See. Figs. 9 bottom, 12 and 13). Even here, the free edge required a boundary stiffening, a curved beam reminiscent of Giorgini's work who first proposed it [3]. In the expanded surfaces, this will require assimilation of portions of undulated ripples in fig.15.

2
A larger domical surface (2006), 8ft diameter, showed that metal sheets could be stitched together and form seamlessly into a continuous surface. This was not obvious and worked on the first attempt.

3
A larger elliptical dome (2006), 16ft by 8ft, provided a further proof-of-concept that 4ft wide metal sheets could be tailored together to make smooth architectural surfaces.

4
Steps in the fabrication of one
asymmetric large piece within the factory
using 3-directional expanded metal
(2006-2007).

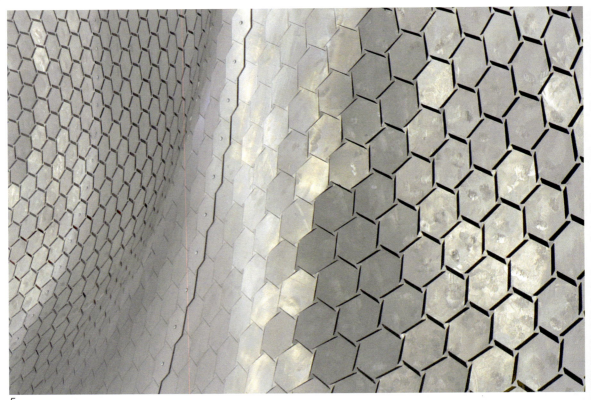

5
Close-up and finished piece from fig.4.

6
Close-up of another sculpture *SUNBURST* (2009) with a multi-directional expansion.

Coding, Shaping, Making

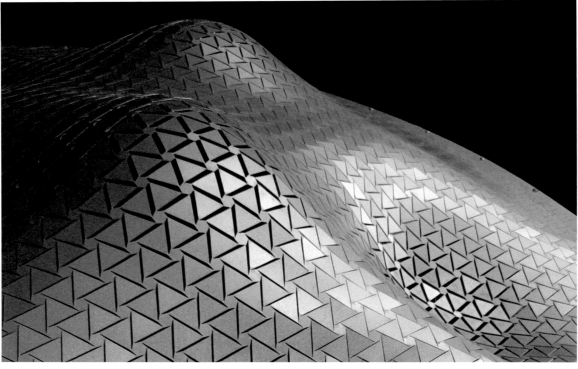

7
Close-up of two sculptures EROS and AMUN (2008), both with different geometries of a 3-way expansion.

Form Follows Force 141

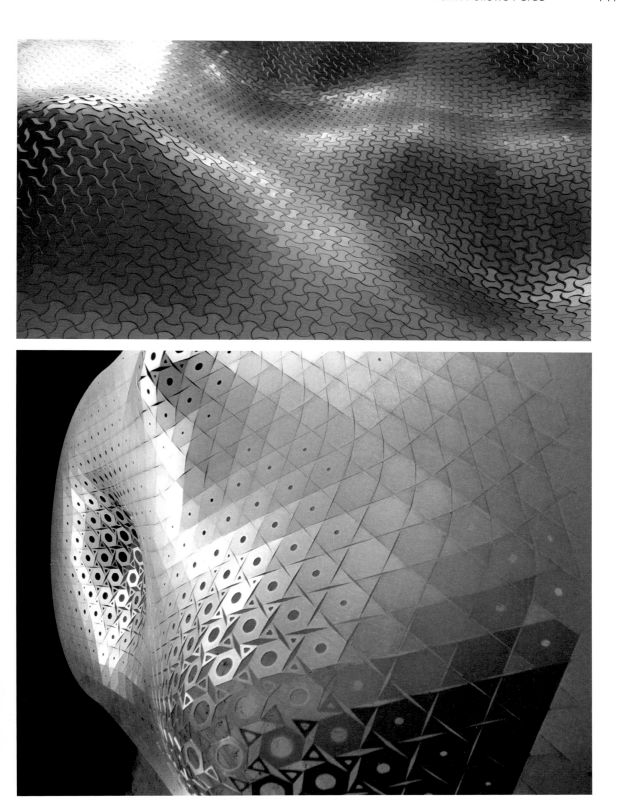

8
Close-up of two additional sculptures, PHOBOS (2008), and large wall piece (2013).

circular pieces, except the last one (**fig.8, bottom**) which is a maquette of a 20 feet wide wall sculpture. The sculptural surfaces in the last two images were the most revealing, also the most surprising. We had discovered a way to make curved asymmetric surfaces using force controlled by the artist, reminiscent of how Jackson Pollack invented his drip paintings by directing a natural process.

There were several additional experiments, all directed towards enlarging the formal vocabulary of physically shaping spaces and surfaces and making structures from small to large architectural scales based on new forming methods. Panelization of expanded surfaces to make architectural products in sheet metal (wall or ceiling panels, for example) required additional innovations especially in the design of panel edges which ranged from straight to curved (**fig.9**). A more standard ceiling system, with 2 feet wide drop-in panels within a standard suspended grid, was one application for producing undulated ceiling surfaces (**fig.10**). Panels with scalloped cross-sections (**fig.11**) to design stronger structures from the same amount of material were achieved. Not surprisingly, they began to resemble fruit skins[9] and carapaces[10] in nature where material economy is presumed to be a guiding principle. This was biomimicry in reverse, the possible discovery of how nature builds. The fundamental struggle between strength and size was in play. The principle of similitude, which according to Thompson determined how big the dinosaurs could be [29], posed a constraint due to the exponential differences in size (length which determines surface area, hence strength) and weight (which determines volume, hence weight). This suggested there would be an upper limit to the sizes we could build.[11]

Surfaces that define or enclose space were the most difficult ones to realize physically. Not many experiments were undertaken due to the more complex fabrication requirements except for some proof-of-concept studies with simple closed or semi-closed forms (**figs 12**). A simple example of an expandable minimal surface that can be extended to make a continuous branched surface was possible and a single module was constructed (**fig.13**). These and other experiments not shown here suggested that larger structures, building facades, skins, bridges and structural envelopes for architecture could be possible.

The opportunity of building these "live" in the presence of an audience did not present itself. When one sees these structures being built in real-time, one realizes the "magic" of a hard material such as metal behaving as if it were a fluid, changing its shape continuously into unexpected curved surfaces, and rigidizing itself in the process. Magic is here used to describe any unexplained technology to paraphrase the late futurist Arthur C. Clarke [30].

No 3D digital models were needed, there was no theory to guide us, it was all empirical work guided by intuition and combined with the joy of making new structures which pushed the bounds of the building art. All it took was some flat

---

[9] Bumpy fruits like custard apple (*Annona squamosa*), Osage orange (*Maclura pomifera*) or raspberry, and vegetables like gourds, pumpkins and squash are "spherical" versions of the "planar" bumpy surfaces in fig.11.

[10] For example, beetle shell e.g. *Brachygnathus angusticollis* or *Dhysores rhodesianus*, both with curved "vaults" oriented along its length similar to fig.11 (top), or the bumps on the carapace of the Red Rock crab which is a smoother version of fig.11 (bottom).

[11] No structural analysis of any of these structures has been done to know how well they perform compared to others. They appear to be light and strong, and it has challenged us to make larger-span structures with greater thickness of material. This remains to be tested.

Form Follows Force   143

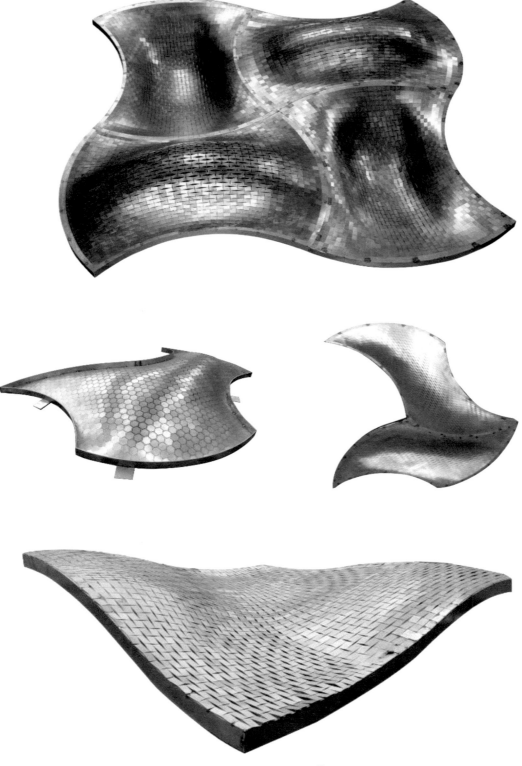

9
Various panel systems (2006-08), different expanded modules with curved boundaries.

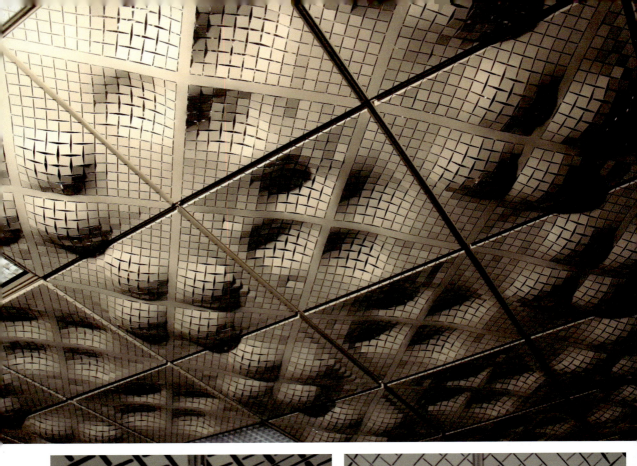
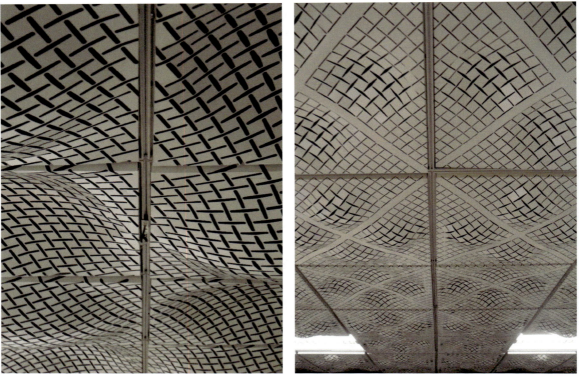

10
Different ceiling systems installed at Milgo-Bufkin (2009).

11
Various expanded surfaces with scalloped sections (2008-09).

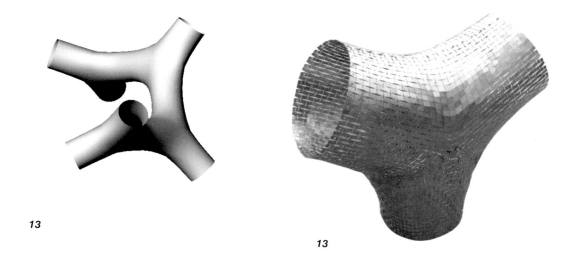
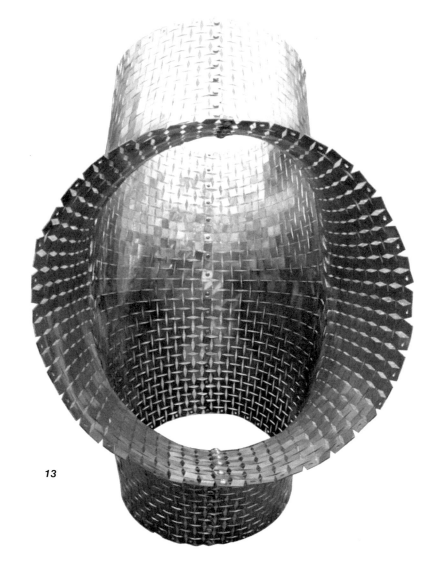

12 (*previous page*) Examples of an expanded hyperboloid (2003), unduloid (2008) and an expanded sphere (2007-08).

13 A unit of an infinitely extendible branched minimal surface (2007).

sheets of metal, force, and a group of adventurous metal workers[12] with considerable expertise, confidence and willingness to experiment with the author within a factory environment where this type of open-ended experimentation was even possible.[13] Seventeen years of author's collaboration (1997-2014) with Milgo is probably unprecedented in the history of artists, designers, architects working with an industry partner. For those involved, it produced memorable works.

Gravitational Rotational Forming

Among the many forming methods we tried in making these structures, one particularly interesting one is described here. It relates to Einstein's theory of general relativity according to which gravity bends (curves) physical space[14] [31]. One of the most popular visualization of bending physical space is by placing Earth in the center of a flexible 2D fabric type grid which bends it into a continuously curved grid. This curved grid represents a slice through 3D space curved by Earth's mass. In one of our experiments in the factory, we started with our standard 4 feet circular diameter flat steel sheet, 1/32nd inch thick in this instance, and rolled a bowling ball starting from the outer edge. The ball founds its way to the center and after repeated cycles the surface became increasingly domical (**fig.14**). This was the invention of *gravitational rotational forming*, a new way to shape surfaces using gravity as a dynamic force, different from the way Gaudi and Otto had used. A weakened metal surface had gradually become stronger and rigid by a repeated action of continually moving weight. We were seeing a rheologic[15] process in real-time, in a material like metal which doesn't flow without being heated. It was also a new method of cold-forming sheet metal.

14
Gravitational Rotational Forming (2007): a rolling bowling ball forms a metal dome (**video 13**). This is a new forming process related to Einstein's general theory of relativity according to which gravity bends space.

 video 13

Many other surfaces, like a rippled wave reminiscent of gravitational waves (**fig.15**) [32], surfaces patterned with lines of forces and shaped like tensile architectural surfaces (**fig.16**) and others were considered using different methods

---

[12] Alex Kveton, Wayne Lapierre, Robert Warzen, Boris Umansky, Tony Pukulinski, Fahan Mohammad, Pauli Mussi, Scott Krissow, Frank Alicea.
[13] Bruce Gitlin's openness to experimentation guided the spirit. It was he who challenged the author in our early days of collaboration to come up with an economical way to make compound surfaces in metal which led to this invention. He dared to increase the thickness to ¼" thick metal sheet accelerating my incremental approach, he also introduced the term "green forming" describing our new methods of building, and the term "gravitational rotational forming" in fig.14.
[14] Artists have used optical and visual tools to represent curved space. See, for example, Escher's use of spherical and hyperbolic spaces for mapping his celebrated day-and-night patterns [33], or Dick Termes use of a 6-point perspective as the visual-spatial structure of his *Termespheres* [34].
[15] *Rheology* has a Greek root, 'rheo' means flow. It is the study of phenomena that embody flow of fluids (water, air, fire, smoke, clouds, viscous liquids like oil or honey or paint, granular substances like sand or grain, molten materials like lava, molten glass during glass-blowing, metal and plastics during casting, etc.).

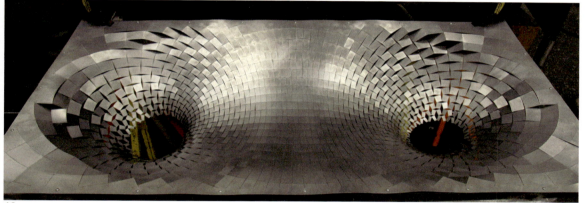

15
A flat surface shaped into an undulating rippled surface (2007).

16
Expanded surfaces following lines of forces (2008).

of forming, all opening up new possibilities. The connection with Einstein's curved space becomes clearer if we map the movement of the surface with all its facets as they morph from flat to curved expanded states. The facets travel through curved space as they rotate and rise from the flat plane.[16]

In *Self-Shaping* (Chapter 6), these ideas will be taken further to show how expanded metal technology can produce tension membrane forms starting from flat sheets. The examples will show that the resulting structures in this process do not need interior "tent poles" as in fabric architecture. The interior space becomes liberated from visible structural elements which obstruct views. In addition, the boundary cables, a requirement in tensioned membrane surfaces, are also gone to free the structures from anchorage to the ground. These two innovations liberate the visible physical elements of tensile fabric architecture but retain its tensioned form.

[16] Tracking the movement of each individual facet as it goes from a flat position to a curved surface and recording the path of each facet as it rotates in 3D space, would provide a trace of the movement of each facet in time. The collective trace forms of all facets would define a curved space. This is the invisible space being traversed as our flat steel sheets become curved. A neat example of curved space-time in action.

Adaptive Morphologies

This body of work began with one material, metal, in the context of a metal fabrication factory as the laboratory for experimentation. New inventions in manipulating sheets of metal advanced the technology of making and also resulted in a new aesthetics of surfaces. However, one obvious feature of expanding surfaces, a dynamically changing form, was prohibited in sheet metal due to its brittleness. Alternatives like pliable and stretched materials require an external structure and are a departure from surface as structure, a harder challenge. A deployable solution, comprising individual rigid parts connected by flexible joints needs further exploration. A flexible but rigid structural surface that can expand and contract, switching from hard to soft reversibly and repeatedly, requires a material we haven't found.[17] Some design concepts for which physical solutions (materials, joints, control, etc.) need to be developed are presented (**figs.17-19**).

Tubular constructions in expanded metal (**fig.17**, *top*) suggested a dynamically undulating structure. This led to the *PERI-TUBE* (**fig.17**, *bottom*) shown in a sequence of images from an animation. It is an actuated 3D conveyor system based on peristaltic motion and provides a new way to transport matter at different scales. Possible applications include transporting grain or fluids, delivering drugs inside a body, a prosthetic intestine, or vehicular transport for humans. It can transport objects larger than the unexpanded diameter of tube. Antoine de Saint-Exupéry's image of the boa constrictor swallowing the elephant in *The Little Prince* [35] comes to mind. Undulating surfaces, with coordinated actuated joints combined with directed movement, may bring us closer to the fairy tale "magic carpet" in another childhood story *Alladin* [36] (**fig.18**, *top*). A pulsating icosahedron-becoming-a-sphere (bottom), a homage to Fuller's *Cloud Nine* spheres [22] that drift in space with internal temperature changes, could rise or fall as it expands or contracts and changes its volume without changing its weight.

Self-rigidization to Self-shaping

The built examples shown introduced the principle of *self-rigidization* as an important feature of large-scale rigid surface structures constructed from flat metal sheets using force. In nature, large sizes are achieved by mass. The tallest redwood (*Sequioa sempervirens*) or the longest blue whale (*Balaenoptera musculus*) have large masses and are not surface structures. Tallest bamboo (*Dendrocalamus giganteus*) is hollow but segmented and thus strictly not a surface structure. Seedpods (shells of nuts, for example) are examples of self-rigidized surfaces but are small. There are no large-scale precedents of self-rigidized surfaces in nature. The imperative of large-scale in architecture and design continues to force us to invent solutions beyond nature, at least in size if not strength and, arguably, design strategy. This appreciation was a driver for these experiments.

A different forming principle revealed itself while making the sculptures in **figs.7** and **8**. The shapes of the metal surfaces were also *self-adjusting* during the iterative forming process. Here, the shape-shifting aspect was enabled by the artist-designer controlling the force. This brought it closer to *self-shaping* where

---

[17] A relatively rigid plastic which can be modified at the connections to be reversibly pliable similar to a "live hinge" is a possibility but it will require some stabilizing element (or elements) for structural integrity as it changes its form. The other possibilities are new advancements in design of metamaterials.

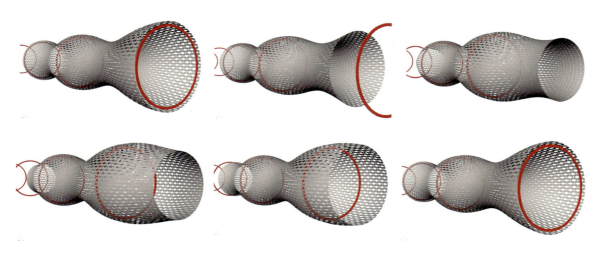

17
Variable tubular constructions (prototype, top (2008). Lower image sequence (digital image of *PERI-TUBE*, 2017; **video 14**) based on peristaltic motion to transport materials/vehicles or objects (represented abstractly by the moving red ring) that are larger than the unexpanded diameter of tube.

video 14

154   Coding, Shaping, Making

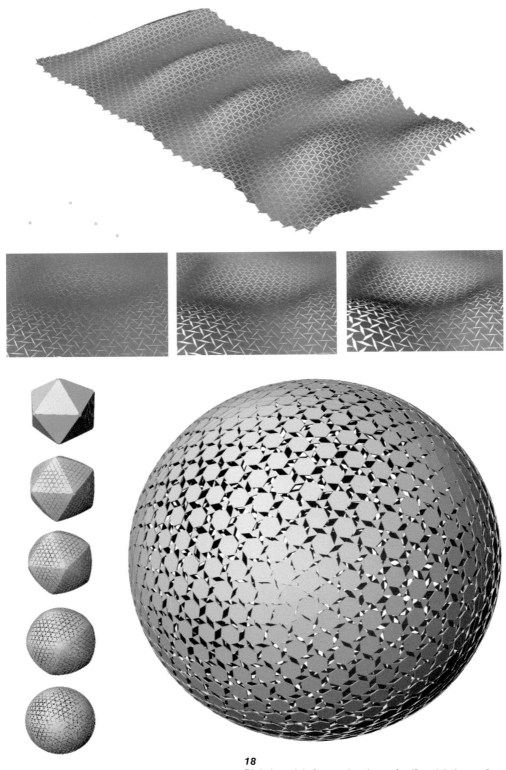

 video 15   video 16

18
Digital models from animations of self-undulating surfaces (2016-17), one inspired by the "magic carpet" (top, **video 15**), the other, a pulsating sphere as a homage to Fuller's Cloud Nine spheres (**video 16**).

the designer is removed from the process and structures shape themselves. The challenges ahead will be to achieve this type of self-shapeshifting on a large-scale, in a controllable manner, or self-controlled in smart versions where shape-shifting is interactive and structures are machines that learn and adapt. With each new experiment we did, our structures were becoming more autonomous. This was heading in the direction of *self-architecture* mentioned in *Meta Architecture* (Chapter 1), a step closer to examples in nature while at the same time removing the designer from the process.

As we head into a future when biology increasingly becomes our technology, new materials and new processes involving encoded matter will be needed to direct assembly, shape, size, organization and material at varying scales. This will require greater collaboration between design artists (artist, designers, architects) and nano-technologists, molecular biologists, synthetic chemists and synthetic biologists. A new field combining STEM fields with Design Arts will be needed.[18] This lateral field is a place where new collaborative discoveries and inventions can rapidly be made to advance the art and science of shaping and making at all scales in all fields.

Binary Continuum

The flexible-rigid continuum exhibited in these structures is part of the broader physical phenomena of soft-hard rheologic continuum exhibited in living and non-living structures. In living systems, structural components grow from soft materials into hardened shells or skeletal parts[19] while non-living natural systems like metamorphic rocks capture flow and movement on a geologic time scale.[20] These two phenomena are similar but have different time scales. Hardened fluid forms in biology are similar to sped up versions of geologic phenomena and geologic flow forms appear to be slow motion growth processes in biology.[21] Our experiments in expanded metal are realized much faster, some structures were built in less than a minute while others took longer. This aspect will be addressed further in *Self-Shaping* (Chapter 6) in the context of "instant architecture".[22]

During these experiments, it became increasingly clear that metal was behaving like a growing material reminiscent of living surfaces in nature. The re-arrangement of the same amount of matter and adding expanding over a larger area was simulating a growing surface and producing emergent forms with different properties and new morphologic features by the use of force. The experiments also

---

[18] This is different from STEAM where A for Art (art, design, architecture) is an add-on to STEM and not an equal.
[19] In the rapidly developing science of Evo Devo (evolutionary developmental biology) [37], one of the great surprise has been the discovery of Hox genes that shape "body plans" of different organisms across the animal kingdom (fly, mouse, chicken, snake, human, zebra fish, etc.) using the same tool kit of genes. More interestingly, the same genes also direct materials as soft embryonic cells grow into bone, muscle, tendons and skin in the vertebrates. This provides a prime example of a soft-hard continuum in biology. On a related note, the Evo Devo developments provide a good test case for the validity of a universal morphological genome (lying outside of biology) and will be addressed in Chapter 8.
[20] Examples range from ripple marks preserved in sandstone, swirls in goggotes, lava flows to accretion growth of stromatolites.
[21] See, for example, Theodor Schwenk's collection of beautiful photographs of flowing forms (water, smoke, clouds, fluid jets, etc.) come to mind [38].
[22] The term "instant" is misleading. It refers to one of our structures which, correcting for stops and starts during the forming process, rose to nearly 12ft in just under a minute.

displayed an interplay between bottom up and top down phenomena. Expanded structures, like the ones shown here, exemplify top down emergence based on physical forces in contrast with bottom up emergence as in rule-based systems like cellular automata [39][23] or L-systems [40].[24] In these expanded surfaces, the genetic component is in the shape of the individual units of the surface and how they repeat while the epigenetic component is in the force that is applied to determine the overall form.

More importantly, this one example shows a continuum between two binary phenomena. A continuum between a binary pair is represented by a line joining 0 and 1 with the line subdivided into decimal places (continuous real numbers) between 0 and 1 [41, 42].[25]

This representation of a continuum works for all binary pairs involved in physical form: rigid-flexible, soft-hard, light-heavy, transparent-opaque, quick-slow, void-solid, wet-dry, hot-cold, smooth-rough, and so on. All binary continua can be overlaid in a composite diagram, a star with outer points defined by the vertexes of a simplex, a higher dimensional version of a triangle [43].[26]

This diagram is the seed of an epigenetic continuum, a universe of all calibrated states of epigenetic phenomena. It can be extended in the same way form is extended as a continuum in the *Morphoverse* (Chapter 8). Genetics of a continuum of form combined with a continuum of physical processes defined by the binary pairs in epigenetic phenomena capture physical morphogenesis. This combines Form and Process within one conceptual space. Adding Material completes the Form-Process-Material triad which underpins physical morphogenesis.

This chapter has shown new experimental structures in sheet metal by deploying force as a new forming technology to build surface structures at an architectural scale. How large can these become remains an open question. The forming process of weakening with laser-cut slits and expanding to make stronger curved surfaces through one forming process exemplifies a rheological continuum from flexible to rigid. A weaker surface gradually becomes harder as it is shaped at room temperature without changing the state of the material. In addition, the material increases its area and covered volume. The increasing area of the expanding surface models a growing surface in living systems and heads towards self-shaping which will be described in the next chapter. In addition, the continuous process from a flexible surface to a rigid one is also a model for a continuum between two binary states. It provides a special case which can be applied to all binary continua involved in making and shaping of physical structures. These can be mapped in an epigenetic universe which combined with the morphological universe defines the space of physical form in natural and human made constructions.

---

[23] Wolfram's cellular automata Rule 30 generates the irregular pattern similar to pigmentation on the sea shell *Conus textile*.

[24] See the work by P. Prusinkiewicz [40] and colleagues on Algorithmic Botany at University of Calgary using L-systems which are named after Lindenmeyer's pioneering work [7].

[25] In the author's work [12], structures were indicated by binary codes (combinations of 0's and 1's) and placed on vertices of higher dimensional cubes or hypercubes. Their continuous transformations along edges and other paths in the hypercubes were suggested by a continuum between 0's and 1's.

[26] This is a higher dimensional analog of a triangle overlaid with a reversed triangle in a Star of David formation. To be more precise, there are two overlaid simplexes in the diagram, one for each of the two opposites.

NOTES/REFERENCES

1. Katavolos, William (1962) *Organics*, [Hilversum] Steendrukkerij de Jong & Co.

2. Giorgini, Vittorio (1995) *Spatiology: The Morphology of the Natural Sciences in Architecture and Design*, L'Arca Edizioni.

3. Doernach, Rudolf, https://de.wikipedia.org/wiki/Rudolf_Doernach

4. Hilbertz, Wolf, https://en.wikipedia.org/wiki/Wolf_Hilbertz

5. Lalvani, Haresh (1973) Towards Automorphogenesis, Building with Bacteria (unpublished).

6. Johansen, John (2002) *Nanoarchitecture: A New Species of Architecture*, Princeton Architectural Press.

7. Lindenmeyer, Aristid (1968) Mathematical Models for Cellular Interactions in Development. I. Filaments with One-sided Inputs, *Journal of Theoretical Biology*, 18, 280.299.

8. Holland, John (1992) Genetic Algorithms, *Scientific American*, Vol. 267, No. 1 (July), pp. 66-73.

9. See, for example, Karl Chu (2006) Metaphysics of Genetic Architecture and Computation, In: *Programming Cultures*, Guest Ed. Michael Silver, *AD*, July/August 2006, Wiley-Academy.

10. Lalvani, Haresh (2010) *A Morphological Genome for Design Applications*, US Patent 7,805,387 B2, Sept 28.

11. Lalvani, Haresh (2018) *Morphological Universe: Genetics and Epigenetics in Form-making*, *Symmetry: Culture and Science*, Vol.19, No.1, Symmetrion.

12. Lalvani, Haresh (1981) *Multi-Dimensional Periodic Arrangements of Transforming Space Structures*, Ph.D. Thesis, University of Pennsylvania; published by University Microfilms, Ann Arbor, Michigan, 1982.

13. Lalvani, Haresh (1982) *Structures on Hyper-Structures*, Lalvani, New York.

14. Ref [11], pages 166-220.

15. Lalvani, Haresh (2021) Expanded Surfaces, Milgo Experiment 2 (1998-2010), In: S.A. Behnejad, G.A.R. Parke, O.A. Samavedi (eds), *Proceedings of the IASS Annual Symposium 2020-21 and the 7th International Conference on Spatial Structures*, 23-27 August, 2021, Guildford, UK, pp.3163-3176.

16. Lalvani, Haresh (2018), *Shaping and Self-Shaping*, Lalvani Studio, NY.

17. Lalvani, Haresh (2011) Multi-Directional And Variably Expanded Sheet Material, U.S.Patent 8,084,117, Dec 27. Filed Nov 29, 2005 https://patents.google.com/patent/US8084117B2/en

18. Bukhari, Ahmed I. and Ljungquist, Eds. (1976) Abstracts, Bacteriophage Meetings, Cold Spring Harbor Laboratory, NY, Aug. 24-29, front and back covers and p.ii.

19. Tomlow, J. (1989), The Model, *IL 34*, Institute for Lightweight Structures, University of Stuttgart.

20. Hennicke, Jurgen. (1974), Grid shells, *IL 10*, Institute for Lightweight Structures, University of Stuttgart.

21. Smolin, Lee (2017/2001), *Three Roads to Quantum Gravity*, Basic Books, 3rd edition.

22. Marks, Robert W. (1973), The Dymaxion World of Buckminster Fuller, Doubleday Anchor Books.

23. Studstill, Kyle (2010) Genomic Architecture Built from Simple Codes of the Natural World, November 8, *PSKF*.

24. Le Ricolais, Robert (1973) In: VIA 2, *Structures, Implicit and Explicit*, Eds. Bryan, J. and Sauer, R., University of Pennsylvania, PA.

25	Fuller, Buckminster R. (1975), *Synergetics: Explorations in the Geometry of Thinking*, MacMillan. https://fullerfuture.files.wordpress.com/2013/01/buckminsterfuller-synergetics.pdf	
26	Le Corbusier (2004, First published in two volumes in 1954 and 1958), *The Modulor: A Harmonious Measure to the Human Scale*. Basel & Boston: Birkhäuser.	
27	Doxiadis, Constantinos (1968), *Ekistics: An Introduction to the Science of Human Settlements*, New York: Oxford University Press.	
28	Mandelbrot, Benoit (1977), *The Fractal Geometry of Nature*, W.H. Freeman.	
29	Thompson, D'Arcy W. (1945) *On Growth and Form,* Cambridge University Press. https://archive.org/details/ongrowthform00thom	
30	Clarke, Arthur C. (1973), *Profiles of the Future*, Harper and Row.	
31	Einstein, Albert (1920), *The Special and General Theory*, Digital Reprint, Elegant Ebooks. https://www.ibiblio.org/ebooks/Einstein/Einstein_Relativity.pdf	
32	*Gravitational Waves Detected 100 Years After Einstein's Prediction*, News Release, February 11, 2016. https://www.ligo.calteChapter edu/news/ligo20160211	
33	Locher, J.L, ed. (1982/1992), *M.C. Escher: His Life and Complete Graphic Work*, Abradale Press/Harry Abrams, NY.	
34	Dick Termes's website, https://termespheres.com/	
35	Saint-Exupery, Antoine de (1943), *The Little Prince*, Reynal & Hitchcock (U.S).	
36	*Alladin,* Walt Disney Feature Animation, Walt Disney Pictures, 1992.	
37	Carroll, Sean B. (2005), *Endless Forms Most Beautiful,* The New Science of Evo Devo, W.W. Norton.	
38	Schwenk, Theodore (1976), *Sensitive Chaos: The Creation of Flowing Forms in Water and Air*, Shocken Books, New York.	
39	Wolfram, Stephen (2002), *A New Kind of Science*, Wolfram Media Inc.	
40	Prusinkiewicz, Przemyslaw, Aristid Lindenmeyer (1990), *The Algorithmic Beauty of Plants*, Springer Verlag.	
41	Lalvani, Haresh (1989) Structure and Meta-Structures, *Symmetry of Structure*, International Society for the Inter-Disciplinary Study of Symmetry (Budapest). See Postcript 2 (fig. A1, p.47)	
42	Ref [11], pages 26-28.	
43	Coxeter, H.S.M. (1973), *Regular Polytopes*, Dover edition.	

POSTSCRIPT 5

The XURF invention traces back to kirigami work in early 1970's. The first few paper models of cutting-and-folding flat sheets were handmade by the author in 1972 and were on display at the School for Morphic Research that year. This development was a spin-off from the one example shown to me earlier by Christopher Castelino in a visit to Montreal. Chris had folded a square net with cut-outs. When I informed him about the examples I had developed since, he was surprised to hear the idea could extend to all tessellations. Among the early folded-nets I made, the most surprising was the discovery of Fuller's octet truss from the skewed hexagonal tiling with hexagons removed. It was also the strongest. When I shared it with Bucky during an office visit in 1976, he acknowledged it immediately. In 1974, Uri Shiran made a few models based on what I had done and an abstract for a joint paper[1] was submitted. In 1975, the work was introduced to students at Pratt and we constructed practically all the derivative folded nets from all combinations of polygons removed from regular and semi-regular tessellations. This body of work has remained unpublished since, the image shown here (**fig.A**) was published last year in (see reference [15] in Chapter 5) and is re-published here. A sheet metal re-construction shows the octet truss configuration folded from a flat sheet (**fig.B**).

Another set of related explorations in kirigami related to the same folded-net concept applied to pentagonal tessellations having rhombic gaps. No cuts were needed, the removal of the gaps by joining the four edges of the rhombus led to 3D folded structures. The work on pentagonal tessellations during 1974 has also not been published. These folded studies were published in 1976 (reference [18] in Chapter 5) on the cover of the abstracts of the Bacteriophage Meetings, Cold Spring Harbor Laboratory, New York, and is reproduced here (**fig.C**). The author is grateful to James D. Watson for inviting me to attend the phage meetings at CSH and also for the cover design. At the meetings, I met practically all the leading structural biologists at the time. This gave me a clearer understanding of how molecular and structural biologists were combining reverse-engineering and detective work in solving some of the most interesting design problems of biological structure at that level. The idea of a genetic code of architecture, using form as a starting point, was born during this mix.

Experiments in different materials (**fig.D**) showed the behavior of the slit surface in the expansion process. The smoother surfaces obtained from the fine steel mesh and cheese cloth produced curved edges and deformed the polygon, the steel wire cloth curved the edges in 3D and produced a bumpy surface, the paper cardboard folded and ripped, and a steel sheet with larger uncut portions resisted expansion and also had surface bumps. These experiments demonstrated that the invention required a careful management of the uncut portion in relation to the cut portion to form the smooth surfaces shown in Chapter 4. It also clarified the joints between polygons were deforming the metal. The bumpy surfaces led to an origami folded surface (**fig.E**) as a solution for deformation-free "joints" of these expanded surfaces, an interesting area for further exploration.

---

[1] Haresh Lalvani and Uri Shiran, 'Folded Polyhedra, and dynamic transformations from 2-dimensions to 3-dimensions'. Abstract (June 18, 1975). Submitted to *IASS World Congress on Space Enclosure*, WCOSE-16, July 4-9, 1976, Monreal.

A
Photographs of physical models of folded nets (made of cardboard, dowels and tape by students at Pratt Institute, 1975) based on author's earlier discoveries. The "jitterbug" tiling composed of hexagons and triangles, related to the expanded metal surfaces shown in Chapters 5 and 6, is in the third row from the bottom.

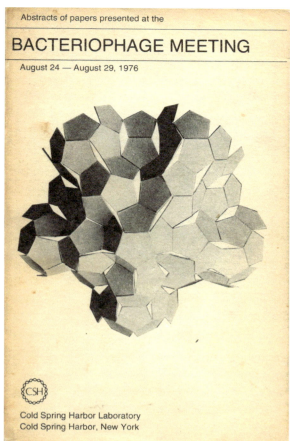

C

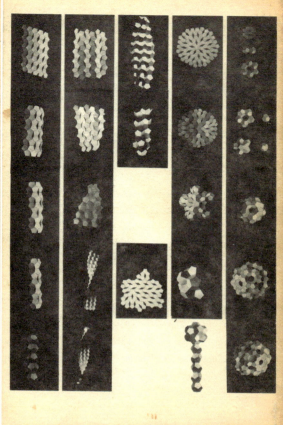

B

D

B
The octet truss folded from a single flat sheet (2007).

C
Front and back covers of the Abstracts, *Bacteriophage Meetings* (1976), CSHL, showing author's models of 3D folded nets from pentagonal tessellations (1972, 1974).

D
Behavior of different sheet materials on a curved surface (2010): *from top, left to right*: fine steel mesh, cheese cloth, steel wire cloth in two views, thin paper cardboard, and steel sheet with longer unslit portions.

E
A metal origami resolution of the unslit portions of the expanded surface (2007).

SELF-SHAPING:
FORM IS PROCESS

"Originality is a return to the origin."
Antonio Gaudi [a] (1852-1926)

The experiment in expanded surfaces described in the essay *Form Follows Force* (Chapter 5) set the stage for the discovery of *self-shaping* where objects began to shape themselves. The direct human involvement in forming processes had begun to recede in these experiments and the intelligence of shaping was gradually shifting from designer to process, culminating in objects shaping themselves under force. The startling results provide a model of morphogenesis of growing surfaces based on *physical emergence* mentioned in the last chapter. With rapid developments in synthetic biology and nano-technology, we are heading in this direction in the future, bringing us a step closer to way nature shapes itself. Besides the huge element of surprise in this discovery while working with metal, passing on the intelligence of design from the designer to objects is philosophically satisfying in view of the unintended impact of our designs on nature. These experiments began with the design of architectural products and gradually morphed into experimental structures and sculptures in order to explore new technologies for building larger structures with less material using economical methods of fabrication.

Two classes of expanded structures are described, the first deals with multi-directional expansion [1] familiar from Chapter 5, the second with parallel expansion familiar from conventional one-directional expanded metal discovered in 1884 by Golding [2] but applied to curved 3D structures. These two classes were named the *XURF*[1] (from eXpanded sURFaces) and *X-STRUCTURES* series, respectively. The publication of works in these two series has been sparse and scattered in catalogs, blogs and reviews of exhibitions [3-13] and were collectively described in [14,15,16] and described further here.

1A-D
PANEL SERIES (2008-09): Three examples (A-C) of expanded metal panels with diagonal restrict bands of un-slit regions; (D) is a side-view of (C) and shows self-shaping, the first example

---

[a] Gaudi, Antonio, https://thequote.art/gaudi-quotes-sagrada-familia/

[1] Milgo-Bufkin's tradename for multi-directional class of expanded metal technologies and products.

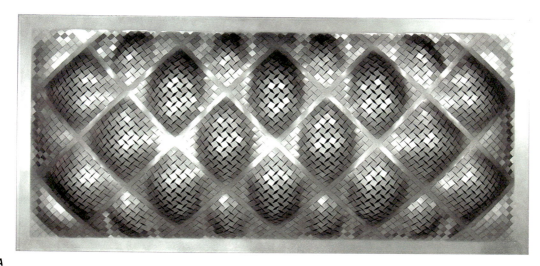

1A

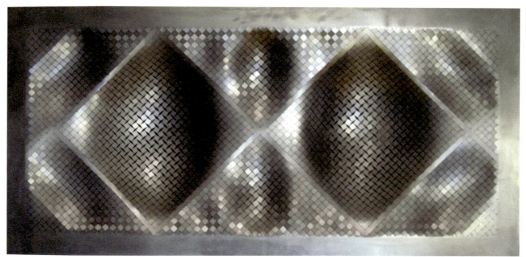

1B

1C

Multi-Directional Expansion (*XURF* Series)

The expansion of space in cosmology and growing systems in biology continue to inspire ideas about 3-dimensional growth in physical systems. The invention of XURF was triggered by the challenge posed to me by Bruce Gitlin who early on asked if I could find a way to make compound curves from sheet metal. The first proof-of-concept of 2-way expansion was developed by the author in 1998 using 1/32 inch steel sheets. Continued development led to the examples described here. While biological growth is entering architecture through use of bacteria, yeast and fungi, in most part as a "living material", the work presented here is a simulation of growth processes and provides a possible model for growth of biological surfaces. Though this remains to be established, it did not start that way. The accidental discovery of *self-shaping* in these experiments steered it in that direction. Some of these discoveries are described through various works on *physical emergence*.

1D

Self-Shaping: Form Is Process

Emergent Folds

PANEL SERIES: A large repertory of architectural panel designs were designed from which three progressive cases are shown in **fig.1A-C**. All examples in this series are rectangular with a square-grid slit pattern turned diagonally; (**A**) has a square grid with half-squares on the boundary, (**B**) has two larger squares, two half-squares on the middle of sides, and portions of squares at corners, all defined by uncut bands, and (**C**) has a large zig-zag restricting band across the panel. The first two (a,b) formed more predictably with bulges within the square regions but the third (c) produced a bumpy surface which is clearer in the side view (**D**). The bumps and valleys in (c,d) were emergent and produced an undulated cross-section. This was not anticipated. No 3D digital model was used here to design the form, the process determined the form. This was a huge surprise since no mold was being used and no heat to soften the metal. A simple cold forming process at factory room temperature using a uniform force applied to the surface. This was the first example of self-shaping. The second set in this series (**fig.2, *top***) had no restricting bands in the surface but had curved boundaries. These too produced a bumpy surface like an undulating wave, also unanticipated. This particular piece evolved into the design of Woo Room ceiling light fixture (**fig.2, *bottom***) where the undulation was confirmed on a scale four times as large (16ft x 8ft). The self-undulation seemed independent of size.

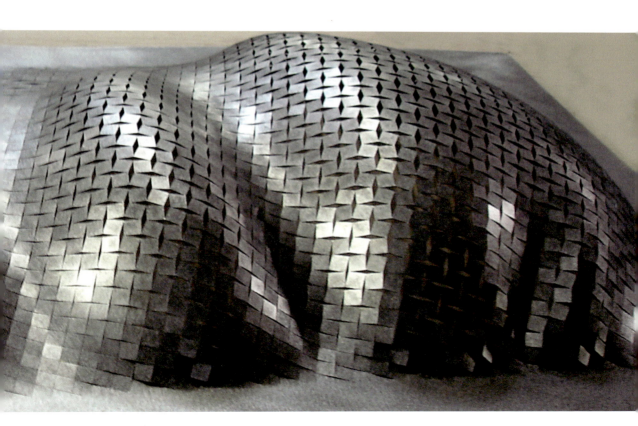

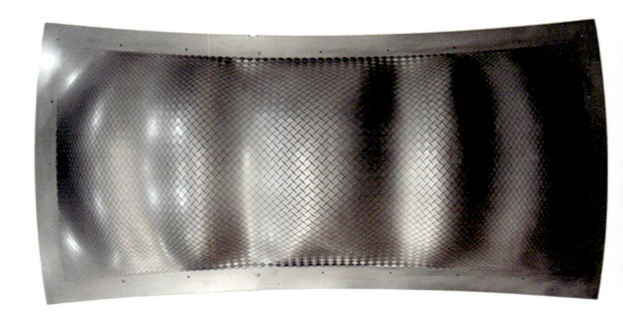

2A

2A-B
Two additional examples of self-shaped surfaces (2008-09), one from the *PANEL SERIES* (2A) without a restricting band, the other (2B) a large ceiling light fixture in the Woo Room, School of Architecture, Pratt Institute (with architect Richard Scherr, 2008-09), based on 2A.

3
CARAPACE SERIES (2008-09). Five more examples of self-shaping with dumbbell and amoeboid-shape plans.

2B

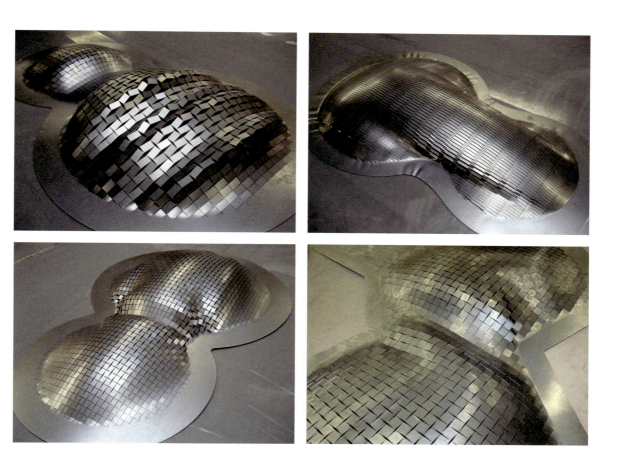

CARAPACE SERIES (**fig.3**): Several designs with dumbbell-shaped and irregular plan geometries were developed in this series. Here too, the appearance of undulations and folds was unanticipated. In contrast with the rectangle-based geometries in fig.1, the concave regions of the dumbbell shape provided a physical constraint which shaped the form. Remarkably, these began to look like beetle shells and carapaces of crabs. In the formed pieces, the curvature of the surface varied according to the location and sizes of openings in the slits. And these openings self-adjusted so that less widths appeared where more strength was needed. The structural elements emerged out of the interplay between expanded and unexpanded regions. This was emergence of structural elements. The unexpanded regions were emergent and were similar to restrict bands in fig.1 which were not needed here. The unexpanded regions were valleys in the undulations and were becoming the "supporting" elements similar to arches and curved beams in architecture. They were emergent structural elements which were denser and stronger than expanded regions which were the "supported" areas between these structural elements. Strength was being produced where it was needed. This was an example of emergent self-stiffening in expanding surfaces.

It was clear that these works displayed *physical emergence*, a counterpoint to *digital emergence* as in rule-based systems like cellular automata or L-systems. The self-undulations were a mystery. Wayne LaPierre, working with the author on the Milgo production end of these pieces at the factory, made a casual remark: "perhaps, the surface has nowhere else to go!" A later experiment, *EMERGENT RIPPLES* (**fig.4**), provided the physical evidence for the self-shaping principle. Wayne's insightful remark was confirmed by this demonstration. Here is the explanation: A vertical pull of an expanding surface produced ripples in a direction perpendicular to the pull. Why? The vertical tensile force initiated the rotation of the facets so the surface began to expand in two directions, horizontal and vertical. Since the horizontal width of the pattern was restricted by top and bottom uncut portions of the metal sheet (the clamped portions), the expansion of surface was restricted in the horizontal direction. Also, a vertical tension produced a horizontal compression across the middle further restricting its expansion. The expanding surface had no choice except to undulate in the third dimension to accommodate the restricted horizontal expansion. Ripples were automatically produced perpendicular to the direction of pulling force.

EMERGENT OPENINGS panel (**fig.5**) displays a different type of self-shaping which produced openings of variable sizes by the same forming process. It has the feel of an organic skin. Openings with varying sizes can correlate with function and structure by enabling variability in rate and amount of matter and energy transported from one side of a surface (inside) to the other (outside).

4
EMERGENT RIPPLES (2009) provides the evidence of self-shaping where folds emerge in a 2-way expanding surface stretched in one direction.

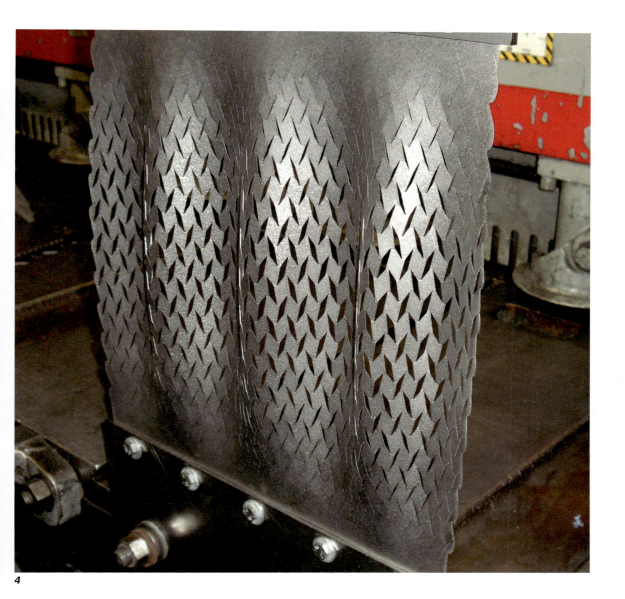

4

Fractured Folds

Another series of works revealed a different class of self-shaping. This was driven by the geometry and topology of the slit pattern which forced the facets to twist unpredictably in 3D space. The topology of the pattern involved odd and even-sided polygons with odd and even numbers of polygons meeting at vertices. This was an example of variable topological valency. An irregular folded surface with openings emerged in *ROUND WHORL* (**figs.6A-B**) and its companion *WHORL*. The base geometry of these two pieces, and some others, is a continuous spiral with evenly spaced points and relates to the *MORPHING PLATTERS* series described earlier (Chapter 4).

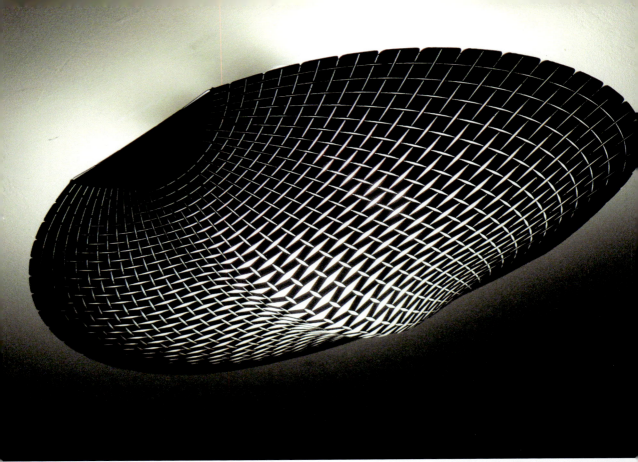

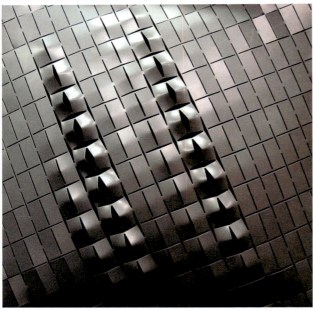

5
EMERGENT OPENINGS panel (2008), another example of self-shaping with variable openings. Top figure is a light fixture with an emergent self-shaped hump.

6A
Four examples, including *ROUND WHORL* (*bottom*, 2010), of a different class of self-shaping where facets twist and turn in 3D space due to the topology of the pattern.

Self-Shaping: Form Is Process 171

6A

6B

6B
ROUND WHORL (2010)
A close-up of the central region in fig.6A.

7
XURF PORTRAITS (2011)
DEBORAH BUCK (*top*) and *RUTH LYNFORD* (*bottom*), an accidental discovery of instant cubism by reflections in a fractured mirror with multiply oriented facets.

Instant Cubism

The dual of Penrose tiling[2] with curved edges in *CALLISTO* (2008) and related mirrored pieces produced a similar twisting and turning of facets as in *WHORL* but, in a mirrored surface, the reflections produced another surprise. It was the discovery of an optical solution to cubism. It fractured an image in a manner reminiscent of cubist artworks as seen in this "Xurf Portrait" of a person standing in front of the artwork (**fig.7**). A transient capture of a fractured image produced by accident. The portraits shown here were taken at author's exhibit *XtraD* [8], Buck House, New York, in front of the piece *HYPERION* (2010) from this series. Though this phenomenon was noticed earlier in the factory, the exhibit setting suggested instant interactive cubist-type portraiture in real time.

---

[2] The Penrose tiling is a non-periodic tiling comprising 36-degree and 72-degree rhombii and named after the mathematical physicist Roger Penrose [17]. In 1981, the author generalized this rhombic tiling as a 2D projection of n-dimensional cubes where the Penrose tiling was n=5 case [18,19]. This will be addressed in Chapter 7. The author also discovered its geometric dual in 1987 [20] by replacing each vertex by a face and each face by a vertex. The CALLISTO and related sculptures are curved variants of this dual tiling.

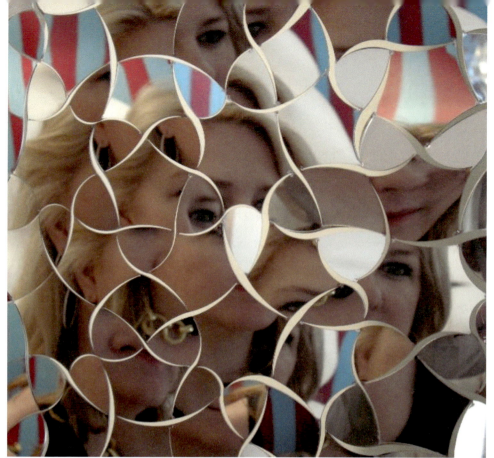

Parallel Expansion (*X-STRUCTURES* Series)

The examples in this section deal with the second class of structures, *X-STRUCTURES* series. *X* represents expanded but also has the double meaning of an unknown quantity as in algebra since none of the *X-STRUCTURES* have been quantitatively analyzed, both in their form and performance. These structures have parallel slit patterns and are closer to the standard expanded metal sheets with the difference that the parallel lines in these are curved. They are either flat surfaces with concentric rings or cylindrical (conical) surfaces with parallel rings in layers. The two are related since concentric rings are also end-caps of closed cylindrical surfaces. The focus on multi-directional expansion had taken precedence in these experiments since it was a generalization of one-directional expansion and was a natural solution to fabricating compound surfaces in a simple and economical way[3] as shown in Chapter 5. From the outset, parallel slitting appeared limited from this standpoint. This was to change with some experiments which revealed different types of self-shaping. They were also relatively easier to fabricate and provided a better opportunity for building larger and lighter structures by re-distributing the same mass over larger areas and covering more 3-dimensional space.

Smooth to Wrinkled, a Continuum

The first experiment with parallel slits began with the *CARAPACE* series shown earlier in **fig.3**, bottom left. It showed emergent folds in the expanding surface emanating from constricted regions of the dumbbell-shaped plan. The second one (**fig.8, *top left***) had concentric rings and a circular plan, 21.5in dia. It quickly grew to a height one and half times its base into a smooth self-shaped tower with a noticeable wrinkle at its apex. This was the birth of the *X-TOWER* series described later. One of these tower structures led to a spontaneous remark from Santiago Calatrava who saw it during a visit to the Milgo factory in 2011: "this could be a building". Though this was an idea that hadn't escaped us, his remark suggests an important aspect of open-ended "pure" explorations where solutions are looking for problems, the opposite of the standard problem-solving method where problems are defined first and solutions follows. The latter is related to the "form follows function" paradigm while an example of the former is a lesson from chemistry where *function follows form.*

The third experiment in this series (**fig.8, *right and bottom***) had a square plan with concentric slit rings that were graded in shape from circle at its center to a square at the boundary. This produced the most startling self-folding in the central region and graded to a smoother surface towards the boundary. This remarkable self-formed feature appeared in several product designs like bowls (**fig.9, *top and bottom left***). The two circular forms (***bottom middle***) displayed a different type of self-shaping. The openings were graded in a spiral formation suggesting the emergence of alternating black

---

[3] Besides the more straightforward solution of a varying cross grid as an exo-skeleton, an interesting spin-off is a deployable version with discrete elements and joints that are activated and stabilized. Or a high-rise version which is progressively jacked up with multiple actuators or erected with the use of an external structure that is raised or lowered.

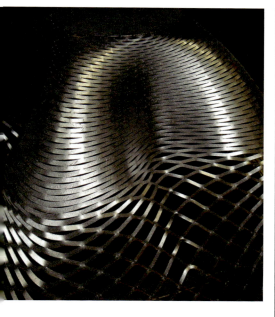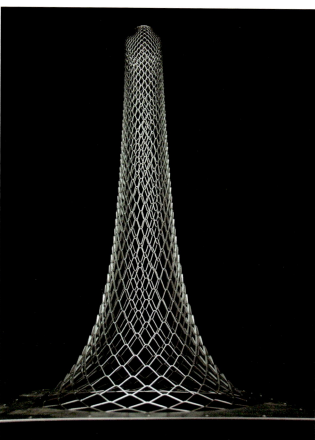

...rallel expansion in circular plans (*top right*) ...f-shaped into a smooth form; in a square plan ...*ght and bottom*) the surface self-folded into ... undulated form which graded from smooth ...rimeter to crumpled center (2009).

and white (lighter-darker, stronger-weaker) spiral bands.[4] This was another surprise not anticipated and revealed only during the forming process. The other two pieces (***bottom right***) showed radial (or circumferential) wrinkling, yet another surprise. These led to a more systematic set of 5 pieces in the *GR FLORA* series (2012) for the Grand Rapids Art Museum exhibit described next.

GR FLORA series (**fig.10**) comprised 5 pieces built using the same amount of material (same area, same 1/8th inch thick stainless steel) but with varying perimeters. Three pieces from this series are shown in **fig.10a**. The series began with a circular boundary (***top***) which gradually morphed continuously towards a square boundary (***bottom***), with the range of intermediate shapes[5] (***middle***). The perimeter increased progressively from the least (circle) to most (square) while the area remained constant. The results were startling. An increase in the perimeter for a fixed plan area led to self-wrinkling (self-folding, self-crumpling) in a surface that was expanding in 3 dimensions. The most wrinkled (**fig.10a, *bottom***) was harder to make, had a lower height and appeared to be much stronger.[6] This indicated a new principle, a surface undulation (crumpling, folding, wrinkling) was produced in an expanding 3D surface that was growing faster than its boundary. This was the opposite of hyperbolic surfaces (corals, some leaves) where the boundary expands faster than area to force the edges to undulate. Though the edges were constrained in a flat plane in these experiments, the principle should apply to non-planar and irregular boundaries. Folds in some leaves and turtle shells came to mind.

Metal as Fluid: The self-wrinkled surface in the *GR FLORA* square piece (**fig.10a, *bottom*; video 17**) shown in a close-up view in **fig.10b *top*** displayed another remarkable phenomenon. A stainless sheet with a slit pattern was behaving like a fluid during the expansion process. Metal at room temperature was flowing under force like a liquid, rapidly crumpling into a complex wavy surface while becoming increasingly rigid. No heat was used to soften the metal or cold to harden it. The sequence of nine images from a video of this piece (**fig.10b, *bottom***) show the emergence of self-folding from a flat sheet. These still sequences are excerpted from real-time forming (**video 18**). The forming process is remarkably beautiful, extremely rapid, the pieces are untouched by human hands. The rheologic process produces the form, *form is process*. In flow of fluids, it is easy to see how process (of forming) and form are the same, but flow is harder to imagine in rigid materials that have not been softened or melted by heat as in this case. This is a step in the direction of how nature designs, where form, function, process and material are one.

---

[4] This suggests the question: Could alternating black-and-white patterns in nature (for example, zebra) be the result of alternating weak and strong regions produced by tension force in an expanding surface or compression force in a shrinking surface?

[5] The intermediates are super-ellipses. The super-ellipse is a transitional shape between a rectangle and an ellipse, the general case of a square transitioning to a circle. Piet Hein coined the term "superellipse" and used it first for a traffic "roundabout" in Stockholm [21].

[6] These are qualitative results. Quantitative results would require laboratory conditions not available in a production environment of a factory.

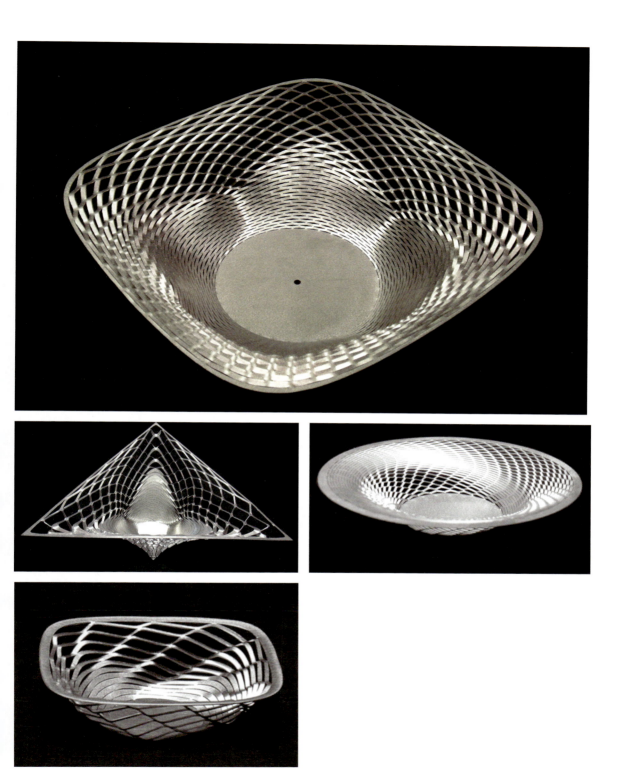

9
Design of self-shaped fruit bowls (2009). The two circular ones (*middle, bottom middle*) with a spiral expansion, and the other two circular ones with an smooth undulated folds (*middle right*).

video 17

10A
GR FLORA Series (2012), 3 of 5 pieces in the series shown. Each piece had the same area but an increasing perimeter. The lowest perimeter was smoothest (*top*), the highest was self-wrinkled (*bottom*) in *GR FLORA* 24 100 2 (**video 17**).

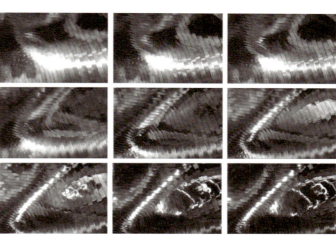

10B
Close-up of emergent self-folding in the
GR FLORA 24 100 2 (*top, middle*) and 9
stills from its real-time forming (*bottom*).
Metal has become fluid at room
temperature (**video 18**).

 video 18

X-TOWERS

The *X-TOWER* series (2009-14) explored the forming of vertical structures from flat sheets by expansion along an axis. Numerous towers from 4ft to 18ft were built to explore formal possibilities and limits of the forming process and material. Our interest in building a 40ft tower from one continuous sheet, albeit stitched together from smaller shaeets due to limited available widths of 4-6ft in sheet metal, hasn't been realized. This was to determine the tallest structure we could build in expanded sheet metal since the increased thickness of metal would present a practical limit to forming tall *X-STRUCTURES* requiring exponentially increasing force. The *X-TOWER* series also demonstrated how self-rigidization of these curved expanded metal structures would eliminate two ubiquitous formal features of tensile architecture built by using cable nets or fabrics, the internal poles (masts) and tie-cables at the ground level or supporting elements. These two opposing ends of fabric surfaces maintain equilibrium, one pushing up and the other pulling down and were not needed in expanded metal structures. The *X-TOWER* series demonstrated that supporting masts in fabric architecture could be eliminated and the self-rigidization would retain the beautiful tension forms. Several examples from this series are shown. Later experiments removed the second feature.

SEQUOIA (18ft tall, 7ft 6in dia, 1/8in thick, 2014) is shown here at Milgo factory along with other tower pieces in the background (**fig.11a**). This is the tallest we have built and rose effortlessly to the highest we could fabricate inside the factory. As the name suggests, the wonder of the Californian redwood trees in this instance, as well as tropical palms and bamboo trees, have been an inspiration for these works. The recent forest fires have posed an existential threat to the majestic redwoods and a tribute to the giant sequoias like General Sherman (Sequoiadendron giganteum), the tallest one at 275ft, is even more timely. *SEQUOIA* was made by seaming 2 flat sheets, but formed flawlessly without wrinkles similar to the larger examples in chapter 5. It rose up to 18ft, expanding to more than three times its surface area. *SEQUOIA 2* at Milgo factory (**fig.11b**) introduced a large opening in the tower for possible entry and exit or provide a way to look inside the tower.

X-TOWER 54.4 (7ft 2in tall, 4ft 6in dia, 2013) is one of the more adventurous pieces in this series and introduced larger openings in expanded surfaces (**fig.12a**). These openings were 3-dimensional and made the structure more cellular in appearance (**fig.12b**). The regions around each opening had multiple bands that were self-formed into a stronger continuous surface. This was another pleasant surprise, a single-band thickness of earlier pieces like *SEQUOIA* had morphed into a multiple-band thickness within one continuous tower surface. The thickened bands were a new way to produce stronger structures from thinner sheets to make 3-dimensional architectural surfaces.

11A
SEQUOIA (2013) at Milgo factory with other tower pieces.

11A

11B
SEQUOIA 2 (2013) at Milgo factory with an opening in the surface.

Self-Shaping: Form Is Process 183

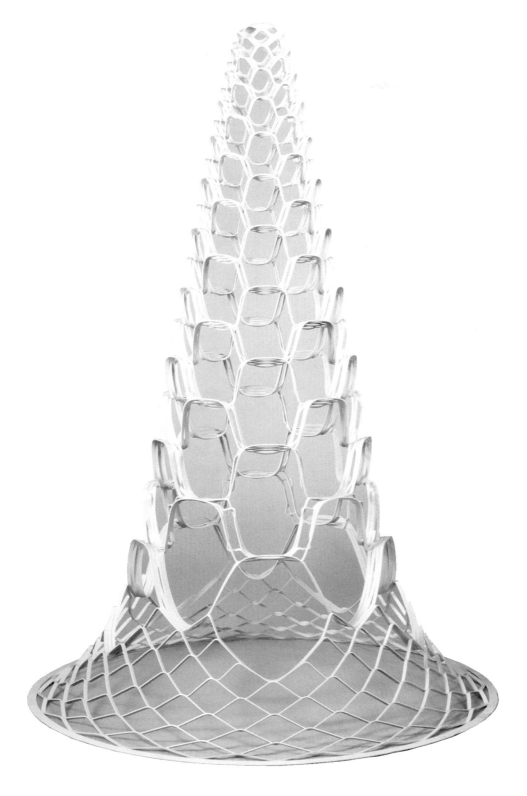

12A
X-TOWER 54.4 (2013), with self-thickened bands around the openings.

12B *(next page)*
Close-up of X-TOWER 54.4.

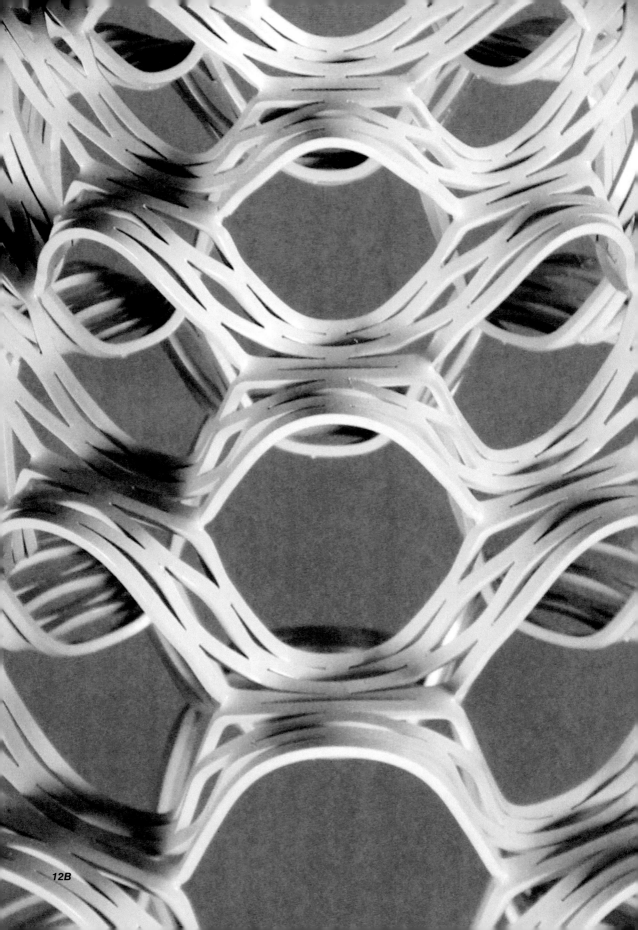

X-TOWER 88.2 (12ft tall, 88in dia, 3/16in thick) commissioned by the Fields Sculpture Park, Omi Arts International Center, Ghent, NY (**fig.13a**) evolved the idea of cellular openings further by exploring their spatial distribution on the tower. Of the many configurations that were explored, continuous spiral openings were selected for this outdoor sculpture. The outdoor setting in a park open to public required a stronger structure and provided the opportunity to work with 3/16in thick stainless steel. With the same footprint as *SEQUOIA*, this tower could only rise to 12 feet within available forming constraints and is the strongest piece in the *X-TOWER* series. The spirally arranged openings are a sequel to the formal features of *X-TOWER 54.4* and the self-shaping of openings is more dramatic and dynamic as seen in the interior view (**fig.13b, right**). The sequence of smaller images (**fig.13b, left**) shows stills of a smaller prototype recording the forming sequence.

The stills from real-time forming of *X-TOWER 88.2* at the factory are shown in **fig.14**. The forming recorded by a Go Pro camera in real-time included all the stops, starts and checks in the forming process which took about 30 min (**video 19**). The video was edited to remove the interruptions to simulate a continuous forming of the tower. It surprised us to learn that the actual forming time for the continuous expansion of this 3/16in steel mesh was under 1 minute. This is extreme rapid-forming and brings us very close to "instant structures" for this size in a hard material like stainless steel.

Additional experiments in this series (**fig.15**) dealt with curved expansion (***top left, top*** and ***middle right***), wavy expansion by joining two curved segments end to end in reverse orientations (***bottom left***), various spiral expansions (***bottom left*** and ***right***). Branched expansions are possible as shown later with some examples. During the process, a collection of *X-TOWERs* (**fig.16**) was made. This family of structures began mirroring the theme of mass customization in Chapters 1-4 and showed a unity in infinitely variable forms obtained from a single process.

Self-Shaping: Form Is Process 187

13B

13A
X-TOWER 88.2 (2014), Fields Sculpture Park, Omi Arts International Center, Ghent, NY.

13B
X-TOWER 88.2, interior view (*left*). Sequence stills of a smaller prototype during the forming process (*right*).

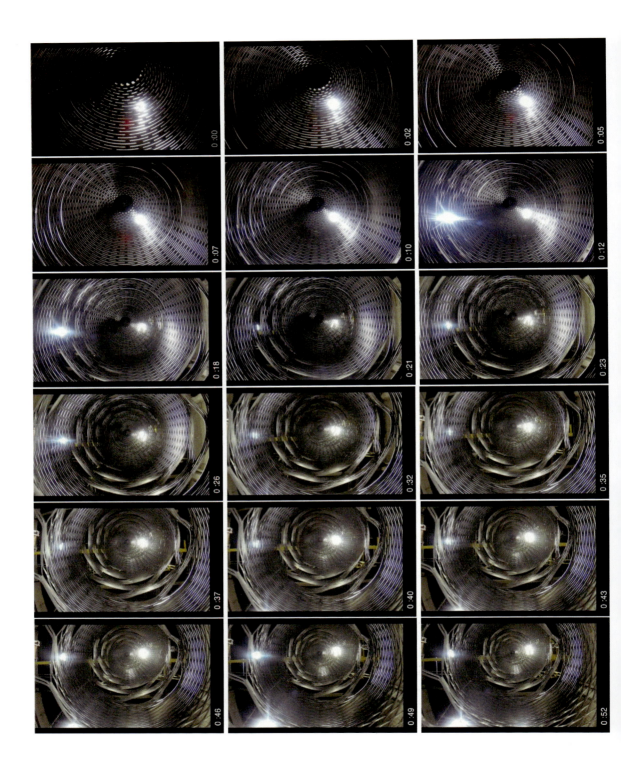

video 19

14
X-TOWER 88.2, 18 stills from real-time forming at Milgo (**video 19**).

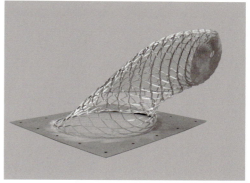

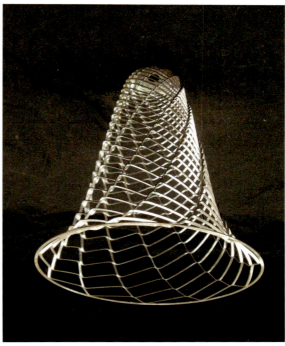

15
Various experiments in curved and spiral expansions (2009).

16 *(next page)*
X-TOWER series (2009-2014) in Milgo's factory.

X-PODS

The still sequence in fig.13B (*left*) showed different stages of an emerging tower from a flat sheet. In the numerous examples we built (**fig.16**), we varied the number and widths of bands and distances between them. Varying these parameters changed the form and a unified morphological model is needed to predict the resulting form. Even in the absence of such a quantitative model, it was clear that the same method of forming was producing surfaces that morphed from flat (planar) to positive (spherical) to negative (hyperbolic) curvature changing from various pod shaped forms to elongated tower-like forms in a continuum. Euclidean and non-Euclidean geometries are a continuum in growing surfaces. A composite photo overlay of various pieces in the *X-STRUCTURE* series (**fig.17**) demonstrates this. The same slit pattern with variation in number, size and location of un-slit regions and widths of slit bands seemed to control curvature of form to produce conical, domical, pseudo-spherical, ellipsoidal and other forms as part of a form continuum. The interplay between spatial distribution of non-expanding components (un-slit parts) and expanding components (slit parts) controlled the curvature of form in these expanding surfaces. Variations in shapes of fruits and subtle shape changes within a single type of fruit came to mind. Pod shaped forms in the *X-POD* series in this shape continuum are described next.

X-POD 138 (7ft tall, 11ft 6in dia, 3/16in thick), (2014) at Architecture Omi, Omi Arts International Center, Ghent, NY (**fig.18**). Bottom right image shows loading at Milgo for transport to Omi. The *X-POD* series was an application of the *X-STRUCTURES* technology to "instant architecture", an idea which lends itself naturally to emergency shelters [12]. Either pre-formed habitats are shipped to the site, or the installation arrives in its 2D state and is formed into 3D on site with the use of special equipment. Pre-formed covering membranes or membranes from stretch materials could be attached before expansion so the finished structure encloses space. *X-POD 138* was conceived as "minimum architecture" which requires one space, one surface and one opening. It is a 3-gene phenomenon, one morphological gene for each of these features (the idea of morphological genes will be described in Chapter 8). If any of these three features is absent we do not have architecture. In the example shown, the ground plane has not been counted as a second surface though mathematicians would call it a dihedron, a 3D shape with two faces like the shape of a pill or a lens, by including it. Conceptually, the ground plane can be seen as a continuation of a single surface, a monohedron (a 3D shape with one face), though practical considerations of fabrication would separate the two.

Several iterations of the *X-POD* were made starting with smaller diameters and scaling up to the full-scale piece shown here. One of the big surprises was the shape of the large opening, a digon, a 2-sided polygon like the shape of our eye. It self-formed when the single elongated slit expanded into an opening and its edge beams at top and bottom emerged during the process (**video 20**). This demonstrated, as in the *CARAPACE* series earlier (**fig.3**), that structural elements like lintels or arches are emergent shape adaptations of boundaries of an expanding opening. This makes you wonder how openings in organisms (like us) emerge from a spherical pre-embryonic stage. A sphere must undergo topological transformations to produce openings that permit the eye, node, ears,

17
X-POD 138 (2014) at Architecture Omi, Omi Arts International Center, Ghent, NY. *Bottom right* image shows loading at Milgo for transport to Omi.

18
A composite photo overlay of various pieces in the *X-STRUCTURES* series that go continuously from flat (Euclidean) to positive (spherical) to negative (hyperbolic) curvature.

Self-Shaping: Form Is Process 193

18

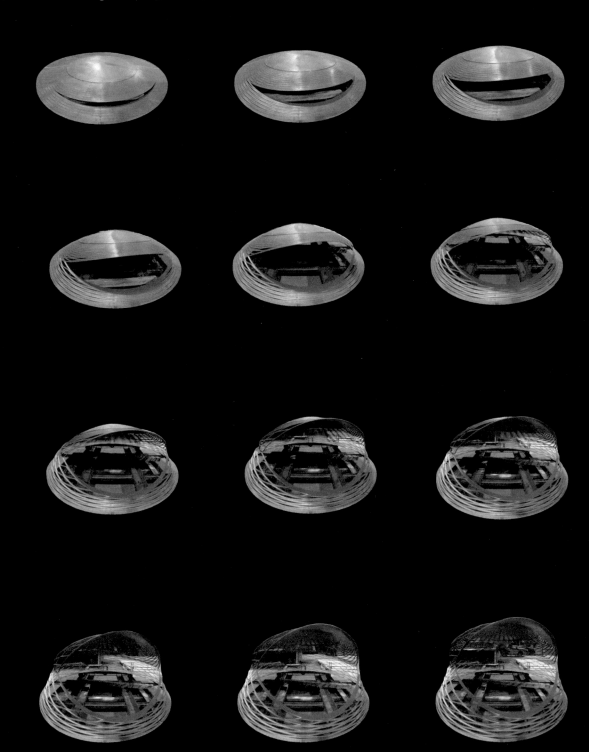

video 20

19
X-POD prototype (2014). 12 stills from real-time forming of a smaller prototype (fig.19; **video 20**). This clearly illustrates the emergence of an opening in the surface by a 3D expansion of a 1D (linear) slit in the surface.

mouth as emergent modifications of a multi-holed donut. Is topological genus an emergent feature in (developmental) biology?

Currently, we are working on a bamboo X-pod structure applying the expansion principles of X-POD138 to an emergency shelter using a regenerative material with a negative carbon footprint.[7] The climate refugee problem caused by hurricanes, coastal flooding and displacement of vulnerable populations from natural or human-made hazards, requires rapid deployment of shelters. Built either on-site or off-site for transport, the concept of "instant structures" mentioned earlier would benefit from rapid forming and deployment. Structures formed by force, in their limit case, form at the speed of force. In gravity-formed structures, this upper limit to forming is the speed of gravity and provides the aspirational goal for ultra-rapid forming of physical structures at an architectural scale.

X-TUBES

The ubiquitous hollow tubes that we see in pipes, furniture, poles, columns and linear elements of larger structures like trusses, grids and space frames, mark an evolutionary shift in building technologies from mass to surface. Mass-active systems (brick, stone, concrete, cast materials, etc.) evolved to surface-active structures (shells, curved surfaces, hollow spheres, cylinders, etc.), leading to higher performance from the same amount of mass re-distributed over a much larger surface or, conversely, reduced mass for the same surface. This theme has been touched on earlier but here we return to it in the context of self-shaped expanded tubular structures which marked a breakthrough in design, construction and fabrication of hollow expanded forms. This is the *X-TUBE* series.

Related to the *X-TOWER* series, *X-TUBEs* have both ends open and appeared to expand the most and shape themselves into stronger concave cylindrical or conical forms. In their vertical orientation, the slits elongated more along the vertical direction making them stronger in compression though no tests were undertaken. This opened up possibilities of spatial and structural modules that could be stacked or joined at angles into lattices, space frames and aggregates. *X-TUBEs*, used as a single element, nested in concentric layers or arrayed as multiples, provide alternatives to rebars in reinforced concrete. On larger scales, they define architectural space. Basic experiments in this series are shown first followed by applications to furniture, architecture and sculpture.

Some of the basic experiments in *X-TUBEs* are shown in **fig.20a-e**. The cylindrical expansions with fixed ends (**fig.20a**) were remarkably easy to expand, reaching about 3-4 times its original height and self-shaped into hyperboloidal forms. Steel was expanding in a way similar to stretch fabric like spandex or lycra. A very elongated version of 20a was the most remarkable, it was the strongest, and suggested that true cylindrical tubes (with constant diameter) could be fabricated this way; its central portion was a uniform

---

[7] This is a project being carried out with students at the Center for Experimental Structures, School of Architecture, Pratt Institute.

columnar form and could be preserved by cutting off the tapered portions. The one in **fig.20b** was another breakthrough. The tie cables in boundaries of membranes as well as interior or exterior supports are gone, a self-rigidized structure can simply sit on the ground and enclose larger open areas. A variant is shown in **fig.20c** and it has self-undulations. These two examples captures the second feature of tension architecture mentioned earlier. A related sinuous cylindrical form (**fig.20d,e**) with two parts joined, each formed from flat sheets like the towers, opened up the repertory of asymmetric curved expanded forms.

Small scale applications to furniture are shown in **figs.21** with three prototypes including a circular table with a hyperboloid base and a wider top (**top**), the slanted *CYCLONE* side table (**middle left**), a lectern in a top view (**middle right**) and two reversed dual formed pieces with a dumbbell plan (**bottom left**) and its close-up (**bottom right**). Digital images of additional designs are shown in **fig.22** and include a continuous minimal surface as an expanded base for a circular table (**top**), two hyperboloids fused in a twist (**bottom left**) and a tripod base for another circular table (**bottom right**). The latter was the only physical experiment in branched expansion. The multiple support prototypes led to arrays of 3D spaces from multiply-expanded columns. Two examples of stills from animations or multi-layered expanding surfaces are shown in **fig.23** using arrays of individual columnar elements (**top**) or it's variant with a continuous surface (**bottom**). These suggested multi-floor building systems (**fig.23**) using expanded mushroom and wine-glass shaped columns which could be expanded on site from flat 2D stacks to 3D architectural structures (**video 21**).

Other examples of applications on a larger scale to architecture are shown in **figs.24** and **25**. A proposal for an outdoor pavilion at the Victoria and Albert Museum, London (**fig.24**), suggests the idea of an "instant architecture" where pre-made component parts are brought from the factory in their 2D state and

20B **20C**

20A-E
X-TUBES self-shape themselves into rigid hyperboloid-like forms (2009-12) shown here with fully constrained ends (*right*), ends constrained at points (*top, right*), and two parts joined to make a tubular structure with an inflected curve (*far right*).

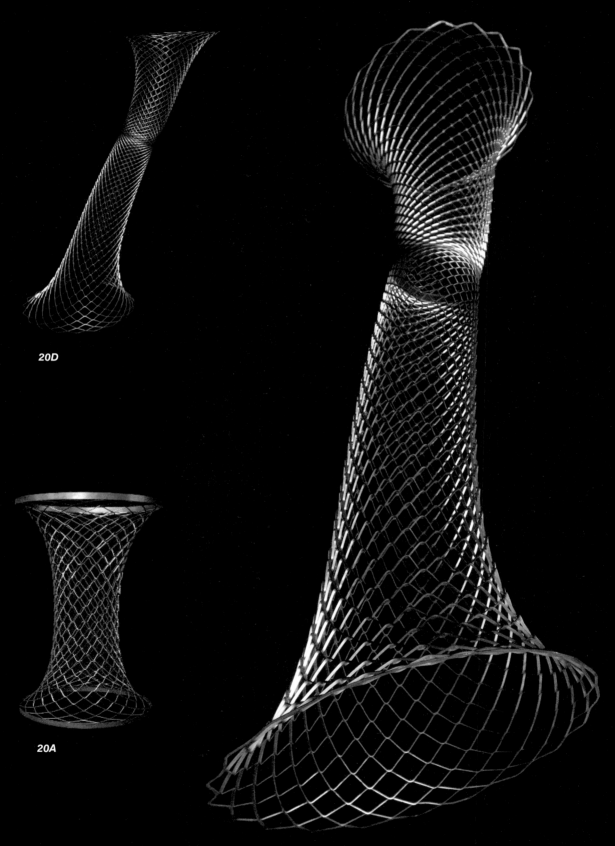

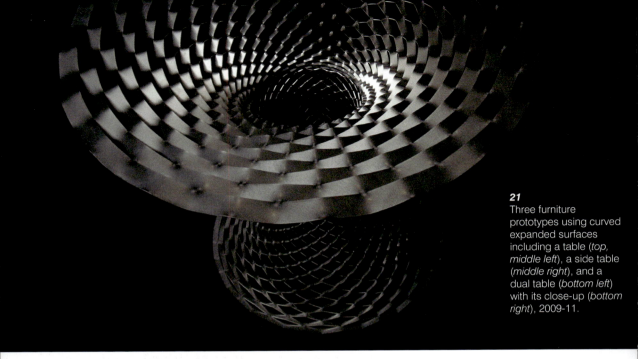
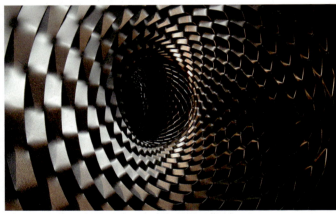

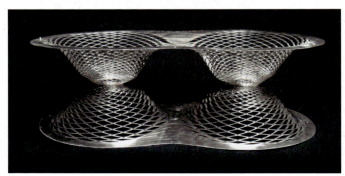

21
Three furniture prototypes using curved expanded surfaces including a table (*top, middle left*), a side table (*middle right*), and a dual table (*bottom left*) with its close-up (*bottom right*), 2009-11.

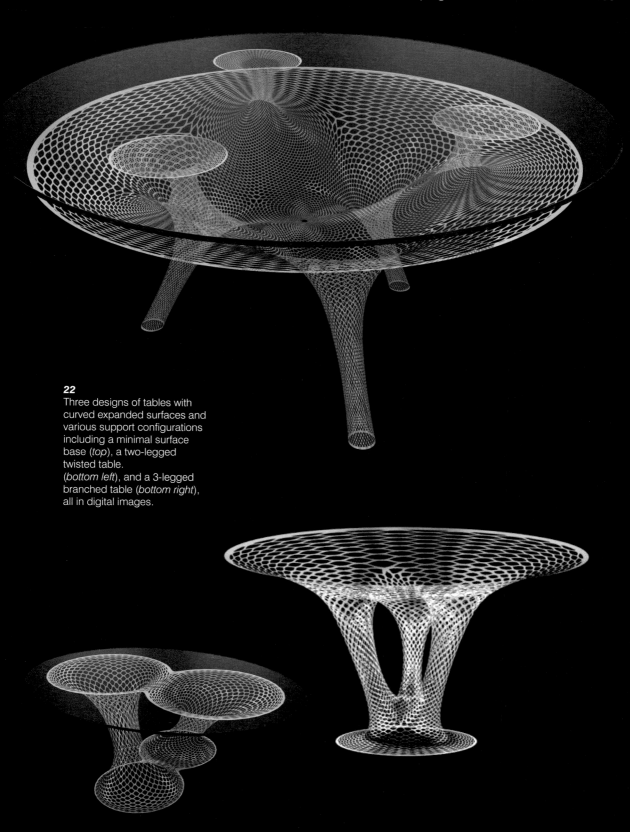

22
Three designs of tables with curved expanded surfaces and various support configurations including a minimal surface base (*top*), a two-legged twisted table.
(*bottom left*), and a 3-legged branched table (*bottom right*), all in digital images.

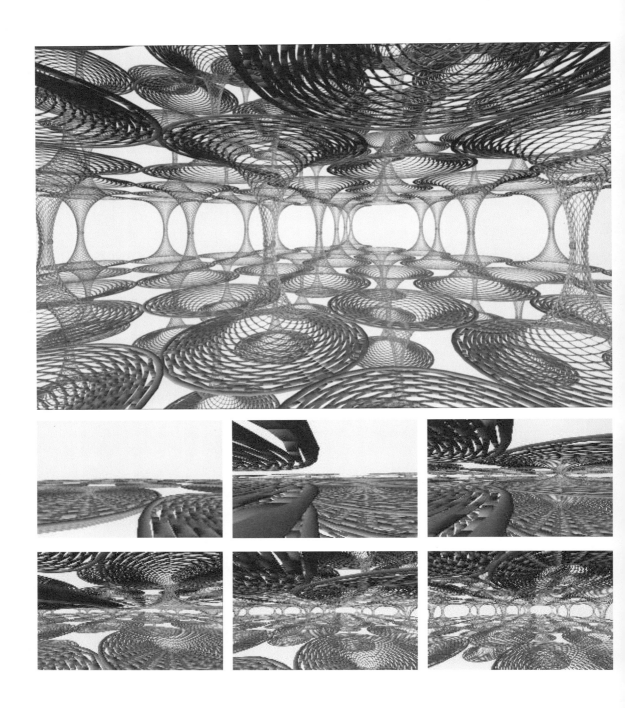

 video 21

23
Digital images of a double-mushroom column system by separating two flat planes using expansion shown in a still (*above*) and a sequence from an animation (*left*) (**video 21**).

Self-Shaping: Form Is Process 201

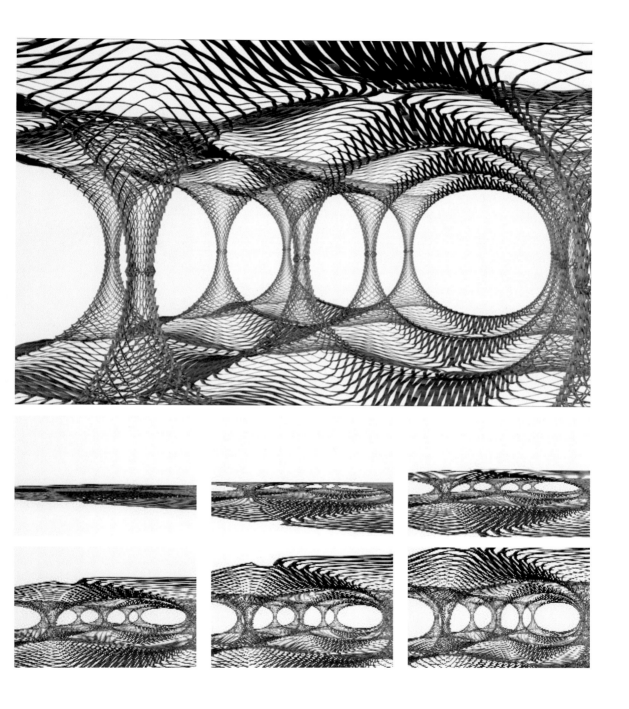

24 *(next page)*
Proposal for V&A Installation (2011),
digital image.

Self-Shaping: Form Is Process

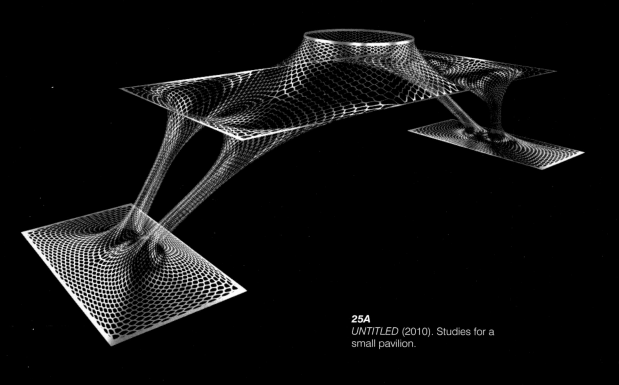

25A
UNTITLED (2010). Studies for a small pavilion.

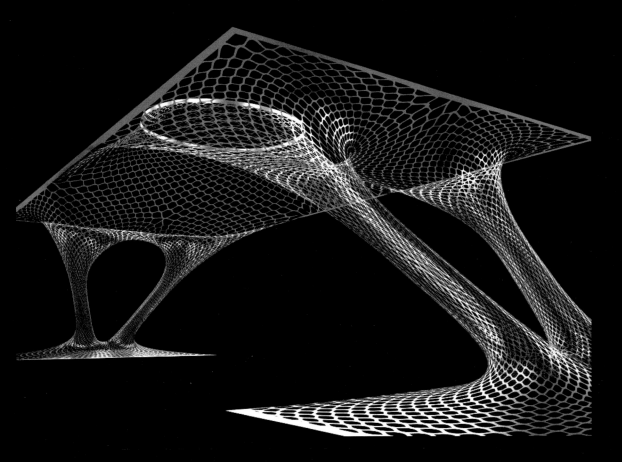

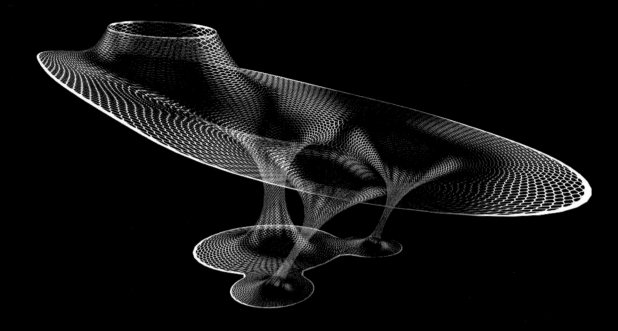

25B
UNTITLED (2010). More studies for a small pavilion.

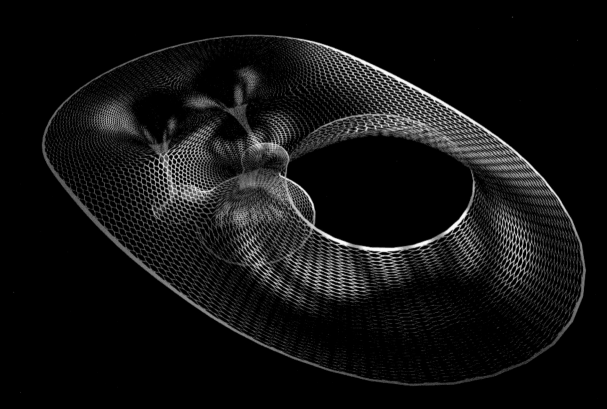

are rapidly expanded into a 3D structure on site. Two examples of studies for small pavilions (**fig.25a,b**) using expanded metal technology to achieve a continuous surface-active structure where supporting elements (columns) and supported elements (roof) have become a seamless whole.

Three examples of large-scale sculptures are shown in **figs 26-28**. *TENT PYRAMID* (**fig.26, *top***), 10ft x 20 ft, is a proposal for a public sculpture where self-shaped tension forms in expanded metal are stacked modules are stacked without internal or external supports. An internal view of a tension-shaped surface without the familiar pole that holds up tents (***bottom left***). A variant with stacked modules constrained by rings on ends of each module (***bottom right***). An example of 5 stacked modules (**fig.27**), each module in one of 10 possible states of rotation encoded by numbers 0 through 9, so 5 combinations of these states produce a distinct combination of 5 numbers corresponding to any zip code of a city. Examples are shown for 15 cities from as many states along with their zip-codes (***bottom***). Top image *ARCH 07201* corresponds Elizabeth, New Jersey, with a zip-code *07201* and is a ceremonial entrance arch at its main New Jersey Transit train station. The proposed installation *WORM HOLE* (**fig.28**) is an expanded hyperboloid form sandwiched between two mirrors and angled to the sky with peek holes on the ground level mirror to capture an infinite tube joining the viewer with the sky.

3D *X-STRUCTURES*

The 3-dimensional version of *X-STRUCTURES*, where the 2D expanded surfaces have become 3D expanded meshes, will be similar to our Algo truss fabricated with Pratt students at Milgo under the NYSTAR project [20] and shown earlier in Chapter 3. A stacked expandable version of this truss, starting from stacked flat layers, will be a good example of a 3D *X-STRUCTURE*. These can be formed in parts and joined, or formed as an instant lattice, a 3D truss with wavy curved elements. Some physical models of these were constructed.

Biomimicry in Reverse

The designs described here and in Chapter 5 produced by expansion processes were highly reminiscent of biological forms and processes. At least four related constructive principles were discovered as part of a broad repertoire of principles underlying morphogenesis of surface forms. These included controlled growth along a preferred direction like an axis, growth of boundary in relation to the interior region of a surface, radial versus circumferential growth, and growth controlled by restrictions at points, along lines or across areas. The results showed that differences in rates of growth in different spatial regions, or in different locations, or along different directions within a growing surface lead to undulations, folds, bumps and wrinkles. Variably expanded metal surfaces were providing a scale-free model for shapes of growing surfaces in living materials. Preliminary unpublished work on these principles suggests that self-shaping morphologies can be mapped in higher dimensional networks in a similar way to the examples described in the Morphoverse in Chapter 8.

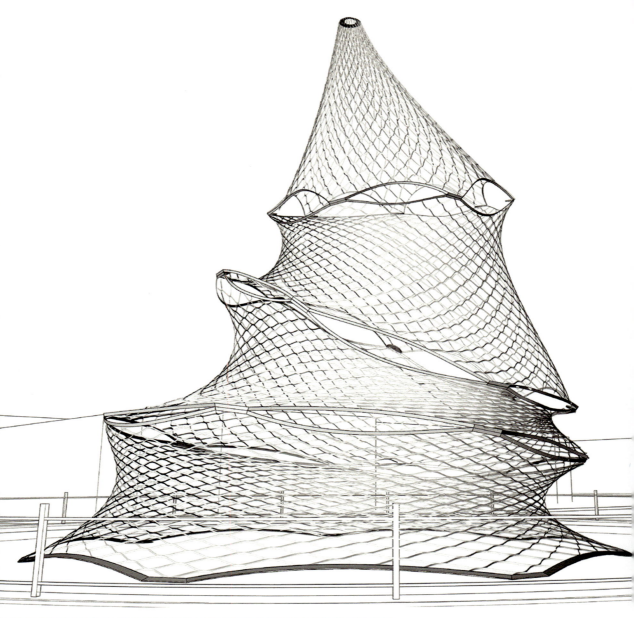

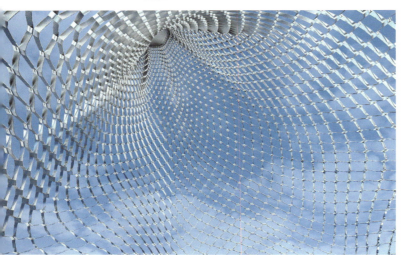

26
Digital images of *TENT PYRAMID* (2016, *top*), interior view of a self-rigid "tent" form in expanded metal without the central pole (*bottom,* 2019).

27 *(next page)*
Digital images of *ZIPCODE ARCHES* (*bottom*), a series of 5-segment structures corresponding to zip codes (**video 22**); *ARCH 07201* (2017, *top*), a proposal for Elizabeth, New Jersey, 07201, for the NJ Transit train station courtyard.

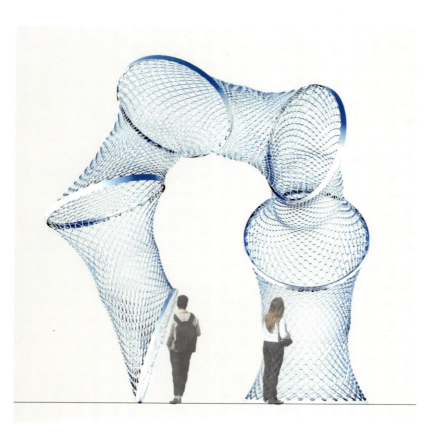

video 22

28
Digital image of *WORM HOLE* (2016), an installation for an infinite tube from multiple reflections which connects our space with the sky.

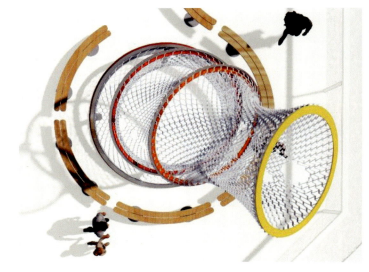

29
A sculpture with stacked and rotated elements, 2019.

The author has been collecting examples of similarities between expanded structures shown and those in nature.[8] Similarities can be deceptive since the same form can arise out of entirely different methods of construction and may be static or dynamic.[9] However, similarities can be powerful since they can provide the starting point for unifying different morphologies. The similarity between feathers, hair and scales [24] is one example. Differences force us to look for common principles and, conversely, similarities propel us to explore differences. The combined knowledge of similarities and differences informs both to complete the whole.

Three examples of self-shaping in expanded surfaces described earlier make me wonder if similar principles are being used in nature. The *EMERGENT RIPPLES* (**fig.4**) where self-folding emerged in the expanding surface, the *CARAPACE* (**fig.3, middle**) where the "supporting" and the "supported" differentiated as the surface expanded, and the *X-POD138* (**fig.11, top**) where the curved lintel and sill around the opening emerged as strength migrated to the boundaries to make "edge beams" where needed. Could nature be using similar constructive principles of self-shaping in growing bio-surfaces like skins of organisms, shells of arthropods, leaves, petals, seed pods and biological membranes?

Any of these principles, if found to exist in nature, would be examples of biomimicry in reverse. In biology, physical processes that shape material into form lie outside genetics and insights into natural form can come from experiments in physical construction. This is generally the domain of architects, designers, artists, engineers and makers, and provides an area of collaboration between design arts (designers, architects, artists, engineers) and biologists in developmental genetics and Evo Devo scientists. These principles are also important for how we in the design arts design our structures and will build growing structures in future.

The two companion essays, *Form Follows Force* (Chapter 5) and *Self-Shaping* (Chapter 6), have tied form to epigenetic phenomena. The formal morphological genes comprising combinations of topologic, spatial and geometric parameters that underpin the *Morphoverse* (Chapter 8) define the genetics of form and was covered in the essays *Meta Architecture* (Chapter 1) and *Genomic Architecture* (Chapter 2). Can the two phenomena be combined? This is an exciting idea and self-shaping provides the first clue.

For the design arts, physical morphologies of continuously growing structures, and physical technologies that enable their construction, will increasingly become more possible and will be realized on larger scales. Laboratory experiments on small-scale prototyping in softer materials will need to be translated into harder or hardening materials and scaled up. It will require the mastery of graded and controlled rheologic processes in relation to form and will need to be tied to manufacturing and fabrication. A marriage

---

[8] These include barks of trees, e.g. birch bark cherry (*Prunus serrula*), Northern red oak (*Quercus rubra*), palms (*Phoenix canaraiensis*), stem of Brussel sprouts (*Brassica oleracea var. gemmifera*), vessels of plants (*Cycas diannanensis*), and others.

[9] This remark results from the communications with Dierke Raabe and Helge-Otto Fabritius on the chitin-protein layer inside the cuticle of the American lobster (*Homarus americanus*) [22]. They pointed out that though there is a visual similarity between the micro-structure in the cuticle and the expanded structures I shared with them, it is a superficial one. The cuticle micro-structure is a static structure (personal communications with Raabe and Fabritius, May 4 and 11, 2021). Fabritius shared a publication on the ice plant seed capsules as an example of a dynamic expanded honeycomb structure in nature [23].

between biological and geologic morphologies as a continuum of form, process and material, will lead new technologies of form-making. This will require us to develop new continuous methods of fabrication (aeolian, fluvial) using gases and liquids to achieve seamlessness in form. Self-shaping achieved by force combined with other physical and chemical phenomena without prescriptive 3D models or use of standard forming methods will continue to advance the cutting-edge of making at different scales. Even though the digital is firmly placed in the design and manufacturing processes, the fundamental divide between digital (software) and physical (hardware) remains. In biology, the two are one. DNA is hardware and software. As we head towards encoded matter, the fundamental technologies for achieving this synthesis will require new breakthroughs in this area and will be addressed in the last essay on *Abiogenesis And The Future Of Architecture* (Chapter 9).

NOTES/REFERENCES

1. Haresh Lalvani (2011) Multi-Directional And Variably Expanded Sheet Material, *U.S.Patent 8,084,117*, Dec 27, (filed Nov 11 2005). https://patents.google.com/patent/US8084117B2/en

2. John French Golding (1894), Machine for Making Expanded Metal, *US Patent 527242A*, Oct 9.

3. Electronic Art and Animation Catalog, Computer Animation Festival, *Siggraph2008*, Los Angeles, 2008.

4. Blaine Brownell ed. (2008), *Transmaterials 2*, Princeton Architectural Press.

5. Steven Mesler (2010), Form Follows Force: Haresh Lalvani, *Huffington Post*, Nov.10.

6. *Haresh Lalvani: 2point5D+*, curated by core.form-ula, De Castellane Gallery, Brooklyn, New York, Nov 2010.

7. Kyle Studstill (2010), Genomic Architecture Built From Simple Codes Of The Natural World, *PSKF*, Nov 16.

8. *XtraD*, An Exhibition of Metal Sculptures and Design by Haresh Lalvani, *ArtFix Daily*, March 21 2011.

9. *PrattFolio*, Spring/Summer 2011, special issue on Innovation, Pratt Institute, NY.

10. William Menking (2014), Sculptures by Dr.Haresh Lalvani on Display in Columbia County, *Architect's Newspaper*, Dec 23.

11. Samuel Hughes (2015), The Shape of Things to Come, *The Pennsylvania Gazette*, Dec. (Cover Story).

12 Zach Mortice (2017), Haresh Lalvani on Biomimicry and Architecture That Designs Itself, *Redshift*, Autodesk, Jan 17. DOI: https://www.autodesk.com/redshift/haresh-lalvani/
Republished by others:
A. Biomimicry with Steel Sheets: Designing "DNA" Into Materials Can Create Architecture that Shapes Itself, *ArchDaily*, Feb 15 2017.
B. Self-shaping Shelters that could Revolutionize Emergency Housing, *Inhabitat*, Feb 16 2017.
C. Haresh Lalvani programs metal sheets to Shape Themselves into 3D Shelters, *DesignBoom*, Feb 17 2017.
D. Cristina Digiacomo, Self-Shaping Shelters Could Revolutionize Housing, *PSKF*, March 2017.

13 Lalvani, Haresh (2018), *Shaping and Self-Shaping*, Lalvani Studio, NY.

14 Lalvani, Haresh (2018), Morphological Universe: Genetics and Epigenetics in Form-making, *Symmetry: Culture and Science*, Vol.19, No.1, p. 240, Symmetrion.

15 Lalvani, Haresh (2021) Expanded Surfaces, Milgo Experiment 2 (1998-2010), In: S.A. Behnejad, G.A.R. Parke, O.A Samavedi (eds), Proceedings of the IASS Annual Symposium 2020-21 and the 7th International Conference on Spatial Structures, 23-27 August, 2021, Guildford, UK, pp.3163-3176.

16 Lalvani, Haresh (2021) X-Structures, Milgo Experiment 3 (2008-2014), In: S.A. Behnejad, G.A.R. Parke, O.A Samavedi (eds), Proceedings of the IASS Annual Symposium 2020-21 and the 7th International Conference on Spatial Structures, 23-27 August, 2021, Guildford, UK, pp.3177-3189.

17 Martin Gardner (1977), Extraordinary Nonperiodic Tiling that Enriches the Theory of Tiles, *Scientific American*, January.

18 Haresh Lalvani (1987/2011), Non-Periodic Space Structures, *International Journal of Space Structures*, 2, 3, 93-108, Elsevier, U.K. DOI: https://doi.org/10.1177/026635118700200204
Republished in: Vol. 26, Issue 3, pp. 139-154, September 1, 2011 DOI: https://doi.org/10.1260/0266-3511.26.3.139

19 Haresh Lalvani, Ref.14, Fig.20.1, p.124.

20 Haresh Lalvani, Morphological Aspects of Space Structures, in: *Studies in Space Structures*, ed. H. Nooshin, Multi-Science Publ., U.K.

21 Martin Gardner (1965), The "superellipse": a curve that lies between the ellipse and the rectangle, Mathematical Games, *Scientific American*, September.

22 D. Raabe, A. Al-Sawalmih, S.B. Yi, H. Fabritius (2007), Preferred crystallographic texture of a-chitin as a microscopic and macroscopic design principle of the exoskeleton of the lobster *Homarus americanus, Acta Biomaterialia 3* 882–895.

23 Guiducci L, Razghandi K, Bertinetti L, Turcaud S, RuÈggeberg M, Weaver JC, et al. (2016) Honeycomb Actuators Inspired by the Unfolding of Ice Plant Seed Capsules. PLoS ONE 11(11): e0163506.

24 Nicolas Di-Poï, Michel C. Milinkovitch (2016), The anatomical placode in reptile scale morphogenesis indicates shared ancestry among skin appendages in amniotes, *Science Advances,* 24 June: Vol. 2, No. 6, e1600708. DOI: 10.1126/sciadv.1600708

HYPER ARCHITECTURE
HYPERSTRUCTURES, HYPERSPACES, HYPERSURFACES

"No man can visualize four dimensions, except mathematically… I think in four dimensions, but only abstractly."

Albert Einstein[a] (1929)

"…we cannot conceptualize what we cannot depict…"

Stephen Jay Gould[b] (1997)

"I see higher dimensions as solutions that are looking for problems."

Author's quote[c] (2021)

The term "hyper" is here used in the mathematical sense as in hyperspace (higher dimensional space) or hypercube (higher dimensional cube). This is a specific definition of the term 'hyper', generally meaning "over", "beyond",[1] since it uses the dimension of space for its definition. Any dimension greater than 3 dimensions is a 'hyper' dimension, and 'hyper architecture' is architecture based on higher dimensions. Hypersurfaces are surfaces, hence 2-dimensional, but projected from higher dimensions. Hyperstructures has two meanings, a mathematically defined higher dimensional structure[2] which is used here for conceptual diagrams [1-3] or physically constructed 2D or 3D structures based on higher dimensions [4]. Since our physical space is 3-dimensional, higher dimensional structures cannot be physically built. However, their representations in lower dimensions provide a possible way to build them in 2D or 3D physical space, albeit with distortions. Such representations are familiar to us from 2D representations (drawings, photographs or images on a computer screen) of a 3D object and extend to 4D and higher dimensions.

[a] Albert Einstein, Interview by George Sylvester Viereck, *Saturday Evening Post*, Oct 26 1929, p.7, 110.
[b] Gould, Stephen Jay, Redrafting the Tree of Life, Proceedings of the American Philosophical Society, Vol. 141, No. 1 (Mar., 1997), pp.30-54
[c] Haresh Lalvani, Hypersurfaces, Milgo Experiment 4, presentation at IASS2020-21/Surrey7, University of Surrey, August 24, 2021.

[1] 'hyper' has a Greek origin 'hupér' and is familiar from the prefixes in hyperactive, hypersonic, hypertext, and other familiar terms.
[2] The term 'hyperstructures' was introduced by the author in the title of [2] as conceptual diagrams and in [4] as physical structures.

Higher dimensions originated in mathematics in the 19th century[3] as extensions of 3-dimensions by adding one more independent spatial dimension, the fourth dimension of space, at 90 degrees to the three orthogonal dimensions of 3D space [5,6]. The '4D space' obtained this way is distinct from '4D *space-time*' introduced after Einstein's work on relativity where *time*, a 1-dimensional physical phenomenon, represents the 4th dimension added to the three dimensions of physical space. Physicists, looking for a physical meaning of hyperspace, soon after introduced 5D space-time with the fifth dimension curled up in an ultra-small circle, an idea that led to more curled up dimensions in hyperspace later [7]. The ideas of Theodor Kaluza and Oskar Klein, who first introduced the 5D theories in physics, led to the speculation of K-K particles (particles named after them) which, if found to exist, would establish the physical existence of higher dimensions at the lowest physical scale, the Planck scale [8]. The Large Hadron Collider, so far, has not detected any evidence of higher dimensions at this ultra-small scale of particles. On the other end of the scale of nature, the cosmological scale, the 4th dimension of space has been theorized as a physical dimension of the overall shape of the universe [9]. Hyperspace theories continue to drive the imagination of physicists [10] as well as science-fiction writers.

Imagining higher dimensions as physical spaces and structures projected in 2D and 3D, and building them, drives the experiments presented here to capture the fascination and inherent beauty of higher dimensions, both physical and conceptual. This chapter deals with the former while Chapters 8 and 9 addresses the latter. Most of the examples described are in Euclidean hyperspace having straight edges and flat planes, and the principles extend to non-Euclidean hyperspace having curved edges, curved planes and curved spaces.[4]

A visual survey of the images in this chapter show higher dimensions as a way to systematically generate a wide repertory of structures with greater spatial and visual complexity. Projected higher dimensional spaces, structures and surfaces in 2D and 3D provide a new way to generate irregular morphologies, both in the individual units or complex aggregates, that are rooted in the underlying order of hyperspace and not arbitrarily designed. The underlying order provides a basis for automation of irregular morphologies that are unique and continue to extend the theme in the first four chapters on mass customization from ordered origins to infinite variations.

Early Experiments

My interest in higher dimensions of space (hyperspace) began in the late 1970's when I discovered they were reference frameworks for morphological possibilities. The high level of order in hyperspace, especially hypercubes (the 4D and higher dimensional analogs of the familiar 3D cube), I found very attractive. A pencil drawing of a 7D cube, executed with a T-square and a pair of drafting triangles to geometrically construct its 448 parallel lines was a high moment. The fact that

---

[3] Coxeter [5] provides a condensed history of higher dimensions, Henderson [6] provides an extended history in relation to the arts.
[4] Each element – line, plane, cell – can exist in 2 states, straight/flat or curved leading to a 3D taxonomy (3x3x3 cube) of combination states for Euclidean and non-Euclidean geometry in one space.

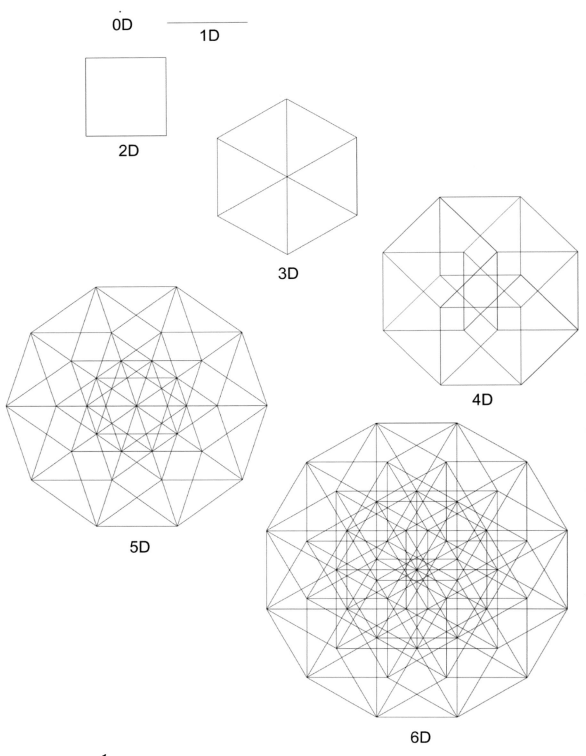

1
Orthographic projections of
n-dimensional cubes for dimensions 0
through 6.

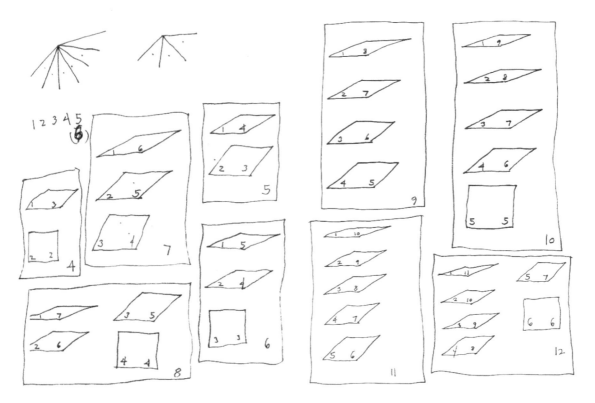

2
Author's drawings (1981) showing the discovery of angle numbers of rhombi that led to the angle-sum rule from projection of *n*-dimensional cubes.

$$1+6 = 2+5 = 3+4 = 7$$

3
Three rhombi (*n*=7) and their angle numbers from 2D projection of a 7D cube.

each point (vertex) of this hypercube with 128 vertices, with each vertex having the same 7 directions (vectors) emanating from it, and all parallel to 7 different directions of 2D space (the plane of paper), I found magical. **Fig.3** shows one half of 7D cube in a later digital image. It was a model example of parallelism that also embodied combinatorial possibilities in a geometric structure, and it was just one case ($n=7$) of an infinite class of n-dimensional structures. I have been using these representations since and the results continue to surprise me. They provide a higher-level system for form exploration and development that is intimately tied to morphogenesis, origin of form, origin of life, and has implications for the future of architecture.

Knowing that it was physically not possible to build these exotic mathematical structures physically, drawings, 3D models, and subsequently digital models, provided the next best option even though they were *representations*. These representations are projections of higher dimensions to lower dimensions, 2D and 3D, and are an example where physically a built 3D structure is a representation. In this example of a representation of representation, the abstract and the physical are one. The 2D cases are described first, followed by the lesser explored 3D cases.

2D Hypersurfaces

Of all known higher dimensional regular structures,[5] hypercubes (higher dimensional versions of the cube) and hyper-cubic lattices have provided the author with the most useful starting point, both as physical structures and meta-structures. Their 2D projections and derivative structures are described followed by 3D projections.

2D Hyper-Tiles from n Dimensions

The most familiar 2D projections of hypercubes, and also most useful, are the orthographic projections which are centrally symmetric as shown in **fig.1** for dimensions 0 through 6 ($n=0,1,2, \ldots 6$). In each, all edges are equal and are parallel to n directions. The outer polygon of these projections is an even-sided regular polygon with $2n$ sides [5]; for example, a 6D cube is bound by a regular 12-sided polygon. Interest in building these 2D projections using physical pieces led to dissections of the diagrams into their smallest units, the 2D building blocks of hypercubes. These building blocks are an infinite class of distinct rhombii. Their listing by dimension led to the discovery of angle numbers of rhombii shown in **fig.2**. Many of author's drawings that follow are taken from various patents during the period 1985-1991 [11-16].

Angle Numbers

The angle number is an important concept in defining the shapes of 2D

[5] The others include the higher dimensional versions of other regular polyhedra, tetrahedron (called a simplex), octahedron (cross-polytope), dodecahedron (120-cell) and icosahedron (600-cell). These have been well-known in mathematics [5] and their visualization as 3d-prints or digital models can be easily accessed on the web.

polygons from higher dimensions. It is a multiple of the central angle of the 2n-sided polygon which is represented by the number 1. As an example, in the 7D cube ($n=7$) illustrated in **fig.3**, this angle is marked with the number 1 at the center. Since the outer polygon is 14-sided, this central angle equals 360/14 degrees. All other angles, convex and concave, derived from a 7D cube are integer multiples of this angle. These can be represented by numbers 1,2,3, ..., here termed *angle numbers*. The surprising result of this representation was the discovery that the pairs of angle numbers of rhombii (shown in the left upper half of the diagram) are the only pairs of numbers that add up to 7. These pairs, 1 and 6, 2 and 5, and 3 and 4, determine the precise shapes of the three rhombii for $n=7$. The angle-pair rule is captured in the equation $1+6 = 2+5 = 3+4 = 7$. Thus only 3 different shapes of rhombii are possible from 2D projection of 7D. Similarly, in the Penrose tiling, with its acute and obtuse rhombii ($n=5$ case), yields the angle-pairs 1-4 and 2-3. Hence, only two tile shapes are possible from 5D. And, similarly for other dimensions as shown in author's drawings in fig.2. These rhombii are the 2D tiling units derived from n-dimensional cubes and their shapes are underpinned by a simple rule: for any dimension n, the only permissible pairs of angle numbers are those that add up to n. Angle number concept provided the clearest case of 'form follows number'.

Periodic Tables of Zonogons

Rhombii, 4-sided polygons ($p=4$) with equal and parallel sides, are the smallest of a general class of p-sided polygons (p even) termed zonogons [5], like hexagons, octagons, decagons, and so on. Zonogons can be convex or non-convex. The convex shapes of these polygons can be organized in nested periodic tables as shown in **fig.4** for $p=4$, 6, and beyond, for $n=4$ through 7. This is an infinite 2D table with n having any value 2 and greater, and p with values 4 and greater. The idea extends to non-convex polygonal shapes as shown with 6-sided and 8-sided examples in **fig.5**. These tables are embedded in higher dimensional periodic tables.

Number Code and Asymmetry

The angle numbers for any polygonal shape, convex or non-convex, symmetric or asymmetric, define a cycle of numbers starting from any one vertex and moving to successive ones in a clockwise (or anti-clockwise) manner till the cycle is complete. This cycle can be represented by a linear number sequence which acts like a unique code for that shape. An assortment of polygonal shapes is shown in **fig.6** with their number codes. For each, its dimension is shown as a single number followed by the angle number sequences. The black dot indicates the start vertex and the sequence is read in an anti-clockwise direction. For each dimension, there is theoretically an infinite number of such sequences since the number of sides of the tile shape can be increased to infinity.

It is easy to see that the number code can be manipulated to change one polygon to another. In **fig.7**, a transformational sequence of morphing shapes for a 10-sided polygon ($n=5$ case) through various intermediates is shown

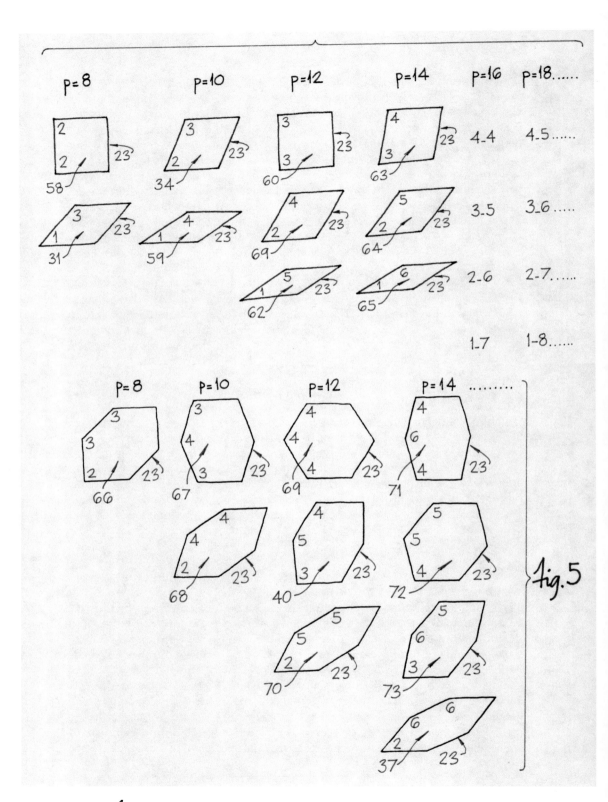

4
Portion of the Periodic Table of Convex *p*-sided Zonogons from Dimension n. Examples of *p*=4 (rhombi) and *p*=6 (hexagons) shown; author's patent drawings, 1987, '91 [14].

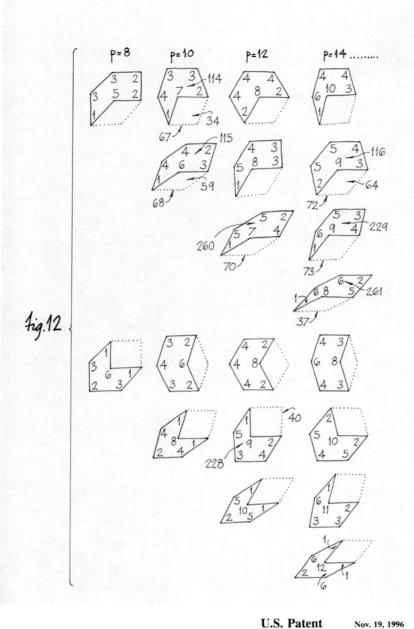

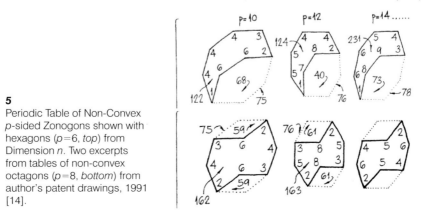

5
Periodic Table of Non-Convex *p*-sided Zonogons shown with hexagons ($p=6$, *top*) from Dimension *n*. Two excerpts from tables of non-convex octagons ($p=8$, *bottom*) from author's patent drawings, 1991 [14].

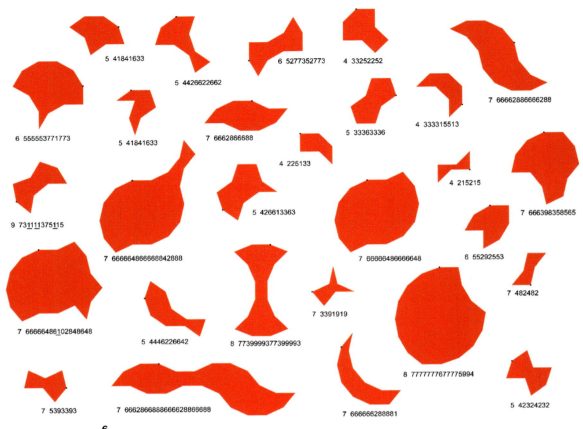

6
Angle number sequences for five non-convex tiles (singly-, doubly- and triply-concave) from one or more 2n-gons fused together (b) for $n=4$, 5 and 7 cases and their corresponding "biomorphic" counterparts (c).

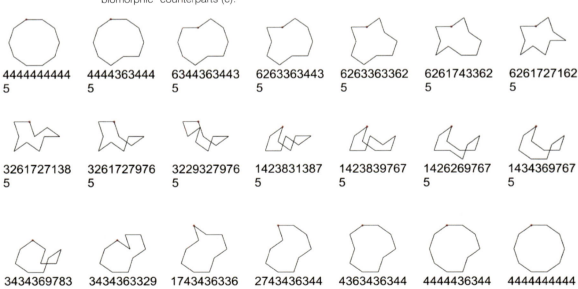

video 26

7
Sequence from an animation of Morphing Decagons (**video 26**). All Map in a 10D Periodic Table of Decagons ($n=5$ case), each indicated with a sequence of 10 angle numbers.

from an animation (**video 26**). In all instances, the sides remain parallel to the 5 directions of 5D space. The morphing admits polygons with cross-over edges which are usually excluded from the standard definition of a polygon. Examples with cross-overs are encountered in knots, weaving and crochet, or in the temporal case of paths of our movements. The morphing sequence shows a continuum between symmetry and asymmetry and also between convex, non-convex and extensions to knotted/woven morphologies.

Tile Shape Rule

The number sequences that close into polygons are determined by a simple rule. The sum of interior angle numbers, angle-sum, of any p-sided polygon – symmetric, asymmetric, convex, non-convex – equals $(p-2)n$ [17]. This simple rule underpins all 2D closed shapes projected from n dimensions.[6] The rule also provides a way to automate number sequences to generate all polygons. The rule also underpins curved polygons derived from these straight-edge counterparts.

Biomorphic Polygons

From the irregular polygonal shapes in **fig.6**, organic polygonal shapes having continuous curved edges and no visible vertices or edges are only one transformation away.

The discrete polygons having well-defined vertices and straight edges of irregular polygons can morph into curved edges of biomorphic polygons by a simple transformation, an iterative procedure, which leads to rounded vertices. This is described in the next chapter (p.321). It is reminiscent of a practical method ofr circling the square by doubling its sides in the binary sequence 4, 8, 16, 32, 64, 128, ..., the method that led to circular domes supported on square walls in Byzantine and early Islamic architecture. The method is similar to what Archimedes used earlier to discover the circle as the limit case of a recursively truncated polygon in an attempt to find the exact value of pi.

Hyper-Tilings

All polygonal shapes, and their biomorphic counterparts, derived from n-dimensions can tile with others within the same dimension or with tiles from related dimensions (factors or multiples). These tilings, termed *hypertilings* here, cover a continuous 2D surface built from units that are projected from higher dimensions and are thus termed *2D hypersurfaces*. The hypertilings are periodic, non-periodic, irregular or random. Individual hypertiles, as we have seen, can be symmetric or asymmetric, and can tile by themselves or in various combinations with other hypertiles. The tiling rule is simple: the angle number sum at any vertex equals n. The shape rule and the tiling rule set the stage for automation of

---

[6] Though the examples shown are for even-sided polygons, the formula also applies to odd-sided polygons. The simplest case of triangles, with angle-sum n, are derived from n-tetrahedra (simplexes) in 2D projections, the way rhombii are derived from n-cubes.

hypertilings. A visual assortment of examples is shown from author's early work (**figs.8** and **9**). Of these, the more interesting cases are *self-tilings* where only one tile shape is used to fill the plane (**fig.10**).

Self-Tilings and Escher

Self-tilings, like the examples in fig.10, show individual hypertiles having sides from 6 to 22. This feat is enabled by allowing 2 tiles to meet at a vertex, a condition that is disallowed in standard tiling designs as a violation of the basic rule of tessellations which require at least 3 tiles to meet at any vertex. In this sense, hypertilings are a different class of tessellations. As the number of sides of hypertiles increase, the tile shapes begin to resemble shapes of creatures as in the examples on the lower right indicated by numbers 10,14,16,18 and 22, leading to irregular tile shapes and resulting aggregates. This is reminiscent of Escher's artworks [18], where periodic arrays of polygons with 3, 4 or 6 sides in most instances, were used to deform the edges of a regular polygon to produce shapes that resembled creatures like fishes, bats, and so on. Escher used topology-preserving transformations to produce his art whereby the connectivity of the base tilings and the number of their sides were preserved while its shape was morphed. Hyper-tilngs, by comparison, are dimension-preserving (dimension of projection, not size). There are (theoretically) infinite shapes of hypertiles since dimension n and number of sides p can both reach infinity. This broadens the options for design and has direct implications for mass customization discussed in Chapter 4.

Pentiles and Froebel

Friedrich Froebel, the originator of kindergarten and a pioneer in early education developed 'Gifts' [19] some of which came to be called Froebel blocks.[7] Frank Lloyd Wright used them as a child and recalled its influence in his autobiography: *"The maple wood blocks . . . are in my fingers to this day"* [20]. Froebel was inspired by his studies in minerology [21] and the crystal forms in nature inspired the idea of open-ended designs from an underlying ordered system. This concept guided his invention of Gifts. Wright saw this immediately: *"Along with the gifts was a system, as a basis for design and the elementary geometry behind all natural birth of 'Form'"* [20]. Among the Gifts, Gift 7 called *Parquetry Blocks*[8] introduced various geometric tile shapes. Froebel's blocks, tiles, and other gifts including the pea-and-toothpick construction, which Buckminster Fuller played with in his kindergarten and recalled years later in his magnum opus *Synergetics* [22], indicates a far wider influence on architecture, design and art [23]. Froebel's inventions also set the

---

[7] See, for example, Froebel's "Sixth Gift" comprising 27 brick-shaped blocks and other "gifts"
[8] Froebel's "Seventh Gift" (or "Gift 7") comprises tile shapes like squares, equilateral triangles, half-squares, half-triangles and one-third triangles; https://www.froebelweb.org/gifts/seventh.html

8
Author's sketches of various hyper-tilings from patents, 1991 [14].

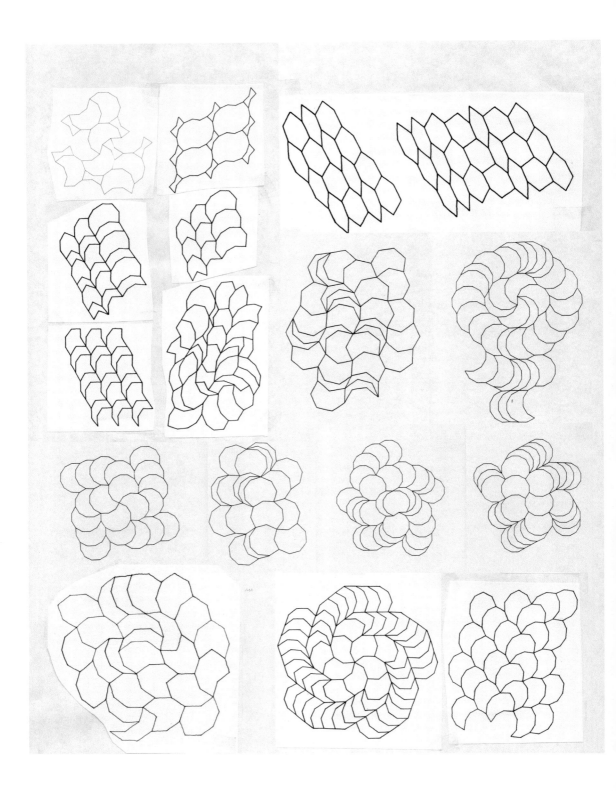

9
Author's sketches of more hyper-tilings, also from patents, 1985,'91 [11,14].

Hyper Architecture 225

10
Author's sketches of hyper-tilings having 6 through 22 sides that tile by themselves; author's sketches from patents, 1987, '91 [14].

226　Coding, Shaping, Making

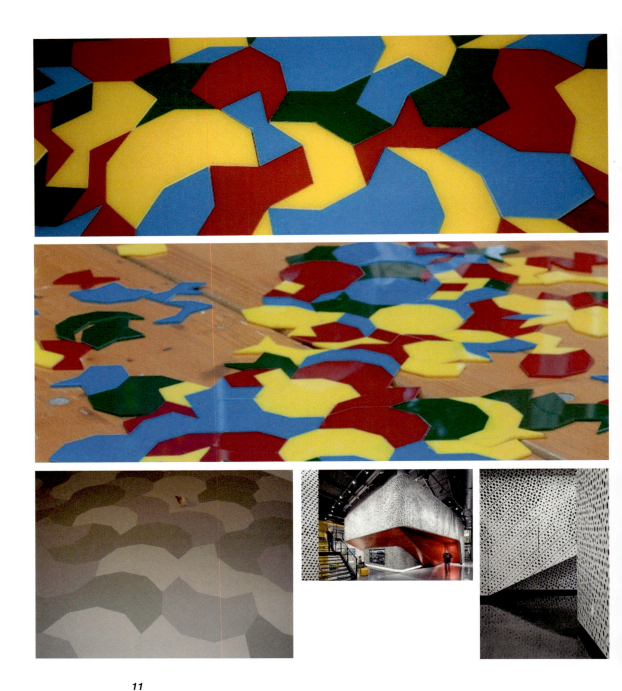

11
The *Pentiles*, based on $n=5$ case, from the Family Day, Symmetry Festival 2016, Karlsplatz, Vienna (*top, middle*), 10-sided rubber Crescent tiles, 1985 [10] on the floor of Lalvani Studio, 2004 (*bottom left*). A continuous morphing of $n=3$, 4, 5 patterns with laser cut-outs on the interior walls of Pratt's Film and Video Department (*bottom middle and right*) for architect Jack Esterson, Think! New York, 2015 [24].

stage for future construction toys.[9]

Pentiles, *n*=5 case of 2D hypertiles (**fig.11**) was developed by the author in the early 90's with beginnings in early 80's. It is a generalization of the "pattern blocks" in educational play kits and also relates to Froebel's Gift 7. It is one of the infinite set of pattern blocks from other values of *n* (*Hextiles, Heptiles, Octiles* and so on). *Pentiles* are currently being used as an educational kit in STEAM workshops by Krtistof Fenyvesi [24]. Later examples in this chapter show the applications of Pentiles to curved hypersurfaces and hyper structures for architecture, experimental structures and sculptures. We are currently looking into their possible application for robotic construction.[10]

Hyperspaces and Hyperstructures
4D Architectonics

Examples of architectonic studies from the 1980's [26-29] are presented in **figs.12-16**. These studies leverage the topology of the 4D cube[11] and explore the formal aspects of its projected 3D geometry. The topological elements of the 4D cube comprise the 24 faces (panels, walls, ceiling-roof planes), the 8 cells (spaces and how they inter-connect), and the 4 axes (spatial directions) of 4D space. In some studies (**figs.12-15**), the regular cube can be seen in its upright position and defines the vertical axis and the wall planes, the other 3D cells are tilted or squished. These compositions are gravity-bound and are thus spatially orientable. They also have an implied ground plane or a split level.

Fig.12. The 4D space frame (***top right***) can be built in 3D from the 12 wall planes, each made from 4-sided rigid frames that define the space of 6 interlocking cells (each a 3D cube in 4D space); the remaining 2 cells lie at the top and bottom to complete the 8 cells of the 4D cube. The 4D spatial composition (***bottom***) is derived by converting each wall plane into an L-shaped rigid panel in different orientations and the floor plane has a split level. In the composition on the middle right, the 6 cells have level floor and ceiling planes, as in a more conventional spatial layout, though the cell spaces interlock.

Fig.13. Two 4D suspended structures (***left***) are anchored to a central core. In one (***top***), a tilted 3D cell below is hung by cables from a similar tilted cell above. The other (***bottom***) uses more conventional cantilevered frame on top attached to the central core by cables from which the bottom frame is hung. The offset structure (***right***) uses the 4D cubic frame as a cantilevered "joint" between two vertical "cubic" towers which are offset from another. The 4D

---

[9] Construction toys like Unit Blocks, Tinker Toys, Lincoln Logs, Pattern Blocks, Tangrams, Lego, Mechano, Synestructics, Zometool, K'nex, Hoberman Sphere, and the more recent ones like 4D Frame, Itsphun and Lux Blox, are lineages and improvements of early educational toys and include new inventions. These continue our childhood fascination with designing and building physical constructions from a kit of parts, a concept Froebel adopted from crystals.

[10] At the Center for Experimental Structures, School of Architecture, Pratt Institute, jointly with Robinson Strong and Ahmad Tabbakh.

[11] The numbers of toplogical elements - vertices (V), edges (E), faces (F) and cells (C) - in the 4D cube is governed by the Euler-Schläfli equation $V-E+F-C=1$. The 4D cube has 16 vertices, 32 edges, 24 faces and 8 cells. In comparison, the cube has 8 vertcies, 12 edges, 6 faces and 1 cell. The 4D cube has 4 spatial directions compared to the 3 (*x-y-z*) in the familiar cube.

12
The 4D cube (*top left*, line drawing) is converted into a 4D space frame (*top right*) constructed from 12 4-sided frames which are joined edge-to-edge as wall planes. A derived 4D composition (*bottom*) showing 6 interlocking 3D cells, each a distinct "cube" in 4D space. Another composition (*middle right*) where the 6 interlocking cells have co-planar floor (and ceiling) planes; author's drawings (1989).

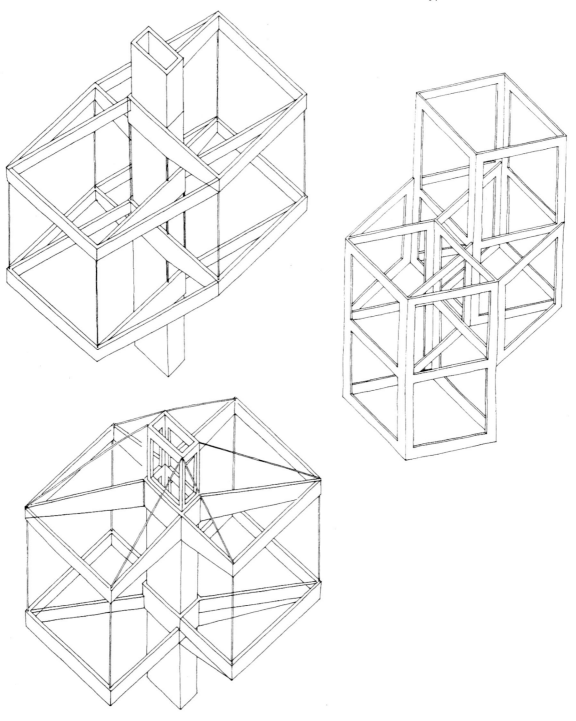

13
Two examples of a 4D cube suspended from a central core by cables (*left*, *top* and *bottom*). The drawing on right shows a way to extend the hypercube space frame of fig.14; author's drawings, 1991.

Coding, Shaping, Making

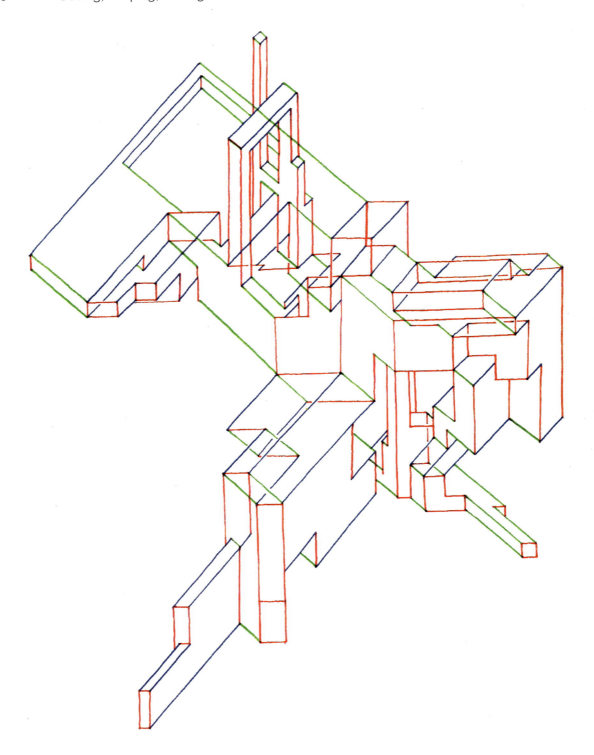

14
A 4D composition defined by the 4 colored axes that define the sides of various 3D elements in fused or inter-locking states and juxtaposed with others in different ways, author's drawings, 1986.

Hyper Architecture 231

15
A 4D composition of planes where the 4th dimension is coplanar to two others; author's drawing, 1990.

232   Coding, Shaping, Making

16
A composition of the 6 colored planes of 4D space (*top*), a composition of 3D L-shaped blocks in 4D space (*bottom*), and a 4D composition of 3D blocks and 3D linear elements (*middle right*), author's drawings, 1989.

cube acts like a spatial-structural transition in a vertical structure to provide a spatial device that potentially accommodates view, light, air, zoning and other considerations.

Fig.14. A spatial composition is defined by the 4 axes of 4D space, each in a different color, representing the edges of various 2D planes – squares, rectangles, parallelograms, L-planes, and planes with other shapes – in different proportions. These planes define more complex 3D block-like elements which appear to be fused, cut-out, or inter-locked. All are juxtaposed with others in different ways along the four spatial directions.

Fig.15. A 4D composition of vertical and inclined planes where the 4th dimension is coplanar with two others. This condition makes 3 planes of the 4D space coplanar. These planes are shown with one color (red) which, along with the remaining 3 planes (yellow, blue, green), make a spatial composition of only 4 planes.

Fig.16. Abstract 4D compositions which are spatially non-orientable. Not having a fixed vertical axis, liberates them from gravity. The designs can be viewed in different orientations by turning the page at different angles. The 6-plane composition (***top***), with each plane in a different color, shows the 2D design elements for 3D spaces. A composition of 3D L-shaped blocks in different orientations in 4D space (***bottom***), and a 4D composition of 3D blocks with 3D linear elements (***middle right***) are additional examples of early hyper-architectonic studies.

4D Modulor

The spatial and formal architectonic explorations were followed by the development of various higher dimensional proportion systems [28, 29]. The table of rhombi in **Fig.4** provided a starting point as higher dimensional analogs of the square-based systems, where each rhombus is a square in hyperspace. The method followed Le Corbusier's *Modulor* [30] which, in the hyper-system is the $n=2$ case. This led to the invention of the *4D Modulor*, $n=4$ case, as a 2D building kit[12] (**fig.17**) as one example from higher dimensional proportion systems. *4D Modulor* permits 4D compositions that appear to be 3D but are 2D. It has 6 colors – three primary pigment colors, red, yellow and blue and white, black and grey – one for each plane in a 4D cube. This linked Le Corbusier's *Modulor* with the works of Piet Mondrian and the De Stijl group, notably, Theo van Doesberg [31]. The flat tiles of the *4D Modulor* can easily extended to 3D blocks.

The 4D compositions in **fig.17** can be rotated through 45 degrees in 8 different positions, each marking the 4 axes in up or down positions, similar to the 6 orientations of a 3D cube. Re-orientable morphologies of projected hyperspace are an interesting feature with implications for more adventurous architectural layouts. 4D compositions can be constructed from 3D volumetric blocks or as 2D panel systems for architecture.

---

[12] A tile kit prototype comprising 48 magnetic foam tiles was prepared for the company Nova Design Group (and the Orb Factory) for introducing the *4D Modulor* as a Design Kit (1997).

17
'4D Modulor' ($n=4$ case) as an extension of Le Corbusier's Modulor ($n=2$ case), 1994, 1997. Diffferent orientations produce different upright 3D blocks.

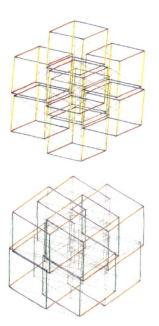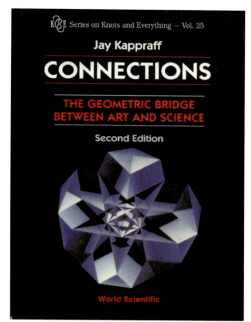
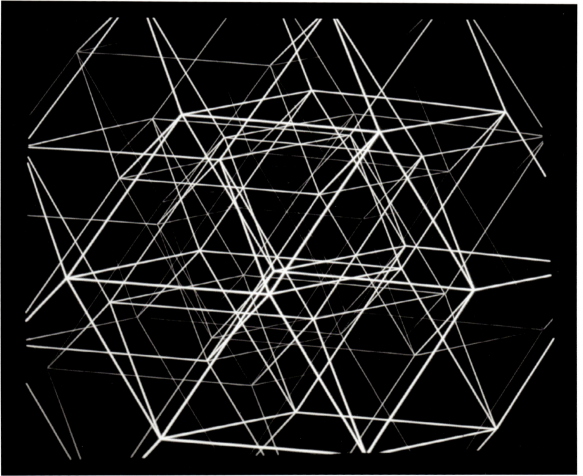

18
Author's drawings (*top left*, 1981) and digital models of 6D cube (1985). The cover of Jay Kappraff's *Connections* (1990) showing one image from serial sections of a 6D cube and a screenshot of its vector model (*below*). Both digital models built by the author at CGL, NYIT.

6D Space and Quasicrystals

Two early drawings of a 6D cube by the author[13] in **fig.18** (***top left***) show the use of 3 primary and 3 secondary colors for its six edges. These drawings came in handy a few years later while modelling the 6D cube with icosahedral symmetry shown here in two versions. A raster model from 1985 (***top right***) is one image from a sequence of 11 serial images, all parallel MRI-type sections first introduced in scientific digital visualization in the 1970's. It was published by Jay Kappraff on the cover of his book *Connections* [32]. A raster version of the corresponding vector model of the 6D cube (***bottom***) shown in a close-up screenshot. The model was enabled by the special real-time interactive software written by Patrick Hanrahan at the Computer Graphics Laboratory (CGL), NYIT, Old Westbury campus. This was part of author's independent work in 2D and 3D projections of higher dimensions on non-periodic structures which became important with Schectman's discovery of quasi-crystals in nature [33]. These crystals were based on pentagonal symmetry forbidden by classical crystallography and have an interesting history of parallel developments from different fields[14] including design arts (Baer [34], Miyazaki [35,36], Lalvani [37,38], Robbin [39,40]). Though quasi-crystals originated in the laboratory, their occurrence in nature was recently traced by Steinhardt [41].

The author presented his discoveries in non-periodic geometries to the scientific community at an invited conference in Germany in 1985 organized by the mathematician Andrea Dress,[15] two papers on non-periodic structures were published in 1986/87 [37,38], a joint paper on quasicrystals with the physicist Peter Kramer at Tubingen University was published [42], and several patents were filed [43-48] in addition to the ones mentioned earlier. Drawings in subsequent sections are taken from these patents and others from the period 1981-1991.

Hyperspace Frames (Figs.19-22)

The simplest cases of hyperspace frames are hyper-cubic space frames (**fig.19**), the higher dimensional versions of the 3D cubic frames familiar to us from rigid frameworks of buildings. The images show arrays of a 4D-cube (***top row***), different 4-pronged tetrahedral nodes needed for 4D lattices (***middle left***), and a hypercube frame with a hypercubic node (***middle right***) as the beginning of a fractal 4D-cube space frame system [26]. This leads to a non-periodic hyper-cubic space frame (**fig.20**) [26]. Simpler non-periodic space frames, $n=7$ case (**fig.21**, ***left*** and ***top right*** [13]) and a single-layer non-periodic frame, $n=5$ (***bottom right***), are possible to build using prismatic nodes. **Fig.22** shows single and multi-layered space frames, $n=5$, that permit zig-zagging along the vertical axis [13].

---

[13] First presented in author's lecture at Arthur Loeb's *Design Science 3*, Harvard University, 1983.
[14] Roger Penrose, Nicholas De Bruijn in mathematics; Alan Mackay in crystallography; Paul Steinhardt, Peter Kramer in physics; Steve Baer, Koji Miyazaki, Haresh Lalvani and Tony Robbin in architecture and sculpture.
[15] 'Geometric Problems in Crystallography' organized by the mathematician Andreas Dress, University of Bielefeld, Bielefeld, August 1985.

Hyper Architecture

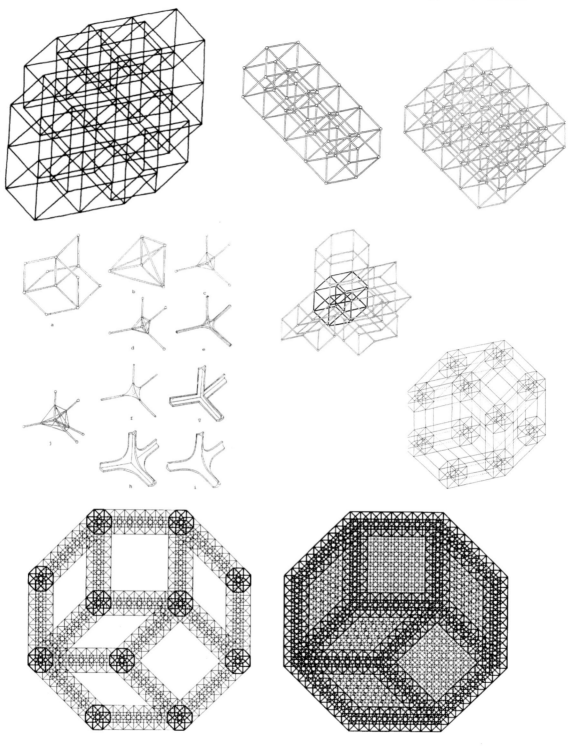

19
Hyper-cubic space frames from the 4D-cube and 4D-cubic lattice (*top* and *middle*, author's drawings, 1993; each 4D space frame comprises different 4-pronged tetrahedral nodes. The bottom two examples (digital images) illustrate a fractal 4D-cube space frame, 1993.

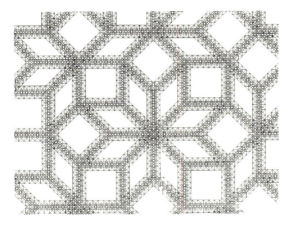 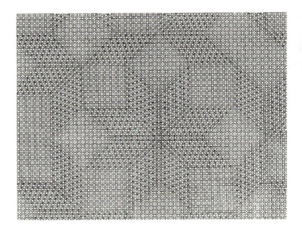

20
A non-periodic hyper-cubic space frame (*left*) based on the fractal space frame in fig.18, and the same with infill lattices (*right*); digital images, 1993.

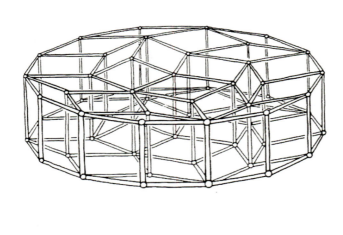

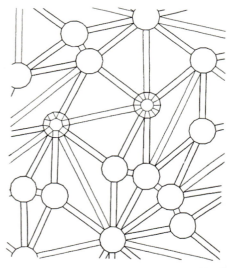

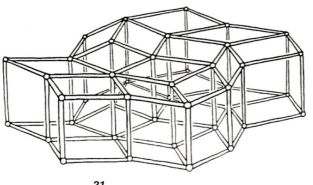

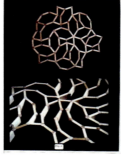 

21
Non-periodic space frames, $n=7$ case (*left* and *top right*, author's drawings (1991) and physical mock-up of a single-layer frames based on $n=5$ having pentagonal nodes (1987).

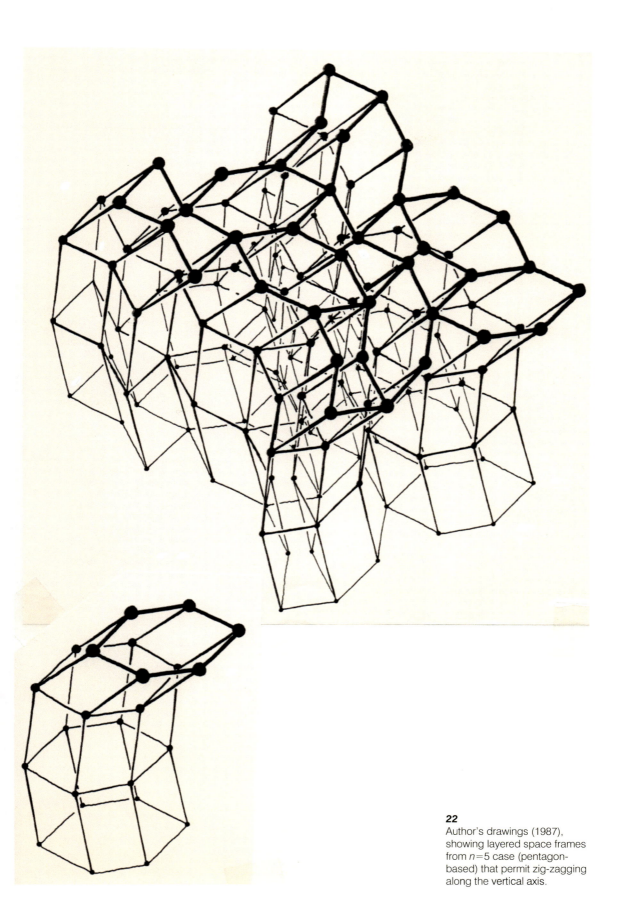

22
Author's drawings (1987), showing layered space frames from $n=5$ case (pentagon-based) that permit zig-zagging along the vertical axis.

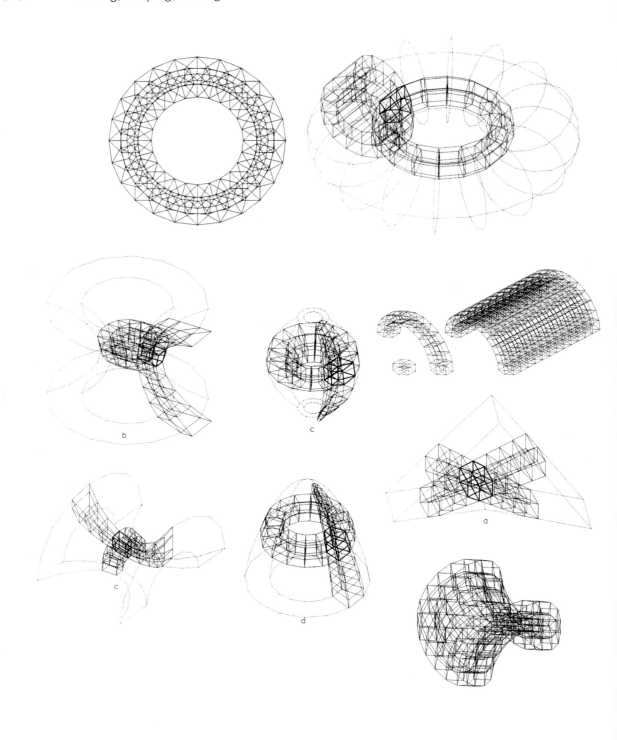

23
Digital images of curved hyper-cubic space frames derived by translation and rotations of linear arrays of hypercubes along straight or curved lines corresponding to curved surfaces obtained by a similar method, 1993.

Curved HyperSpaces

Curved hyperspaces derived by translation and rotations of linear arrays of hypercubes along straight or curved lines in various combinations, corresponding to their lower dimensional counterparts, are shown in **fig.23**. All are shown as curved hypercubic lattices in various morphologies - ring, torus, hyperboloid, cylinder, hyperbolic paraboloid, paraboloid, a saddle with straight or curved lines, and so on [28,49]. These are non-Euclidean counterparts of the examples shown so far. Symmetric, asymmetric, periodic, non-periodic and irregular versions of tilings and space-fillings extend the formal vocabulary of curved hyperstructures.

Curved Hypertiles (Figs.24-27)

Hypertiles with straight edges and flat faces shown thus far provide the starting point for generating a variety of curved structures in 2D and 3D by geometric transformations on elements of structures. These produce local curvatures on the tiles from four possible combinations of the binary states of edges or faces, each existing in straight (0) or curved (1) states. Adding a negative curvature adds the third state (-1), leading to 9 combinations. In the examples in **fig.23**, individual cells have flat faces and straight edges. The examples in **figs.24-27** show curved faces with straight and curved edges [38]. The straight and curved elements exhibit the combinatorial continuum between binary states seen earlier in the context of pigmentation patterns in seashells and insect carapaces in Chapter 4.

Examples in the upper two sets of **fig.24** illustrate a variety of saddle polygons derived from even-sided zonogonal prisms[16] having 4, 6, 8 or 10 sides based on hypertiles (n=5). The saddles are derived from edges or face diagonals of these prisms. Some curved states are shown in the *bottom right* column. Curved hypertilings on a plane composed of saddle zonogons (**fig.25**) have plan geometries that are periodic, non-periodic or random tessellations and comprise combinations of even-sided convex and/or non-convex plane hypertiles.

Curved hypertiles in multi-directional spatial orientations lead to an infinite class of saddle zonohedra[17] [44,45] and are shown with some examples in **fig.26**. They are composed of even-sided convex and non-convex saddle polygons and tile with others in periodic, non-periodic or irregular arrays. The two examples on *top right*, both from a zonohedron with 20 faces and pentagonal symmetry (n=5 case) shown on their left, are respectively built from 10-sided and 6-sided saddles. Each has two poles, one at top and bottom. Since n can be any number greater than 2, this infinite class yields structures from a single 3D saddle hyper-tile having sides ranging from 6 to $2n$. These are similar to the flat tiles with large number of sides

---

[16] Termed zonogonal prisms which, like prisms, have p-sided polygons on two ends joined by parallelograms. In zonogons, as stated before, p is even.
[17] Zonohedra are a special class of 3D polyhedra with even-sided polygonal faces arranged in parallel pairs and located on opposite sides around the center. Zonohedra define the outer shell of hypercubes and their edges are parallel to a finite set of directions which determines their dimension. Saddle zonohedra are composed of even-sided saddle polygons like those having zig-zag edges and a saddle surface.

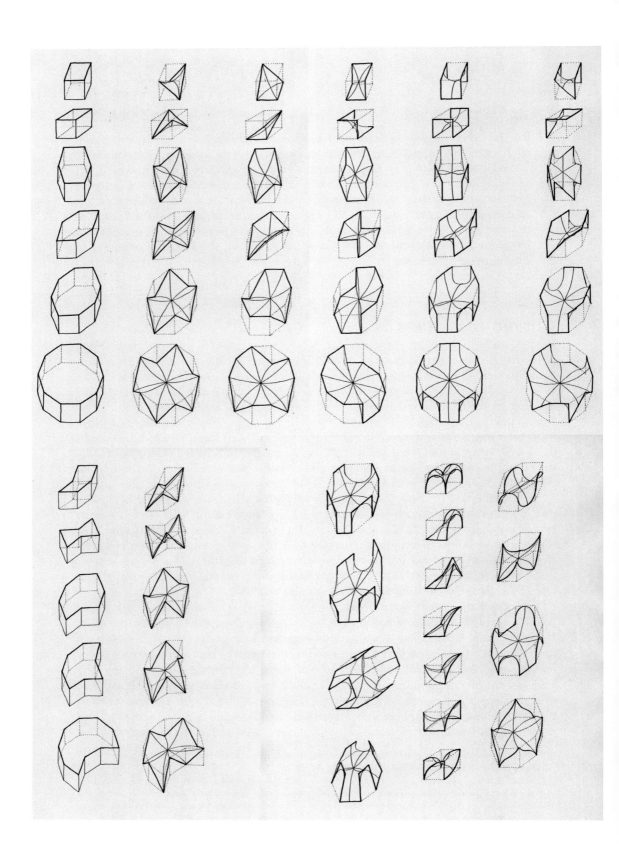

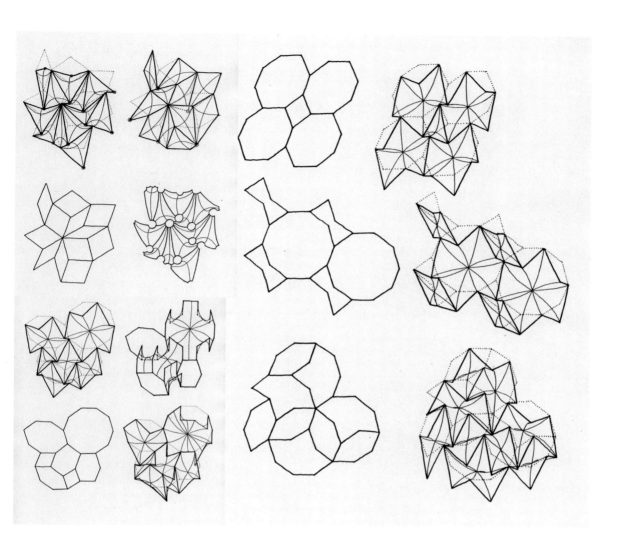

24 *(left)*
Curved hyper-tiles having curved faces and straight edges with a few examples having curved edges, all derived from even-sided prisms and 3D structures with even sided parallel faces called zonohedra; author's drawings (1987, '89).

25 *(top)*
Curved hyper-tilings on a plane having plan geometries which are periodic, non-periodic or random tessellations comprising combinations of even-sided convex and/or non-convex plane hyper-tiles. These are also derived from tilings of zonohedral prisms; author's drawings (1986-87, '89). They lead to membrane structures and fabric architecture.

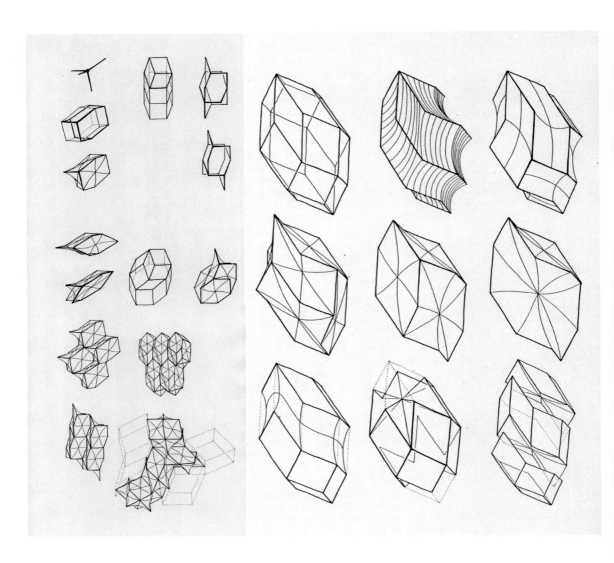

26
Curved hyper-tilings corresponding to saddle zonohedra, comprising saddle polygons that tile to produce open or closed cells array in periodic, non-periodic or irregular spatial configurations; author's drawings (1987, 1989).

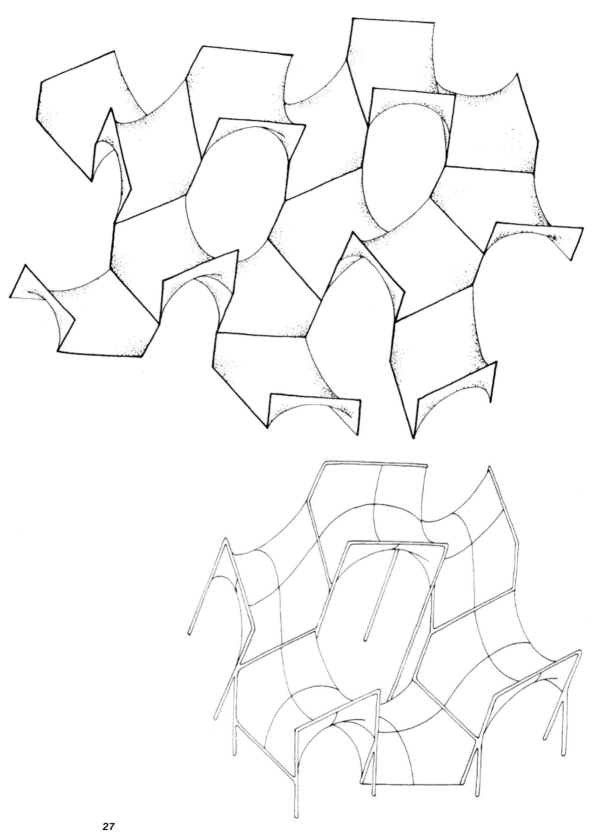

27
An example of curved hyper-tiling composed of 8-sided saddle polygons in a non-periodic array (*top*) and its derivative as a framed membrane structure (*bottom*); author's drawings (1991).

shown in **fig.10** and lead to a broader class of structures from zonohedra and their space-fillings. One example composed of 8-sided saddle polygons, each derived from a tilted parallelepiped is shown in **fig.27** as a portion of a non-periodic minimal surface (***top***) with its derivative framed membrane structure (***bottom***).

Curved HyperSurfaces

Early Developments: In 1986, the author developed a recursive procedure to generate the Penrose tiling which enabled building increasingly larger self-similar triangular regions which could be mirrored and rotated to generate the Penrose tiling recursively.[18] Subsequent work soon after led to the observation that the Penrose tiling (n=5) could be dissected into finite regular pentagons and decagons of incremental sizes that appeared as local regions within the tiling pattern. It followed that other values of n could yield subdivisions (or dissections) of regular n-gons and $2n$-gons using their associated rhombi as shown in the digital images (**fig.28**). These subdivisions comprised radially symmetric as well as non-periodic portions of tilings, all examples of hypertilings that, in principle, could be generated procedurally. Isolating these polygonal regions permitted an interesting spin-off. Irregular hyper-tiling patterns could be obtained by re-arranging the tiles within the polygons without altering their boundaries. These tiled regular polygons could be used as modules to build a variety of plane and curved hypertilings termed 2D hypersurfaces earlier. 2D hypersurfaces can be built from regular polygons, namely, all regular polyhdera, all infinite polyhedra in Burt et al [50], deltahedra, all polyhedral packings, all zonohedra and higher dimensional regular polytopes, and various curved surfaces like the Mobius strip, cylinder, torus, paraboloid, hyperboloid and so on. An assortment of examples of curved hypersurfaces is shown in **figs.29** and **30**. The discoveries of curved hypersurfaces using hypertilings were first presented[19] in Lalvani [28,47,51] and some excerpts were included in [32].

Thin laser-cut cardboard models (2ft dia) of two spherical hypersurfaces were on public display during the mid-1990's at a special bay in the Cathedral of St. John the Divine, New York, where the author was an artist-in-residence.[20] One of these is shown in **fig.30** (*top row, left*). The late Very Reverend Dean James P. Morton thought *"How fitting it would be that Cathedrals which have symbolized architectural geometry in the past, could once again become a repository for the cutting-edge development in geometric researches in the*

---

[18] By dissecting the two golden rhombii ("skinny" and "fat" rhombii with angles 360 and 720; see Fig.4, n=5) into a pair of golden triangles which defined the fundamental region of Penrose tiling. This region could be mirrored and rotated to enlarge the pattern in a self-similar way to generate the Penrose tiling.

[19] Art and Mathematics conference, SUNY, Albany, 1992, organized by Nat Friedman.

[20] In 1994, the author was formally invited by Dean James P. Morton of the Cathedral of St. John the Divine, New York, to be an artist-in-residence at the cathedral to continue advanced explorations in geometry. In the same year, a special Cathedral bay was dedicated in 1993-94 to display selected models showing new work. It was adjacent to the Sacred Geometry Bay initiated by Keith Critchlow, author's predecessor at the cathedral in 1982, and which housed original models of Matila Ghyka as part of Cathedral's Library of Sacred Geometry.

28
Subdivision of various planar regular polygons into symmetric, non-periodic and asymmetric tessellations of rhombi, 1992-93.

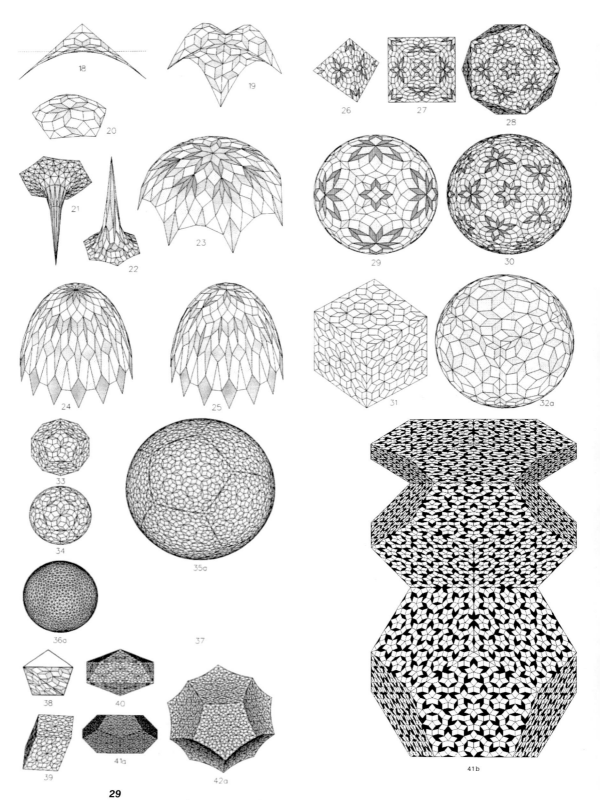

29
Examples of a different hypersurfaces obtained by using the tessellated polygons in fig.24 as units of polyhedral and higher dimensional structures, 1992-93.

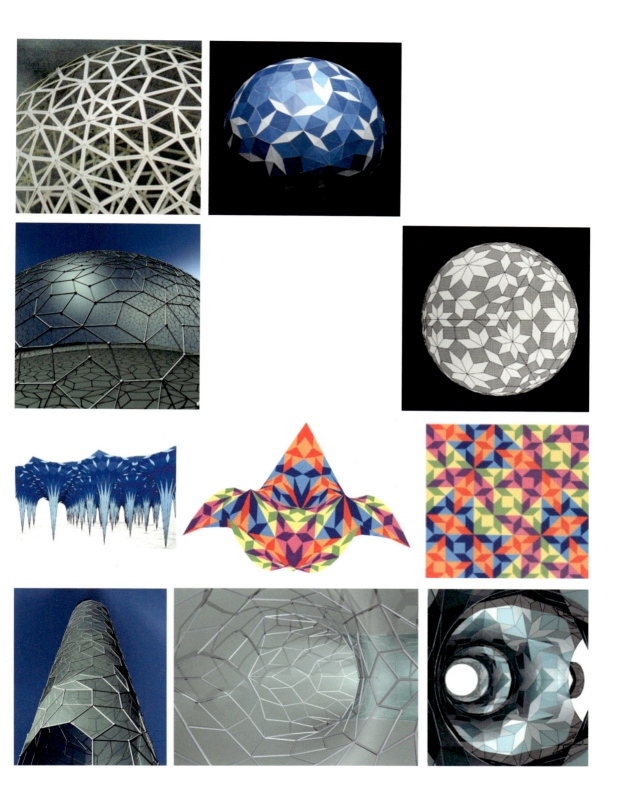

30
Various examples of non-periodic tilings on the surfaces of spherical polyhedra (*top two rows*), a membrane surface with cutting patterns a saddle surface with fractal coloring (*third row*), a cylinder, a torus and the Schwarz surface (*bottom row*), 1992-93, '98.

modern world".[21] Another physical model (*top row, right*) from that period was displayed, also at the Cathedral, at the Bucky Centennial exhibition curated by the author in 1995 to commemorate Buckminster Fuller's 100th birth anniversary.[22] This model was built in cardboard with blue colored triangles in 3 shades, each color representing one of the different rhombi from the $n=6$ case. These early models set the stage for their subsequent extension a decade later into a group of metal sculptures as part of *Milgo Experiments 4* [52].[23]

Hypersurfaces in Sheet Metal: Hypersurfaces in sheet metal at larger scales in a more substantial material than paper were built at Milgo by translating complex digital models into physical structures accurately. The accuracy of assembly became critical with increase in size which increased in thickness of material which, in turn, increased its resistance to folding flat sheets. These were overcome and new methods were invented in realizing the largest built piece *SEED54* in this series, 8ft x 6ft x 4ft. Some built examples are shown in **figs.31** and **32** followed by digital works in **figs.33** and **34**.

In fig.31, the aperiodic elliptic cut-outs (*top left*) correspond to the rhombic cells in the base tiling pattern. 3D and 4D cells projected on the spherical surface can be seen by visualizing circles on faces of cubes which are distorted in their projected states. It led to the triangulated version (*middle left*). The ellipsoidal versions of spheres (*top* and *bottom right*) added another layer of dimensionality to these works. An ellipsoid inscribed in a rhombohedron is a 3D projection of a sphere inscribed in a cube in higher dimensions. The tiling of the ellipsoid with curved hypertiles, also projected from higher dimensions, makes the tiled ellipsoid doubly-hyper.

Non-convex Hypertiles: These pieces paved the way for more adventurous structures with non-convex hypertiles (**fig.32**). In the examples shown, two adjacent rhombi are fused into a non-convex hexagon, a 6-sided crescent shape, and the spheres are tiled with different crescents. The crescents are in two versions, as "voids" (*top row*) or "solids" (*bottom row*). The solid versions using rounded crescents enabled biomorphic tile shapes for a curved hypersurface. The opportunity to scale up these experiments came with *SEED54*.

Fig.33 shows additional examples in digital images and include a sphere with a tiling of solid 6-sided crescents having rounded corners (*top left*) and others with void tiles. An ellipsoid with 8-sided non-convex void polygons (*middle left*) The upper limit to the number of sides of non-convex tiles will depend on the number of rhombi in the original 2D tiling of regular polygons. The two spheres (*top* and *bottom right*) are topologically isomorphic and show

---

[21] Excerpted from Dean Morton's letter to a foundation (Rolex Awards for Enterprise) seeking support for the Pratt Cathedral Center for Morphology to be housed in the Cathedral's south-west tower where *"this most collection of the most scientifically advanced, and at the same time extraordinarily beautiful, geometric models can be displayed in the most inspiring setting to the widest public"*, 25 February 1992.
[22] An expansive international exhibition 'Contemporary Developments in Design Science' comprised drawings and models from the cutting-edge work at that time from more than 75 leading individuals and firms in experimental structures, theoretical and applied morphology within architecture, design science, science and mathematics.
[23] See Introduction, footnote 6, for the others.

Hyper Architecture

31
Various examples of tiled spheres and ellipsoids with voids based on triangles and rhombi in the *Hypersurfaces* series (2007-2011). The bottom two images on the left are from Design Miami 2011 solo exhibition presented by Moss Gallery and the window display at Moss Gallery, New York.

32
Four examples from the Hypersurface series (2007-2011) with 6-sided crescent shaped tiles in two states, "voids" (*top row and bottom*) and "solids" (*middle row*).

Hyper Architecture 253

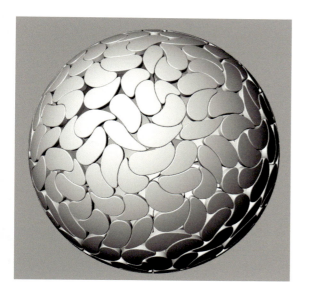
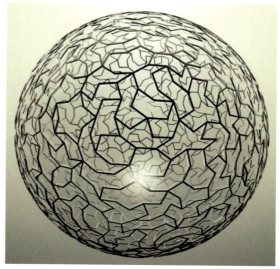
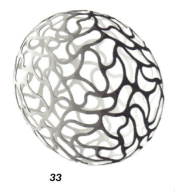
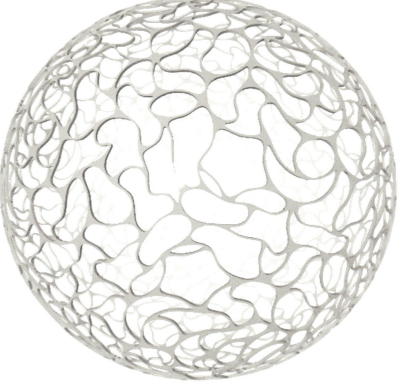

33
Digital images of a spherical tiling with 6-sided "solid" crescents touching point-to-point (*top left*, 2008), an ellipsoid with 8-sided non-convex "void" polygons (*bottom left*, 2017), a sphere with a combination of even-sided non-convex polygons having 4, 6, 8, or 10 sides shown as a frame structure (*top right*, 2006) and its biomorphic version (*bottom right*, 2006).

34 *(next page)*
SEED54 (2012), laser-cut stainless steel, located on the street plaza of 1330 Avenue of the Americas (at 54th Street), New York City.

a combination of even-sided non-convex polygons having 4, 6, 8, or 10 sides with straight edges in its curved biomorphic state.

SEED54: *SEED54* (**fig.34**), a public sculpture for the outdoor plaza of the Emery Roth building at 1330 Avenue of the Americas and 54th Street, Manhattan, was commissioned by RXR Realty for their plaza renovation by Moed De Armas and Shannon Architects [53,54]. Similar to a couple of earlier examples, this too was a tiling of 6-sided crescents on the surface of an ellipsoid. Tiling the surface of a sphere with only hexagons is denied mathematically [55]; the hexagons are implied to be convex, as in a beehive. However, *SEED54* provided a tongue-in-cheek solution to tile a sphere with only non-convex hexagons. As in polygons with high number of sides we saw in **fig.10**, the definition of polygons in a tiling is used here is non-standard. It alters the standard tiling rule which requires at least 3 polygons meeting at a vertex. Tilings with non-convex tiles have 2 polygons meeting at a vertex at many locations, a condition excluded in standard tiling designs.

The smaller pieces in the hypersurface series ranged between 1ft and 4ft diameter with thicknesses between 1/32nd and 3/32nd inches. *SEED54* at 8ft tall was in 1/4in thick stainless steel, something we hadn't done before. It was surprising that the thickness of sheet metal did not inhibit its bending or distort the resulting geometry due to the high-level fabrication skills at the factory that guaranteed the precise dihedral angles needed for a closed faceted surface. The fabrication technique, kept visible in the final form, was part of the sculpture's aesthetic.

SEED54 dealt with a small portion of possible hypersurfaces that can be built from sheet metal. We look forward to further experiments that push the bounds of building with sheet metal while we continue to explore higher dimensions as a source for physically built structures using other building systems and green materials.

SEED54, and other seed sculptures shown, are a homage to seeds, the carriers of life. With the continuing destruction of our forests, seeds are threatened. The interest in their designs has applications to design of human habitats and also to the design of future seeds.

Hypersurface Blobs: The principle of fusing 2D cells (rhombi) to generate convex and non-convex tiles is extended to 3D cells (rhombohedra) which are fused to get non-convex shells (non-convex zonohedra) as exterior surfaces of a single interior space. The shared interior walls of the 3D cells are removed. In their rounded states, this generates non-convex curved shells shaped like blobs, amoebas, squiggly forms and a variety of irregular morphologies. These irregular curved forms, hyperblobs, are an example of *cellular biomorphism* based on an underlying order derived from higher-dimensions.

Two examples of multi-cell closed hypersurfaces are shown in **fig.35**. A 3D kidney-shaped sculpture (*top images*) is defined by the tiled outer faces of 2 fused rhombohedra. and projected on to a smooth curved surface which blends the two ellipsoids. As in the spheres and ellipsoids shown earlier, the surface tiling can comprise convex or non-convex tiles in solid or void states. The pieces shown here have elliptic holes. The bottom image shows another piece derived

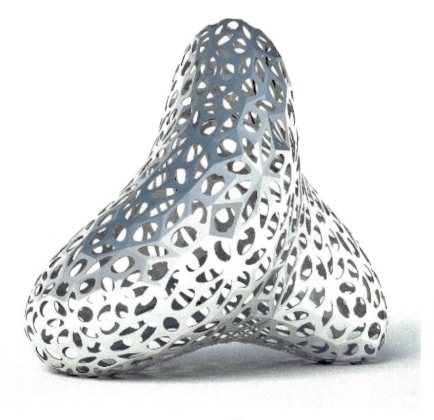

35
Digital images of two hyper-blobs. A 3D kidney shaped sculpture composed of rhombi with elliptic holes and derived from the outer surface of 2 fused rhombohedra in front and side views (*top*, 2012) and another piece derived from three fused rhombohedra tiled with rhombi having elliptic holes (*bottom*, 2018).

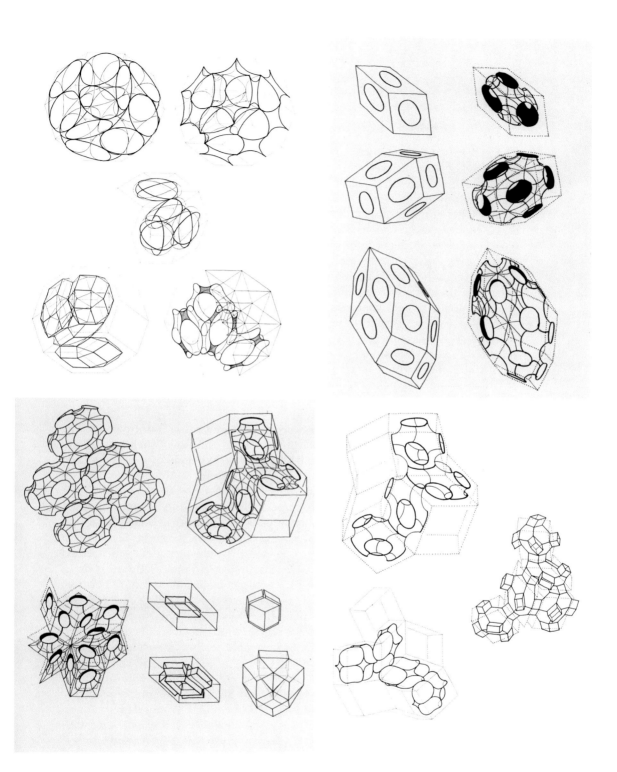

36
Multi-cells embedded in hypercubes and portions of non-periodic configurations; author's drawings, 1986-87, '94.

from the outer surface of 3 fused rhombohedra tiled with rhombi; it's curved surface blends three corresponding ellipsoids into a single continuous surface.

Multi-Cell Hypersurfaces

Multi-cell hypersurfaces comprise hypercells nested without intersections and embedded in hypercubes [35,38,42]. Five examples from a larger inventory are shown in *top left* cluster of images in **fig.36**, all derived from a 5D cube, all shown with non-intersecting cells. One has cells with flat faces and straight edges (*bottom left*), two have a combination of flat and curved faces with curved edges (*top*), and one has curved edges and curved faces (*middle, bottom right*). Continuous periodic, non-periodic or irregular surfaces can be derived systematically by simple geometric operations on the base cells. The top right cluster of images shows 3 cells on left — outer shells of 3D, 4D and 5D cells — and corresponding hypercells with saddle polygons and holes derived from them. These pack together to fill space as shown with two examples (*top*) in the *bottom left* cluster of images which are portions of a periodic and a non-periodic minimal surface. Additional examples in this cluster include a minimal surface from the Penrose tiling (*bottom left*), and cells of a 6D non-periodic infinite polyhedron with straight edges and flat faces based on icosahedral symmetry [38]. The *bottom right* cluster of images shows additional examples with multi-cells as portions of non-periodic configurations.

A special class of multi-cell hypersurfaces are analogs of infinite polyhedra mentioned earlier. Their curved counterparts are minimal surfaces [56-59]. shown with some examples in **figs.37** and **38** as portions of a non-periodic minimal surface derived from the well-known quasicrystal composed of two space-filling golden rhombohedra. Each rhombohedron is a skeleton of a curved 3D cell of the well-known triply-periodic surface, the Schwarz surface. **fig.37** shows three examples of pairs of cells of minimal hypersurfaces (*top row*) and their corresponding clusters in serial sections below. The one on left comprises curved 3D cells of a projected 6D Schwarz surface described further in **fig.38** (*top row*) which shows the projected hyper-Schwarz cells in a sequence of increased dimensions proceeding from left to right. The two pairs of 3D cells on left, a 4D cell comprising four such 3D cells next, a 5D cell comprising ten such 3D cells and a 6D cell comprising twenty hyper-Schwarz 3D cells on the right. All cells are shown within their skeletal cells indicated by lines. Extending beyond the 6D cell,[24] additional 3D cells can be added to make an aperiodic minimal surface.

Four studies of sculptures based on a small portion of this aperiodic minimal surface are also shown in **figs.38** and **39**. In **fig.38**, an open cluster of open 3D hyper-Schwarz cells (*bottom left*), a cellular triangulated version and with graded holes to decrease weight with height (*bottom right*), one with uniform 6-sided crescents as in earlier spherical pieces (*above left*), and another cellular one having graded rhombic holes in **fig.39**.

---

[24] The outer skeletal form corresponds to a rhombic triacontahedron, a multi-faceted form with 30 rhombi.

Hyper Architecture

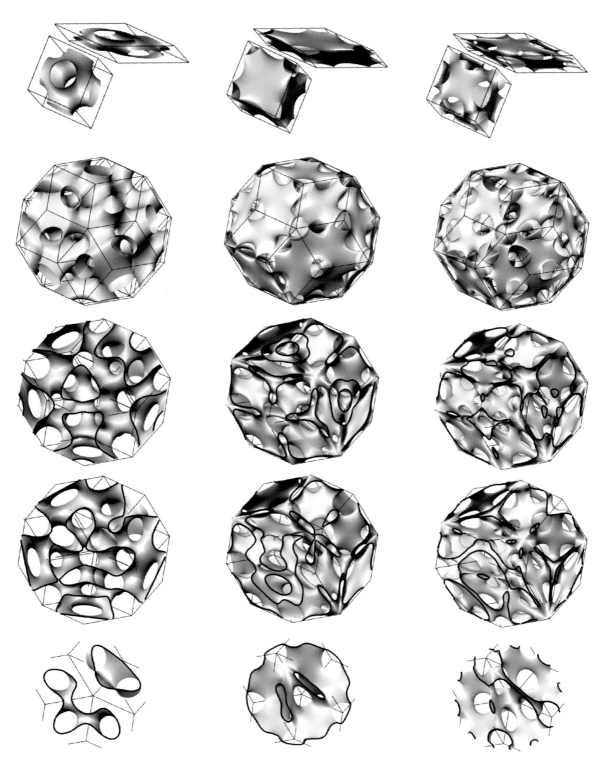

37
Three examples of non-periodic minimal surfaces with icosahedral quasi-symmetry shown with their 2 cells in an exploded view (*top row*), and corresponding serial sections of clusters below.

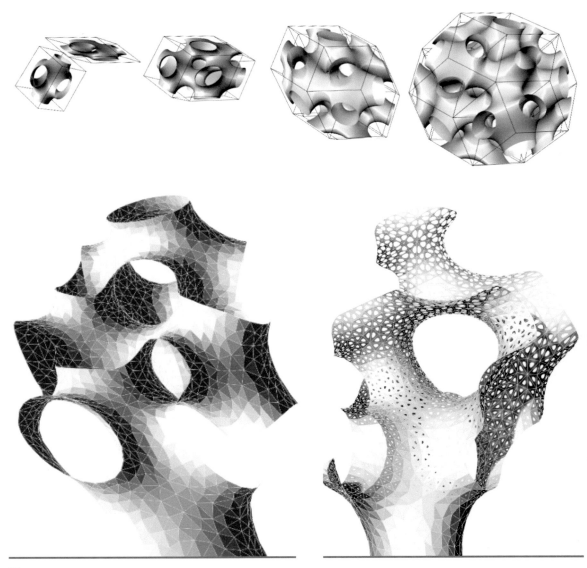

38

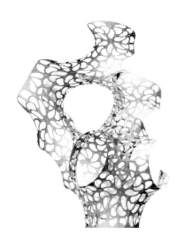

38
Top row shows a sequence of two 3D cells, a 4D cell, a 5D cell and a 6D cell composed of golden rhombohedra with derived Schwarz surface cells within each 3D cell as the building blocks of a non-periodic minimal surface that fills space. Various sculpture studies (2017) based on the hyper-Schwarz surface (*below*).

39 (*next page*)
Another sculpture (2017) based on the hyper-Schwarz surface tiled non-periodically and having graded elliptic holes to reduce mass with increasing height.

Hyper Architecture 261

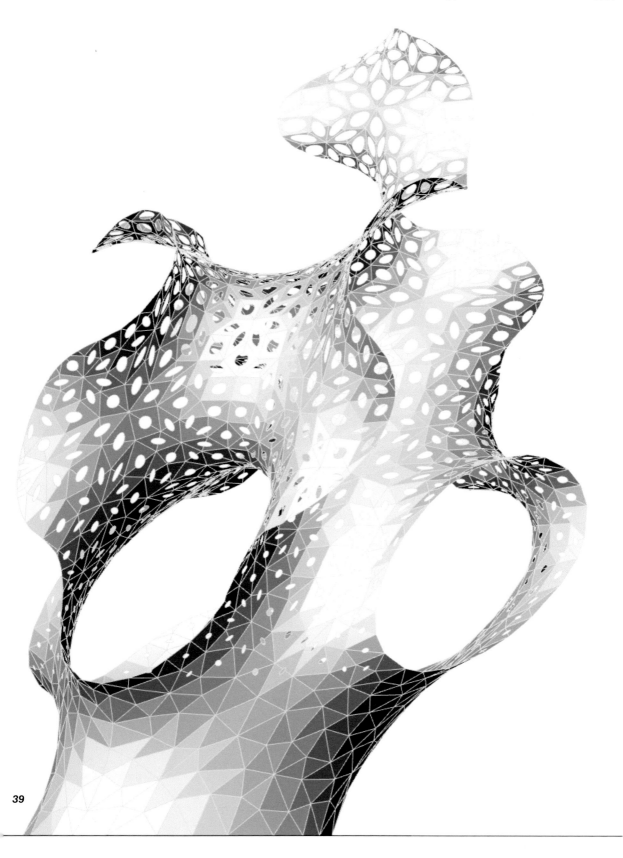

39

Multi-Cell Hypersurfaces with Interior Elements: An interesting class of multi-cell hypersurfaces are those with interior walls or other interior elements. This is a very large class of hyperstructures which can also be derived combinatorially. These are different from the minimal surfaces or infinite polyhedra mentioned earlier. Two different examples are shown in **figs.40** and **41**. The design of the inaugural NYCx Design Pavilion (**fig.40**) in 2015 at Astor Plaza[25] provided an opportunity to apply hypersurfaces to architecture. Hypersurfaces embedded in 5D space, based on n=4 non-periodic tiling (in plan) plus one vertical dimension, was chosen to design the layout for a semi-open public display and gathering space, indoor spaces for exhibition and presentations for this plaza. The proposed system used an inventory of curved pipes and PTFE membrane surfaces constructed from Teflon coated woven fiberglass material.

One-sided Hyper Minimal Surface: Hypersurface embedded in 8D space ($n=7$ tiling in plan plus a vertical dimension) was applied to an outdoor structure *3D Hypersurface* originally designed for Architecture OMI (**fig.41**). It is an example of a one-sided hyper minimal surface with three fused cells. It is 'one-sided' like the Mobius strip which has no inside or outside but ours is a complex 3D surface without a self-intersection. 'Hyper' because it is embedded in higher dimensions, a 4D cube portion of the 8D space. 'Minimal', because the surface is conceived as a tensioned membrane having least surface with well-defined boundaries similar to soap films within wire frames. This provided the starting point for a structure that has no inside or outside. It is being built as a structural surface from an aluminium laminate. The twist in the surface helps it withstand the uplift from strong winds and designed for its original outdoor setting, an open field in upstate New York. This project is in production at the Center for Experimental Structures (CES), School of Architecture, Pratt Institute, with the active involvement of students, interns and guidance from experts.

The surface relaxation of the original cells led to the surprising "emergence" of the column as part of the continuous surface. This hints at the origin of micro-columnar structures in nature (e.g. trabeculae in bone or reticulates in minerals) which could emerge from minimizing flat cell walls to become tubular networks, or in an industrial process for making cellular foam materials. Cellular aggregates (composed of only faces) can thus morph continuously into biomorphic lattices (grids, frames) with smooth saddle polyhedral joints.

The topology of *3D Hypersurface* questions the basic premise of architecture as well as living organisms both of which require a controlled "interior" separated from the "exterior" environment. This requires both to be built as "two-sided" surfaces, one side facing the outside and the other facing the inside. In the box-shaped rooms most of us live in, the walls are two-sided. One-sided surfaces like the Mobius strip or *3D Hypersurface* makes one wonder if one-sided buildings or organisms are possible? Chemists have proposed Mobius strip molecules including one from graphene, a single-atom thick surface made from carbon [60]. Will synthetic biology follow?

---

[25] The author was invited by Ilene Shaw, an independent producer, who wanted a distinct pavilion for the annual NYC event.

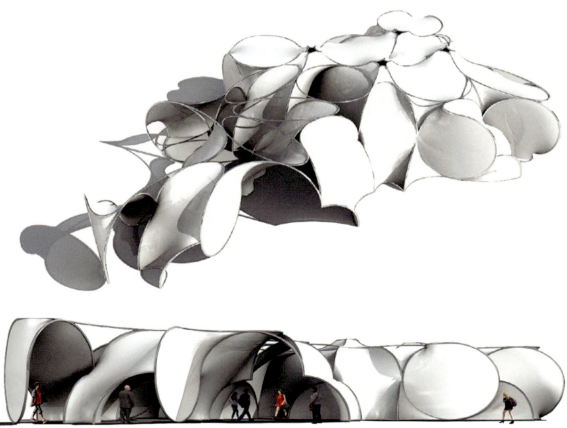

40
Hypersurfaces applied to NYCx Design
Pavilion 2015, Astor Place, New York
(2014).

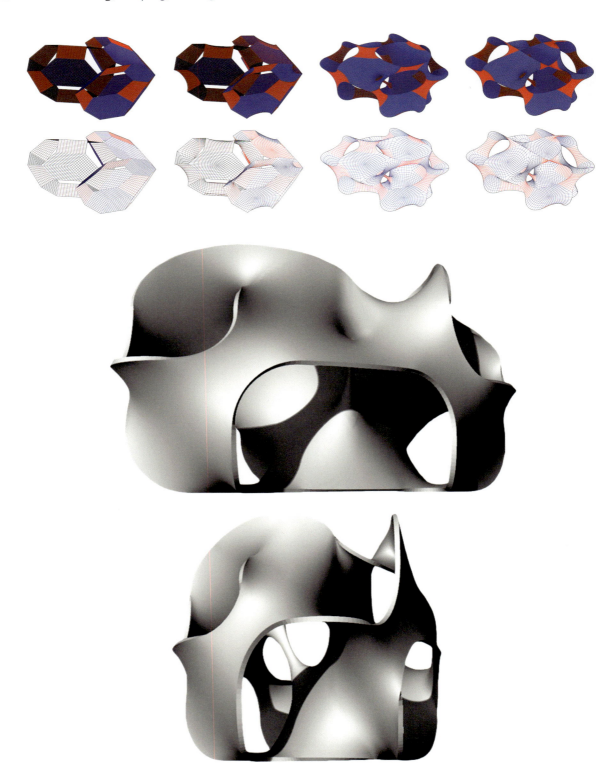

41
Hypersurfaces (*top*) applied to an outdoor pavilion-type structure *3D Hypersurface* (*below*), 2014-15.

Performance of Hyperstructures: The experiments presented in this chapter are focused on the rich morphologic possibilities of higher dimensional structures as an open-ended resource for design. They need to be combined with material, constructive, programmatic and environmental aspects of built structures at different scales. As physical structures, projected hyperstructures can be built using familiar or emerging building technologies and materials which are heading towards biology as the driving technology. Their formal and technical advantages, besides a different look and spatial experience, need to be firmly established. For example, projected hyperstructures pack more structure (mass) or surface area per unit volume of space and have greater connectivity. These inherent features of hyperspace may prove advantageous in designing structures with a higher density, strength, spatial porosity which impact how designed objects interface with the environment. These are important criteria as we address planetary climate issues that impact design of spatial envelopes of buildings and other structures in different environments especially extreme climate conditions, or physical structures that interface land and water in coastal regions, or underwater structures like artificial reefs that restore the ecosystem. The performative promise of projected higher dimensional structures remains to be explored.

High Dimensions as Metastructures: The important use of higher dimensions in organizing information is the subject of Chapters 8 and 9 and display its significant conceptual value as meta-structures, structures underlying structures.

REFERENCES

1. Lalvani, H., *Multi-dimensional Periodic Arrangements of Transforming Space Structures*, PhD Dissertation, University of Pennsylvania, Published by University Microfilms, Ann Arbor, 1981.

2. Lalvani, H. *Structures on Hyper-Structures*, (self-published), Lalvani, New York, 1982; a slightly revised version of [1].

3. Lalvani, H. Structures on Hyper-Structures, *Structural Topology*, University of Montreal, Quebec, 1982.

4. Lalvani, H. Hyper-Structures, In: *Engineering a New Architecture*, Conference Proceedings, ed. Dombernowsky, P. and Wester, T., May 26-28, 1998, Aarhus School of Architecture, Denmark, 261-271.

5. Coxeter, H.S.M. *Regular Polytopes*, Courier Corporation, 1973.

6. Henderson, L. D. *The Fourth Dimension and Non-Euclidean Geometry in Modern Art*, MIT Press, 2013.

7. Greene, B. *The Elegant Universe: Superstrings, Hidden Dimensions, and the Quest for the Ultimate Theory*, W. W. Norton & Company, 2010.

8. Randall, L. *Warped Passages: Unraveling the Mysteries of the Universe's Hidden Dimensions*, Ecco, 2006.

9. Luminet, J-P., Weeks, J.R, Riazuelo, A., Lehoucq, R., Jean-Phillippe Uzan, J-P. Dodecahedral space topology as an explanation for weak wide-angle temperature correlations in the cosmic microwave background, *Nature* 425, 593–595 (2003).

10. Kaku, M. *Hyperspace: A Scientific Odyssey Through Parallel Universes, and the 10th Dimension*, Anchor, 1995.

11. Crescent Tiles, In: Best Design in America, ID's Annual Design Review, *Industrial Design*, N.Y., July-August 1985, 129.

12. Lalvani, H. Crescent-shaped Polygonal Tiles, *U.S. Patent 4,620,998*, Nov 4 1986 (filed Aug. 15 1985).

13 Lalvani, H. Non-periodic and Periodic Layered Space Frames Having Prismatic Nodes, *U.S. Patent 5,007,220*, April 16 1991 (filed Dec 2 1988, cont. of Apr 9 1987).

14 Lalvani, H. Node Shapes of Prismatic Symmetry For a Space Frame Building System, *U.S. Patent 5,265,395*, Nov 30 1993 (filed Mar 4 1991, cont. in part of [12]).

15 Lalvani, H. Periodic and Non-Periodic Tilings and Building Blocks from Prismatic Nodes, *U.S. Patent 5, 575, 125*, November 19 1996 (filed April 15 1991, cont. of [12], cont. of Apr. 9 1987).

16 Lalvani, H. Non-convex and Convex Tiling Kits and Blocks from Prismatic Nodes, *U.S. Patent 5,775,040*, July 7 1998 (filed November 18 1996, cont. in part of [14], cont. of [12], cont. of Apr 9 1987).

17 Lalvani, H. Pentiles: Discovery and Concepts, Symmetry: Culture and Science, Vol. 32, No. 2, 2021, 129-132.

18 Locher,J.L. The Magic of *M.C. Escher*, Thames and Husdon, 2013

19 Jarvis, J. (trans.), *Friedrich Froebel's Pedagogics of the Kindergarten*, D. Appleton & Company, New York, 1896.

20 Wright, F.L. *A Testament*, Horizon Press, 1957.

21 Kahr, B. Crystal Engineering in Kindergarten, *Crystal Growth & Design*, Vol. 4, No. 1, 2004. Bart cites Rubin's book *Intimate Triangle: Architecture of Crystals, Frank Lloyd Wright, and the Froebel Kindergarten*, Polycrystal Book Service, Huntsville, Alabama, 2003, for a crystallographic view of Froebel's work.

22 Fuller, R.B. *Synergetics*, McMillan, New York, 1975.

23 Bosterman, N. *Inventing Kindergarten*, Harry N. Barms, New York, 1997.

24 Fenyvesi, K. Lalvani's Pentiles Tessellation System At Experience Workshop's Educational Experience, *Symmetry: Culture and Science* Vol. 32, No. 2, 273-276, 2021.

25 Bernstein, F.A. Jack Esterson Designs Film and Video Department for Pratt, *Interior Design*, September 29, 2015. https://www.interiordesign.net/projects/10900-jack-esterson-designs-film-and-video-department-for-pratt/

26 Lalvani, H., Towards *n*-Dimensional Architecture, In Structural Morphology, *Proceedings of the First International Seminar on Structural Morphology*, pp. 63-74, Montpellier, France, 1992.

27 Lalvani, H. Architectural Spaces and Structures Based on the Hypercube, Sept 1993, Patent application, abandoned.

28 Lalvani, H. *Higher Dimensions and Architecture*, Manuscript, Graham Foundation For Advanced Studies in the Fine Arts, Chicago, 1994.

29 Lalvani, H., n-Dimensional Proportion Systems for Architecture, *HyperSpace*, Vol. 3, No. 1, pp.8-24, Japan Institute for HyperSpace Science, 1994.

30 Le Corbusier, *The Modulor: A Harmonious Measure to the Human Scale*. Basel & Boston: Birkhäuser, 2004; Two volumes published earlier in 1954 and 1958.

31 Jaffe, H. L.C. *de Stijl*, Abrams, 1971.

32 Kappraff, J. *Connections: The Geometric Bridge Between Art and Science*, McGraw-Hill, N.Y. 1990. The second edition, World Scientific, includes a section S.5 'New Morphological Discoveries by Haresh Lalvani', 482-6, 2001.

33 Shechtman, D.; Blech, I.; Gratias, D.; Cahn, J. "Metallic Phase with Long-Range Orientational Order and No Translational Symmetry". *Physical Review Letters, 53 (20): 1951–1953, 1984*.

34 Baer, S. *The Zome Primer*, Zomewroks Corp., Albuquerque, New Mexico, 1970.

35 Miyazaki, K. and Takada, I. Uniform ant-hills in the world of golden isozonohedra. *Structural Topology*, 4, 1980, pp.21-30.

36 Miyazaki, K. *An Adventure in Multi-Dimensional Space*, Wiley & Sons, NY, 1986.

37 Lalvani, H. (1986/87) Non-periodic Space Structures, *Space Structures* 2 (1986/87) 93-108, Elsevier, UK.; re-published in *Int'l Journal of Space Structures*, Vol. 26, No. 3, September 2011, pp.139-154.

38 Lalvani, H. Non-periodic Space-fillings with Golden Polyhedra, in *LSA'86 Proceedings, First International Conference on Lightweight Structures in Architecture*, pp. 202-211, Univ. of New South Wales, Sydney, Australia, 1986.

39 Robbin, T. *Shadows of Reality: The Fourth Dimension in Relativity, Cubism, and Modern Thought*, Yale University Press, 2006.

40 Peterson, I, Shadows and Symmetries, *Science News*, Dec 21 1991.

41 Bindi, L., Steinhardt, P. J., Yao, N., Lu, P. J. *"Natural Quasicrystals"*. Science 324 (5932): 1306–9, 2009.

42 R.Haase, Kramer, L., Kramer, P. and Lalvani, H. Three Families of Cells for Quasi-lattices Associated with the Icosahedral Group, *Acta. Cryst*, a 43 (1987) 574-587.

43 Lalvani, H. Building Structures Based on Polygonal Members and Icosahedral Symmetry, *U.S. Patent 4,723,382*, February 2 1988 (filed Aug 15 1986).

44 Lalvani, H. A Building System using Saddle Zonogons and Saddle Zonohedra, *U.S. Patent 5,036,635*, August 6 1991 (filed Oct 26 1989, cont. of Aug 24 1987).

45 Lalvani, H. Building Systems Using Saddle Polygons and Saddle Zonohedra Based on Polyhedral Stars, *U.S. Patent 5,155,951*, October 20 1992 (filed June 1991, div. of [38]).

46 Lalvani, H. Building Systems with Non-regular Polyhedral Nodes, *U.S. Patent 5,505,035*, April 9 1996 (Filed June 24 1992).

47 Lalvani, H. Space Structures with Non-Periodic Subdivisions of Polygonal Faces, *U.S. Patent 5,524,396*, June 11 1996 (Filed Jun 10 1993).

48 Lalvani, H. Building Systems with Non-Regular Polyhedra Based on Subdivisions of Zonohedra, *U.S. Patent 5,623,790*, April 29, 1997 (filed April 25 1994, cont. of Aug 5 1991, cont. in part of [38], cont. of Mar 6 1989, cont of Aug 24 1987).

49 Lalvani, H., The Architectural Promise of Curved Hyper-Spaces, *Proceedings, Second International Seminar on Structural Morphology*, University of Stuttgart, 1994.

50 Burt, M., M. Kleinman, and A. Wachman, *Infinite Polyhedra*, Technion, Haifa, Israel, 1974.

51 Lalvani, H., Hyper-Geodesic Structures: Excerpts From a Visual Catalog, In: *Spatial, Lattice and Tension Structures*, Eds. John F. Abel et al., IASS-ASCE Symposium '94, Atlanta, GA, 1994. Republished in HyperSpace, VoL. 3, No. 3, 1994, 37-50, Japan Institute of HyperSpace Science.

52 Lalvani, H. Hypersurfaces, Milgo Experiment 4 (2006-2012), *Proceedings of the IASS Annula Symposium 2020/21 and the 7th International Conference on Spatial Structures, Inspiring the Next Generation*, Eds. S.A.Behnejad, G.A.R.Parke and O.A. Samavati, 23-27 August 2021, Guildford, UK, 3190-3202.

53 Fountain, H. A Play on Nature's Patterns, *New York Times*, Science Times, Feb 13, 2012.

54 Tamarin, N. Sphere of Influence (centrefold), *Interior Design*, April 2013.

55 Thompson, D'Arcy W. *On Growth and Form*, Cambridge, 1942, 708.

56 Burt, M. Spatial Arrangements and Polyhedra with Curved Surfaces and their Architectural Applications, Technion, Haifa, 1966.

57 Pearce, P. *Structure in Nature is a Strategy for Design*, MIT Press, Cambridge, MA, 1978.

58 Schoen, A. Infinite Periodic Minimal Surfaces Without Self-Intersections, *NASA Technical Note* D-5541, May 1970.

59 Lalvani, H. Visual Morphology of Space Labyrinths: A Source for Architecture and Design, In: *Beyond the Cube, The Architecture of Space Frames* by Francois Gabriel, 406-26 and color plates (6-14), John Wiley, 1998.

60 Lalvani, H. Morphological Universe: Genetics and Epigenetics in Form-Making, *Symmetry: Science and Culture*, Vol. 29, No. 1. 240pp, January 2018.

61 Lalvani, H. [61], Introduction by Gyuri Darvas, 5-6.

MORPHOVERSE
AND THE UNIVERSAL MORPH TOOL KIT

"God made the integers; all else is the work of man."
Leopold Kronecker[a] (1886)

"...from so simple a beginning, endless forms most beautiful and most wonderful have been, and are being evolved."
Charles Darwin[b] (1859)

"I saw in a dream a table where all the elements fell into place as required."
Dmitri Mendeleev[c] (1869)

The concept of the *Morphoverse*, a morphological universe, was introduced in Chapter 2 (pp.29-35) and in some parts of Chapters 3 and 4. This chapter presents a journey through parts of the morphoverse mentioned earlier as the (theoretical) *"universe of all possible morphologies"* (p.29). This chapter also presents parts of a universal *morph tool kit* as the generator of space-time transformations of form. This toolkit enables the wide variety of structures inhabiting the morphoverse. Though the examples presented here are relatively simple forms, and include many symmetric cases from the work during the 1980's-90's [1-7], they have provided the basis for an extension to asymmetric and irregular morphologies. The organization of form into families, transformations *within* and *between* families, some *discrete* and others *continuous*, are applicable to complex, irregular and emergent morphologies. The concept of continuous transformations is linked to kinetic and growing morphologies. Some examples of emergence were presented in Chapter 3 (p.75), and Chapter 4 (pp.110-11), and non-periodic and order-based irregular geometries from 2D and 3D projections of higher dimensions in Chapter 7. High

---

[a] Mathematician Leopold Kronicker (1823-1891), https://en.wikipedia.org/wiki/Leopold_Kronecker; this quotation was adopted by Stephen Hawking in his book *God Created the Integers: The Mathematical Breakthroughs that Changed History*, Running Press Adult, 2007.

[b] Darwin, Charles, *The Origin of Species*, 1859; extended quote adopted from Sean Carroll's book title[8].

[c] Popova, Maria, How Mendeleev Invented His Periodic Table in a Dream, https://www.themarginalian.org/2016/02/08/mendeleev-periodic-table-dream/

Morphoverse

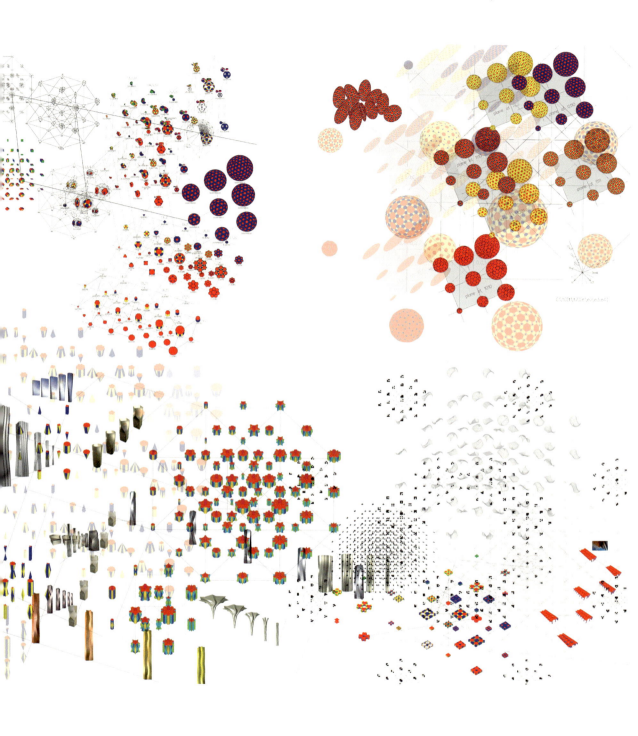

1
A very small portion of the *Morphoverse* (morphological universe) is shown as a visual montage of various smaller portions juxtaposed in this image from 2004 (**video 23**).

video 23

symmetry and irregularity in form, lying on the two extremes of the complexity slider, appear to be united through an underlying *meta-symmetry* captured in the diagrams shown (see, for example, **figs.47** and **48** of this chapter).

The term "morph tool kit" in the title of this chapter is adapted from the term "gene tool kit" in Evo Devo [8], the field of evolutionary developmental biology mentioned earlier (p.128). Though it is a new term in the author's work, it is an apt description of the concept that has been in continuous development in this work over the years as a form-generation tool using a wide range of geometric and topologic transformations. Since space-time-mass is common to form in the living, non-living and human-made worlds, the designs within these three different worlds are related and constrained by the same rules of geometry, topology and physical structure [9-12]. This is a work in progress and the fully developed morph tool kit should have applications across these worlds. The focus of the work so far has been mostly to form as a space-time phenomenon.

Several windows into the *morphoverse*, overlaid in an early montage (**fig.1**, **video 23**), are extended in this chapter. These windows are higher dimensional, each a microverse of related morphologies or families of form (structure, shape) selected from the "infinite infinities" (p.72) within the *morphoverse*. The visually inclined readers can skip the technical aspects of the work and flip through the images to get a sense of the visual logic in the diagrams. This will help to see how the universe of form is inter-connected, open-ended and can expand by adding new formal features, how symmetry-preservation and symmetry-breaking are related processes at different scales of complexity and, more importantly, how *unity* and *diversity* are twin processes intertwined in the ever-expanding morphoverse.

Transformations and Form Continuum

The term *structure* is used here for a topological structure and *transformation* for topologic and geometric transformations of the structure. The transformations extend each structure into a *family of derivative structures* which transform to each other *within* the family (*intra-transformations*). This implies a genealogy. In addition, a structure is part of a larger family of structures where the transformations are *between* the structures (*inter-transformations*). The transformations are either *continuous* or *discrete* with clear instances where these two binary states are in a continuum. Continuous transformations between structures, or within their derivatives, demonstrates a *form continuum*, an idea central to this work. Though the examples shown are for higher-order symmetric structures, the transformations are general and apply to asymmetric structures. Two important classes of transformations, *topology-preserving* and *topology-changing* transformations, are central to the examples that follow.

Structures and Meta-Structures

Structures are mapped in higher dimensional space and represented in diagrams (networks) which show the relationship between structures. When all diagrams are inter-connected, they define the *morphoverse*. The diagrams are underlying structures or *meta-structures* which have a topological structure as well. This makes the structures (form) and meta-structures (meta-form) self-similar. This chapter is about the morphology of morphology[1] and has been in continuous development since its first introduction in [1].

Hyper-Structures

The diagrams used here are higher dimensional structures or *hyper-structures*. The hyper-structures include n-dimensional cubes, n-simplexes, n-cubic lattices and associated higher dimensional polytopes. They are shapes of spaces (meta-spaces, meta-structures, meta-forms, meta-shapes, meta-patterns, meta-networks) that define the *morphoverse*. The hyper-structures embed periodic tables of form, ad infinitum, and are a map of all interactions (combinations, permutations, point mutations and continuous transformations) between the parameters of form, hence between the generated forms. The periodic tables also include periodic tables of families of forms. This leads to periodic tables of periodic tables, demonstrating a self-similarity between levels of organization of knowledge. Though the examples of hyper-structures shown here are in Euclidean space, the extension of the underlying networks to curved non-Euclidean spaces is possible based on some unpublished examples developed by the author.

Morph Code

The mapping in higher dimensions is enabled by associating each *independent* generative parameter of any structure (form, shape, configuration, graph, network) with a point location within its meta-space and specified by a coordinate system, something like a higher-dimensional GPS of form. In the examples used here, namely hypercubes, hyper-cubic lattices, and simplexes in Euclidean space, these point locations are specified by the vertices of hyper-structures by their higher dimensional Cartesian coordinates. The coordinates provide a precise location for each structure expressed in a sequence of numbers. In the diagrams, these coordinates are the numbers or symbols within brackets ( ). Consecutive brackets ( )( )( )... indicate several hyper-structures are involved, sometimes connected, otherwise in parallel and awaiting a future connectivity. These number sequences

---

[1] The term meta-morphology, as a description of this work on "morphology of morphology", was first suggested to me by Ted Goranson who described it as a "new science" (1986, personal communication). Goranson's term came from his background in computer science and he immediately saw the application of this work to artificial intelligence. Anne Tyng coined the term "metamorphology" earlier in the context of 'cycles' in her work. It was derived from metamorphosis, changes in cycles, as the egg-caterpillar-pupa-butterfly transformation [13].

provide a unique "genetic code of form" or *morph code* (p. 30). Each group of numbers within a bracket ( ) represents an independent family of related geometric or topologic parameters and defines an independent *morph gene* (morphological gene, p.72). A collection of morph genes ( )( )( )... defines a *morph genome* (morphological genome, p.28). This permits morph genes to be added, subtracted or compounded, in the same way as independent parameters. The hierarchical self-similarity gives the morphoverse a fractal structure. The number of independent parameters may be very large, some finite and others infinite, but it appears that the number of morph genes may be finite and much smaller in number. This makes identifying and mapping the morph genes more manageable. A collection of morph genes comprises the morph tool kit. Details of the morph code appear in [14,15]. Some of the images presented here show the generating stars (vector-stars) comprising the parameters of relevant morph codes.

Morphogenetic Pathways

The network representations in higher dimensional diagrams (*hyper-networks*) permit continuous transformations of form in a *form continuum*. These changes are represented by the edges of the networks and join one vertex of the network to its neighbors. Continuous changes of form within the network provide all possible cycles (closed loops) or open paths (pathways) of transformations within one diagram by connecting *any* vertex to *any* other vertex in *any* sequence. These paths are *morphogenetic pathways* that guide design at different scales from origin to development (*ontogenic design*) to evolution (*phylogenetic design*). The history of design, architecture and art are selected pathways within maps of all possible pathways. The latter provides paths for alternate "histories" and also a strategy for future designs.

Not all examples shown here display continuous transformations due to the divide between *discrete* and *continuous* phenomena. However, restricting to space-time phenomena, at least some morphologies that resist the continuum between discrete and continuous appear to give way when the binary states 0 and 1 are converted to continuous states by introducing real numbers (numbers with decimal places) between 0 and 1 [16,17].

Conservation Principles

Several conservation principles of form are observed in these hyper-cubic represetations of form dealing with topology (defined by integers like number of sides, number of neighbors, etc.) and geometry (defined by integers and real numbers like lengths, radii, etc.) [1-3]. These principles deal with complementarity introduced by William Katavolos with his visual mathematics,[2] conservation in metrics as well as topology with the observation that the net sum of transformations equals zero,[3] recently described as *zero-cyclic sum* in connection with chemical elements and some of their properties [18,19]. The principles underpin all hyper-cubic networks, apply to form and its transformations within these networks, and extend to hyper-cubic lattices. Selected applications to other disciplines are included in Chapter 9.

Variable Topologies

Variable topologies are described first (**figs.2-31**), variable geometries next (**figs.32-38**). This separation between the two is not a strict one since topology and geometry are inter-connected. This is followed by examples of fruits, sea-shells and Gaudi, and a revisit of one example of emergence dealing with cellular automata.

VEFC Space (figs.2-9)

The VEFC space is an inclusive 4D space for the number of 3D topological structures (forms, shapes, structures, spaces) based on the count of topological elements comprising them – vertices V, edges E, faces F, cells C. In the design arts, the equivalent terms are point, line, place and "volume" or "solid"; the term "cell", a biological term adopted in mathematics, is used here. This number is the Euler count,[4] introduced by the 18th century mathematician Leonhard Euler with the equation V-E+F=2 for single-cell structures (polyhedra). Single-cell structures are mapped in the 3D VEF space, and extend to multicellular 3D structures in the 4D VEFC space, and in higher dimensions (5D space for 4D structures, nD space for (n-1)-dimensional structures). The taxonomy of structures in these spaces is solely

---

[2] The author is indebted to William Katavolos for teaching him De Morgan's Law – unions, intersections and complementarity - using his primary colored cubes and blocks (personal communication, 1975). He is the first one to introduce 2D and 3D matrices in design. I was exposed to his unpublished work, with his associate Stanley Wysocki, in 1975. At the time, I had been working on continous trasformations of polyedra and tessellations and was organizing them into similar families. The exact organization scheme wasn't clear though I was able to inter-relate all transformations and had worked out the different families. When I discovered the relationship between their "heights" [46, pp.53-60] within the families, the relationhips matched Katavolos's visual organization of DeMorgan's law into a cube. This convinced me that the model of organization of polyhedra was a cube. Later, when I discussed it with him, he immediately (and graciously) acknowledged my contributions of continuous transformations, higher-dimensional organization schemes (beyond the cube), and generalized topologies and geometries (also beyond the cube) as original.

[3] In [1, p.41], topological conservation was stated as "sum of all transformations equals zero". Since I have found this "principle" to apply to many different examples - topology, geometry, numbers, diagrams of change (morphogenesis) - it leads me to think it is the inherent property of networks in which edges are vectors. Zaran Lalvani has provided a proof by induction to show why this result is "obvious", mathematically speaking (personal communication, October 9 2017) [19]; see Supplement, pp.393-4.

[4] The term "Euler count", adopted from aerospace engineers at NASA-Langley during my sabbatical there in 1989-90, was being used in the design of structures for outer space in counting the numbers of components in a space station structure to track the costs of the payload aboard the shuttle and in-space time by astronauts to assemble the structure.

based on topology where the metrics of lengths, angles and curvature are irrelevant. A small portion of this space which includes singularity structures is shown (**figs.2-9**). Eulerian structures, based on Euler's formula and their extensions, lie within a smaller subset of this space.[5] The transformations between structures in this space are discrete and change the topology.

Singularity Structures

Singularity structures, where edges, faces, cells emanate from a single vertex, are important in the origin of form. These exist in 2D, 3D and higher dimensions. Examples shown include singularity structures with more than one vertex as part of an extended family. In the latter structures, singularities are distributed and localized at different vertices.

VEF Space

A zoom into one 3D window of the *Morphoverse*, a VEF cube (**fig.2**), shows topological structures composed of numbers of vertices V (points), edges (lines) E and faces (surfaces) F represented by corresponding 3D VEF coordinates. This cube maps structures that have no 3D cells (C) and includes rosettes and relatives. Here, numbers force shapes to have holes. The important topological property, genus (g, the number of holes) is a by-product. The structures are characterized by the known Euler equation $V-E+F=1-g$.

Euler Characteristic

Parallel triangular planes in the VEF cube reveal topological structures with varying genus g (**figs.3A**, *left page*; **3B**, *right page*). In the Euler equation $V-E+F=1-g=\chi$, the right hand side of the equation is termed *Euler characteristic* and represented by χ (Greek 'chi') varies. The diagram captures polygons, rosettes and pod-like structures.

2
A 3D window into VEF space showing topological structures with numbers of vertices V, edges E and faces F and represented by a 3-digit code. Here, numbers force structures to have holes.

---

[5] The preliminary results indicate that the V-E+F=2 structures, all one-celled, lie on a flat plane in the VEF space, the "Euler plane". This plane extends to infinity and is flanked by two axes that extend to infinity, one with regular p-sided dihedra with VEF coordinates (p,p,2) and diagonal prisms (2,q,q). This anticipates multi-cellular Eulerian structures satisfying the equation V-E+F-C=1 to occupy a 3D 'Euler space' within the VEFC space.

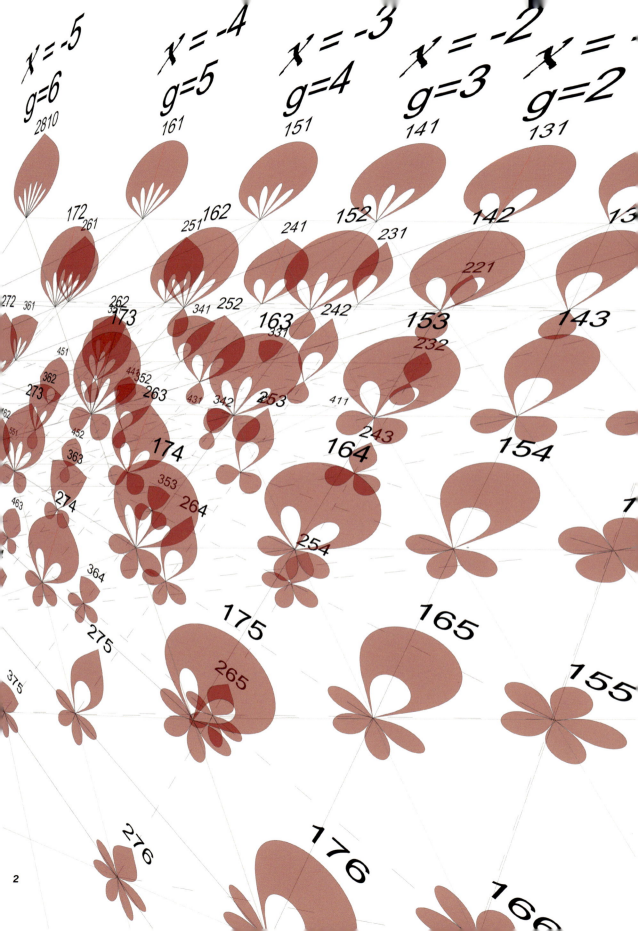

278 Coding, Shaping, Making

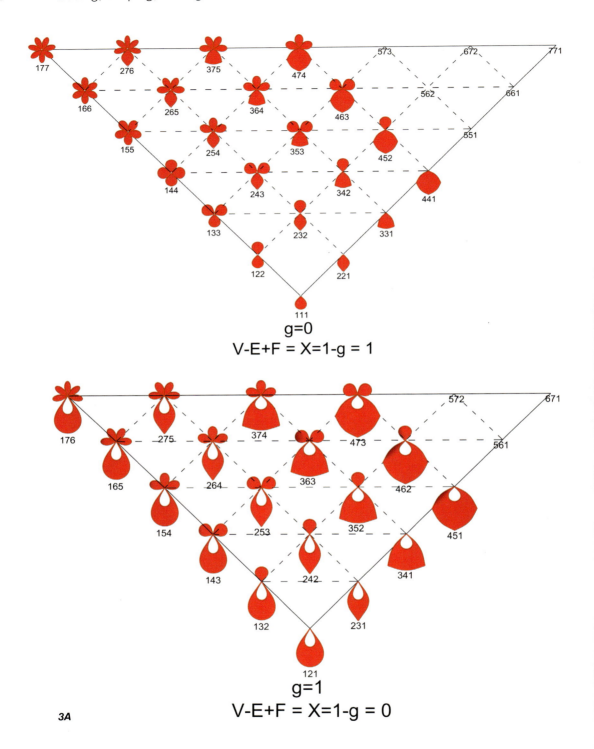

3A

3A, B
Triangular section planes in the VEF cube showing topological structures like polygons, rosettes, pod-like structures having the same number of holes (genus g) within each diagram.

Morphoverse 279

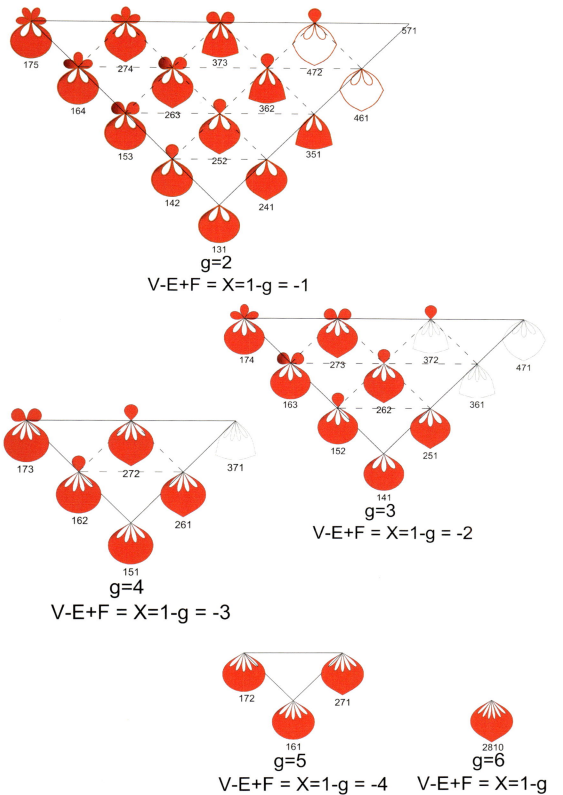

3B

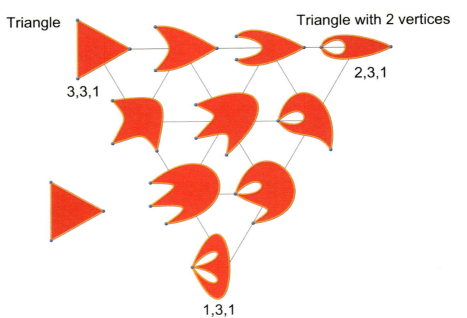

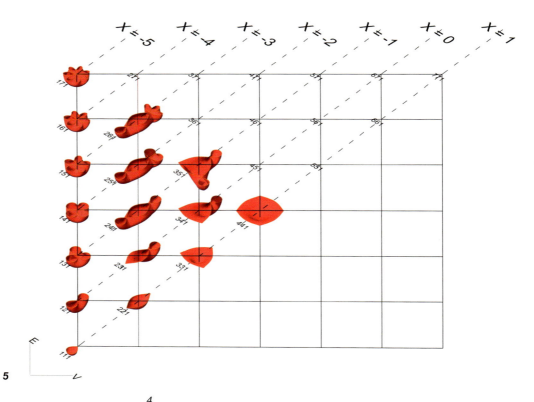

4
Continuous transformations between 3 structures from the VEF space redefine a triangle 331 (*upper left corner*) which morphs to a triangle with 2 vertices (*upper right*) and 1 vertex (*bottom*) (**video 24**).

video 24

5
A new class of polygons having one face (F=1) and no holes (g=0) that violate Euler's equation.

Triangle with 2 Vertices

Continuous transformations between 3 structures from the VEF space redefine a triangle 331 (*upper left corner,* **fig.4**). The morphings permit a triangle with 2 vertices 231 (*upper right corner*) and another with 1 vertex 131 (*bottom*) (**video 24**). All have 3 angles and are thus triangles (tri-angles) by definition.

Departure from Euler

A new class of polygons having one face (F=1) and no holes (g=0) in **fig.5** are examples that violate Euler's equation V-E+F=1-g= χ. Here, χ varies. **Fig.6** shows eight different pentagons (5 angles) with only 2 vertices which also depart from Euler's equation V-E+F=1-g= χ. Here g varies but χ is constant. Examples are shown for χ= -2, and similar examples exist for all polygons and higher structures. Fig.6 also displays an example of isomerism in form (form isomers),[6] those that have the same Euler count but have different forms.

Multi-cellular Dihedra

The 3D VEF space is extended to 4D VEFC space by adding the number of cells C (**fig.7**). Examples are shown with singularity structures having a varying number of cells with two faces (dihedral cells or dihedra). These structures have varying g and satisfy the Euler equation V-E+F-C=1-g= χ.

Tetrahedron with 2 vertices

Continuous morphing of a familiar tetrahedron 464 having 4 vertices (*left,* **fig.8**) to a new tetrahedron 364 with 3 vertices (*middle*) and a tetrahedron 264 with 2 vertices (*right*). Tetrahedron 164 is not shown. The numbers indicate their VEF coordinates. These examples are 3D versions of the triangles with 2 and 1 vertices in **fig.5** though the term tetrahedron (4 faces) holds. The method applies to all deltahedra, 3D structures composed only of triangles. The idea extends to all 2D and 3D structures, including space-filling structures, composed of polygons. The resulting structures depart from Euler's equations.

Spheres to Orbitals

Singularity structures obtained by the continuous morphing of a spherical hexagonal prism resembling a beach ball (*top left,* **fig.9**) to its various states as edges and vertices collapse in a continuum. Each state is shown with its VEFC coordinates and varying χ (the number inside a circle) from the formula V-E+F-C=1-g= χ. All, except three indicated with 1 inside a circle, violate Euler's formula which has χ =1 for the structures with no holes (g=0) like the ones

---

[6] The term is borrowed from Chemistry where the same number and types of atoms are arranged differently.

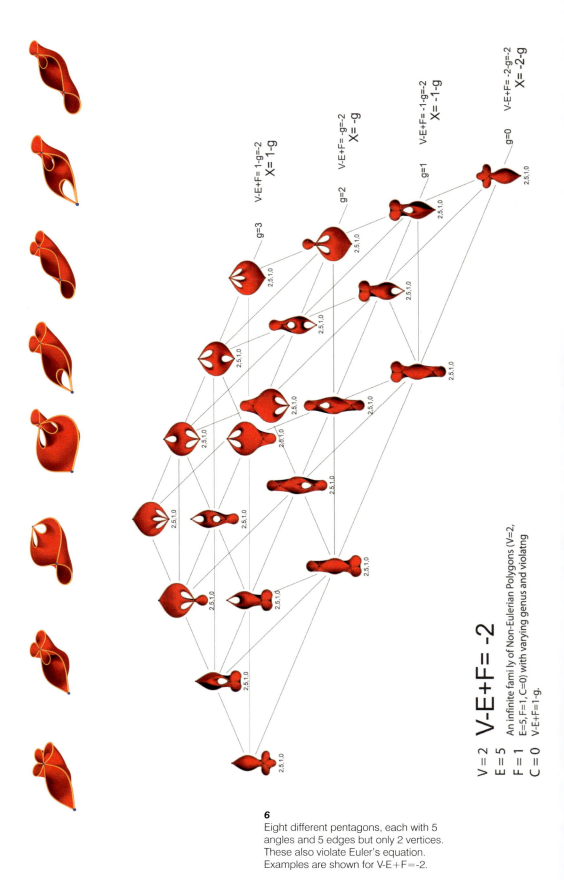

6
Eight different pentagons, each with 5 angles and 5 edges but only 2 vertices. These also violate Euler's equation. Examples are shown for V−E+F=−2.

Morphoverse

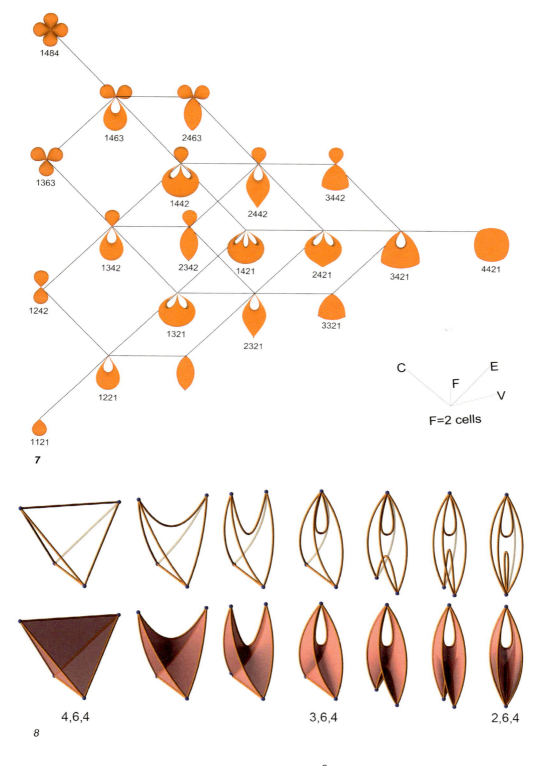

7
The 3D VEF space is extended to 4D VEFC space by adding the number of cells C shown with singularity structures having cells with two faces.

8
Continuous morphing of a tetrahedron 464 (*left*) having 4 vertices to a tetrahedron with 364 having 3 vertices (*middle*) and a tetrahedron 264 with 2 vertices (*right*). Tetrahedron 164 is not shown.

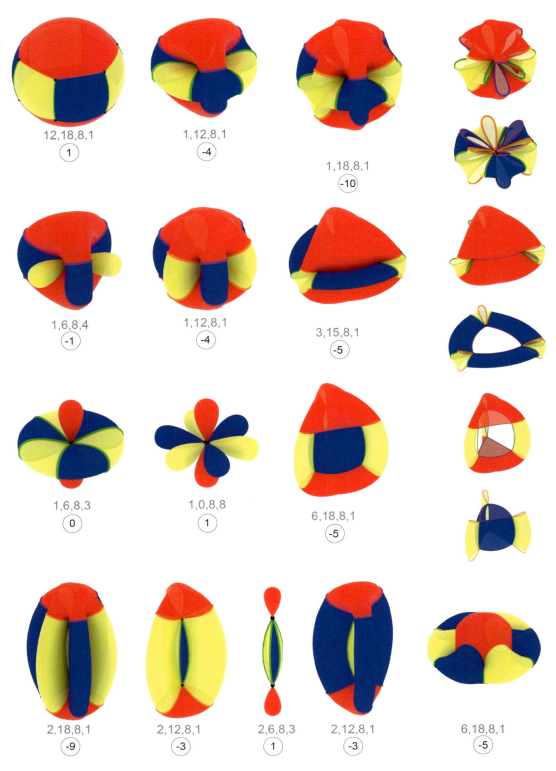

9
Singularity structures from the continuous morphing of a spherical hexagonal prism (*top left*) to its various states as edges morph and vertices collapse in a continuum.

shown here. These add edge-preserving transformations,[7] which conserves the number of edges E, to the inventory of spatial transformations.

Regular Structures and Derivatives (Figs.10-21)

Regular structures are a fundamental class of topological structures with the highest order and have provided the test bed for developing a universal morph tool kit that applies to asymmetric, irregular, and growing structures. Regular structures comprise topological structures with one type of face (polygon) that meet identically at all vertices. They can be generated by placing elements (points, lines, faces, cells) within the fundamental region, the smallest unit that can generate the structures through symmetry operations (reflections, rotations, translations). Starting with these as source structures, other orderly structures can be derived by various transformations applied within the fundamental regions. The derived structures have lower regularity but retain the high order of their source structures. The derivatives shown are a sampling from a larger group obtained by applying three different classes of transformations to the regular ones. They are mapped in 15D space.

15D Space

Shown here (**fig.10**) is a portion of a 15D lattice of regular and semi-regular structures in 2D thru 4D space, their surface subdivisions, and it also includes structures obtained by removing faces from closed cells. The regular and semi-regular structures in region A comprise polygons, prisms, polyhedra (**videos 28,29**), plane and hyperbolic tessellations (in finite portions) in 5D space. Their subdivisions in region B, which include geodesic spheres, expand it to 7D. Region C includes 4D polytopes and extends the space to 9D, and region D adds addition-removal of faces to produce additional structures in extended 15D space. Additional transformations expand the space further.

Regular Polyhedra, Tessellations

This is a primary class of regular structures and is represented by the Schläfli symbol (p,q)[8] named after the mathematician Ludwig Schläfli.[9] The structures are derived from pairs of two topological numbers, p (the number of sides of a polygon) and q (number of polygons at each vertex), each having integer values. They are mapped in an infinite 2D periodic table (**fig.11**) [4]. This space includes all regular polygons $(p,1)$, regular stars $(q,1)$, dihedral prisms $(p,2)$ composed of only 2 faces and termed dihedra, diagonal polyhedra $(2,q)$ composed of only 2-sided polygons termed digons. Higher numbers include 3D polyhedra (all 5 Platonic "solids"),[10] and 2D plane and hyperbolic tessellations.[11]

---

[7] This is one of the 4 classes of topological transformations which include V-preserving, F-preserving and C-preserving cases.
[8] The original symbol in mathematics uses curly brackets {p,q}; see Coxeter [20]
[9] Schläfli was a nineteenth century mathematician and a pioneer in higher dimensional geometry.
[10] The tetrahedron (3,3), octahedron (3,4), icosahedron (3,5), cube (4,3) and dodecahedron (5,3).
[11] The three regular plane tessellations are composed of triangles (3,6), squares (4,4), and hexagons (6,3). The regular hyperbolic tessellations include $(3,q>6)$, $(4,q>4)$, $(5,q>3)$, $(6,q>3)$ and $(p>6,q>2)$.

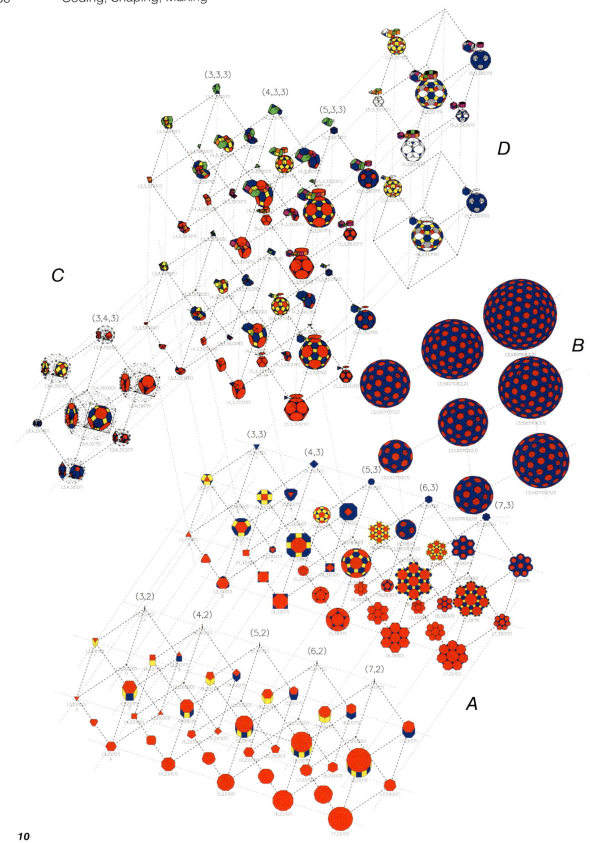

1,∞	2,∞	3,∞	4,∞	5,∞	6,∞	7,∞	8,∞	9,∞	∞,∞
1,9	2,9	3,9	4,9	5,9	6,9	7,9	8,9	9,9	∞,9
1,8	2,8	3,8	4,8	5,8	6,8	7,8	8,8	9,8	∞,8
1,7	2,7	3,7	4,7	5,7	6,7	7,7	8,7	9,7	∞,7
1,6	2,6	3,6	4,6	5,6	6,6	7,6	8,6	9,6	∞,6
1,5	2,5	3,5	4,5	5,5	6,5	7,5	8,5	9,5	∞,5
1,4	2,4	3,4	4,4	5,4	6,4	7,4	8,4	9,4	∞,4
1,3	2,3	3,3	4,3	5,3	6,3	7,3	8,3	9,3	∞,3
1,2	2,2	3,2	4,2	5,2	6,2	7,2	8,2	9,2	∞,2
1,1	2,1	3,1	4,1	5,1	6,1	7,1	8,1	9,1	∞,1

$q \uparrow$, $p \rightarrow$

10 (previous page)
A portion of a 15D lattice of regular and semi-regular structures in 2D, 3D and 4D space, their surface subdivisions, and structures with open cells obtained by removing faces from closed cells (first displayed at the Bucky Centennial exhibition, Cathedral of St. John the Divine, New York, 1995).

11
Regular structures have identical vertices with the symbol (p,q) having integer values for p (the number of sides of a polygon) and q (number of polygons at each vertex).

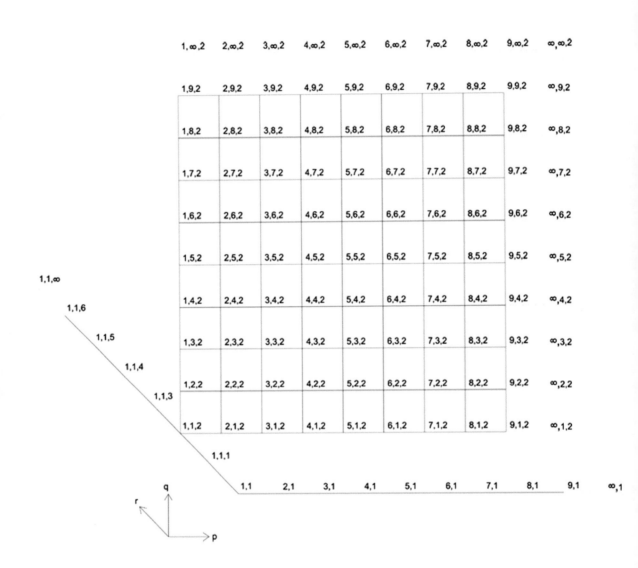

12
Adding a third topological number, r, the number of cells at each edge, leads to an infinite 3D cubic lattice of regular (*p,q,r*) structures which include 4D polytopes in Euclidean and hyperbolic space.

Regular 4D Polytopes

Adding a third topological number *r* (number of cells at an edge) generates the 3D cubic lattice of regular 4D structures (*p,q,r*) [6] from three topological numbers (**fig.12**). These include the 4D polytopes[12] [19] in Euclidean and hyperbolic space. It also includes the 3D cases like the infinite cubic grid (4,3,4) which has 4 cubes (4,3) meeting at each edge.

Polygon with 3.67 sides

The infinite number line of integers (*top*), identical to the (*p*,1) line along the axis in fig.11, translates into *p*-sided regular polygons (*middle*) with one polygon corresponding to each integer. The space between any two integers on the number line (for example, 3 and 4 as shown in **fig.13**, *bottom left*) can be populated with all fractions between two numbers. This leads to a periodic table of stellated polygons mapped on the triangular grid of fractions shown (*bottom right*). Stellated polygons embed convex polygons having sides in real numbers (or fractions), for example 3.67 (11/3 = 3 2/3) sides, as an intermediate polygon lying between the triangle and a square.[13] These examples provide a method so any closed polygon, symmetric or asymmetric, individual or part of a tiling, can "grow" a new edge. This is part of the universal morph tool kit.

Four Edge Transformations

Any regular (*p,q*) structure can be subdivided using four independent edge transformations comprising two pairs of dual edges. This leads to a family of 16 classes of structures for any (*p,q*) and includes regular and semi-regular structures which have at least two different types of faces. **Fig.14** uses three of these transformations (**video 28**), extending the 2D table in Fig.11 to a 5D table by adding a cube of transformations (**video 29**). The faces within each family have 3 primary colors and are analogous across the table. Within each cube, the structures morph to each other in a continuum. In **fig.15**, the fourth edge transformation is added to complete the set. This extends the 5D lattice to 6D shown here with the tetrahedral family (3,3) where the 16 structures are located on the vertices of the 4D cube [4] (**videos 30, 31**). One intermediate on each edge, face, and cell of the 4D cube shows the continuum within this family. This is a general class of 16 transformations that apply to any surface tiling, regular or irregular, on a closed or open surface. These transformations are part of the morph tool kit and come in two pairs of complementary pairs. Their continuum enables the edges to grow or shrink in combinations. Faces grow as a result, sometimes shrinking or changing the number of its sides.

video 28 video 29

---

[12] The term polytope is an extension of the 2D polygon, the 3D polyhedron to 4D and higher dimensional figures with flat faces. (https://en.wikipedia.org/wiki/Polytope).
[13] This suggests extending the concept to the (*p,q*) table and beyond to higher dimensional structures (*p,q,r,s,...*).

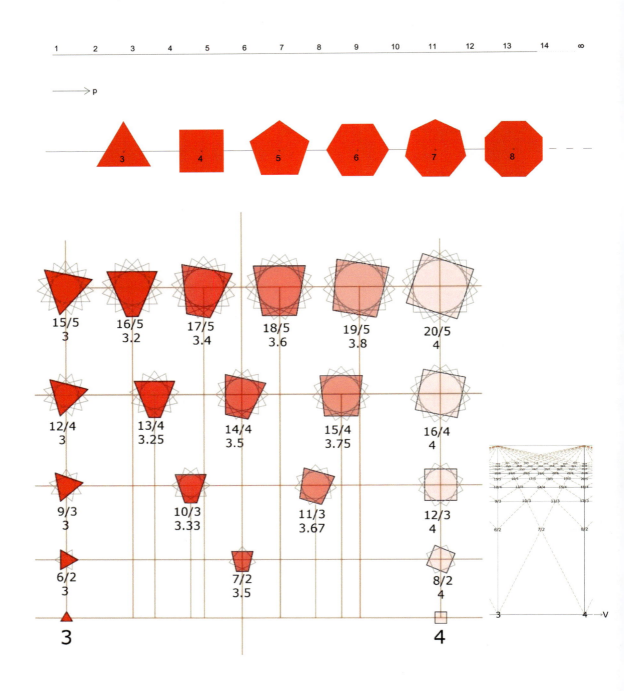

13
The infinite number line of integers (*top*) translates into p-sided regular polygons (*middle*). The space between any two (e.g., 3 and 4 (*bottom left*)) can be populated with stellated polygons based on the triangular grid of fractions (*bottom right*).

14 (*next page*)
A portion of an infinite 5D lattice of regular and semi-regular structures – polygons, prisms, polyhedra, plane and hyperbolic tessellations. This exends the 2D table in Fig.11 by adding three independent transformations.

Morphoverse 291

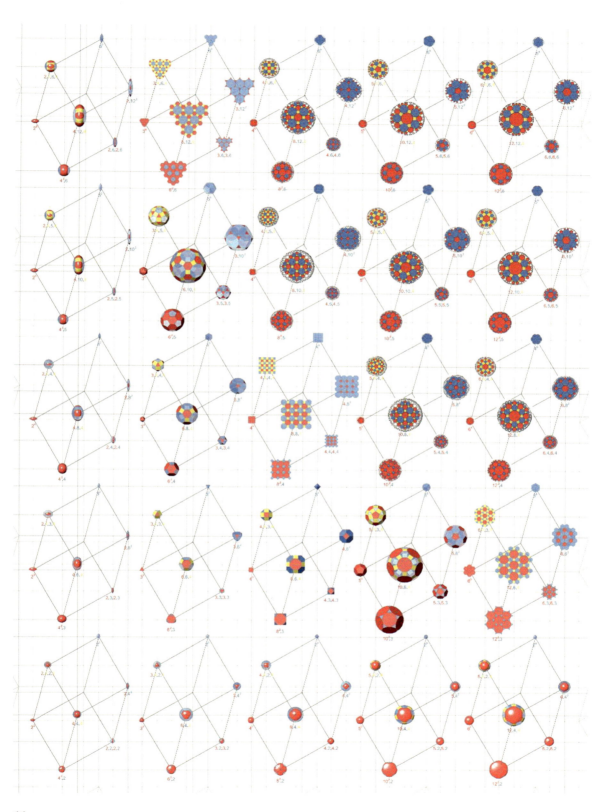

14

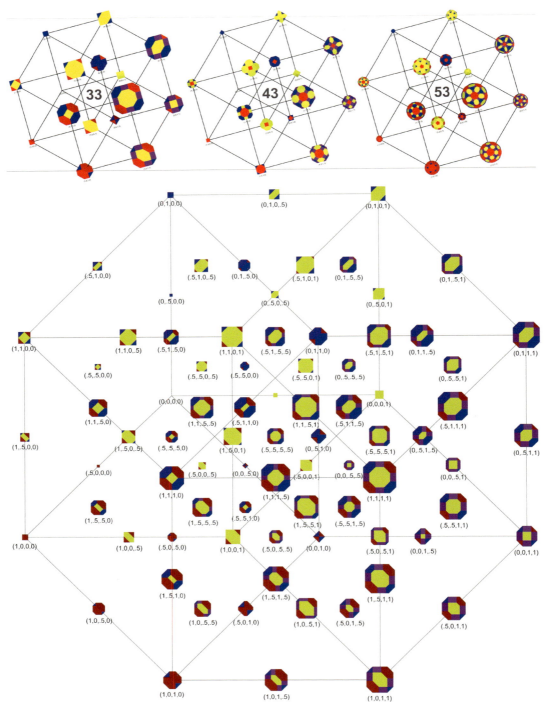

video 30

video 31

15
Adding one more edge transformation to the three in Fig.14, makes each cube into a 4D cube (*bottom*). This extends the space to a 6D microverse comprising 16 classes of surface subdivisions for any (*p,q*) (**videos 30,31**). Three families - (3,3), (4,3), (5,3) - are shown (*top*).

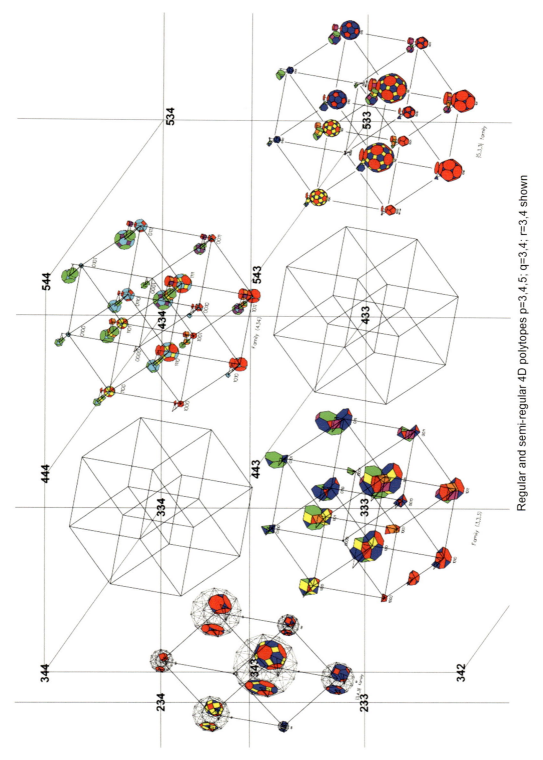

16
Each (*p,q,r*) point location in Fig.12 can split into 4 new directions to make a 7D lattice of regular and semi-regular 4D polytopes shown here with a portion from an infinite 3D array of 4D cubes.

Four Edges in 4D Polytopes

Any point location in the 3D (p,q,r) cube in fig.13 can split into 4 new directions based on 4 different edge types to make a 7D lattice. Each (p,q,r) generates its own family of 16 regular and semi-regular 4D polytopes, each organized on the vertices of analogous 4D cubes [6]. A portion of the 7D lattice space is shown in **fig.16** and the families of 4D structures correspond to the region C in fig.11. Each family has faces in six colors in 3 pairs of primary and secondary colors and their distribution within each family is analogous across the lattice. One family of a 4D tetrahedron (4-simplex) from **fig.16** is shown in **fig.17** in its 16 states of transformations. This is a general group of 16 transformations that apply to 3D multi-cellular structures and are also part of the morph tool kit. The transformations enable continuous growth or shrinkage of its elements and apply to regular and irregular morphologies.

Frequency of Subdivisions

The frequency of subdivision [21] of each face of a regular structure[14] adds two more transformations [5] which extend the 7D space in **fig.16** to 9D. The transformations generate a family of 16 classes of surface subdivisions for any regular polyhedron or cell of any 4D structure (**fig.18**). The subdivisions include mirrored and skewed versions (left-handed or right-handed). The example shows a skewed subdivision of the icosahedron (3,5) with a family of 16 geodesic spheres mapped on the vertices of a 4D cube. These have the same 4 types of edges as shown earlier in **fig.16**. The spheres can morph continuously from one to another as shown with one intermediate earlier (p.32). The subdivisions correspond to the region B in **fig.10**.

Adding-Removing Elements

This is a fundamental class of transformations that deals with adding or removing the four basic elements – vertex, edge, face, cell – that comprise all 3D forms and structures. Their combinations lead to 16 classes of addition-removal (inclusion-exclusion)[15] transformations which can be mapped on the vertices of a 4D cube. Examples are shown with face-removal and its dual edge-removal, each for one family of structures as representative examples. Corresponding examples exist for the other pair of duals, vertex-removal and cell-removal. This group of transformations is part of the morph tool kit and also applies to regular or irregular morphologies. The transformations exist in discrete or continuous states. In the latter, the elements grow or shrink in a continuum.

---

[14] This is known for the square and the triangle from their subdivision into corresponding square and triangular grids. An example for the subdivision of a pentagon was shown in ([4], Fig.A).
[15] Termed inclusion-exclusion in [1], and the 4D map of the 16 transformations was introduced here [2].

Morphoverse 295

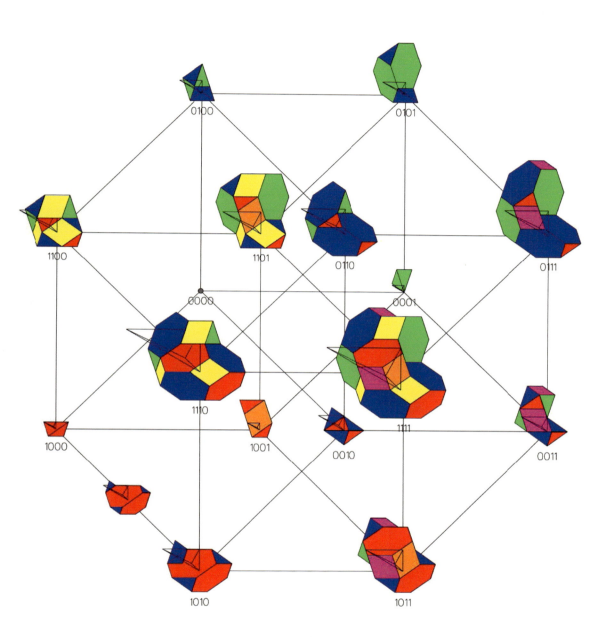

Family (3,3,3)

17
A close-up of one family of 4D polytopes (3,3,3) from Fig.16. This structure, in different 3D projections, permits variable 3D geometries which elongate, shrink, tilt or squish the 3D cells in different ways.

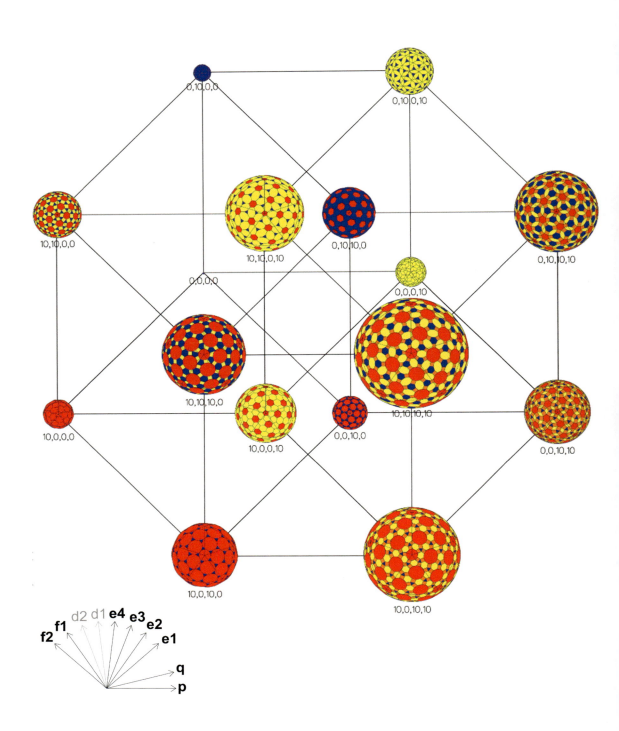

18
Two more transformations, frequency of subdivision, added to the 7D space in Fig. 16 extend it to 9D microverse.

Morphoverse

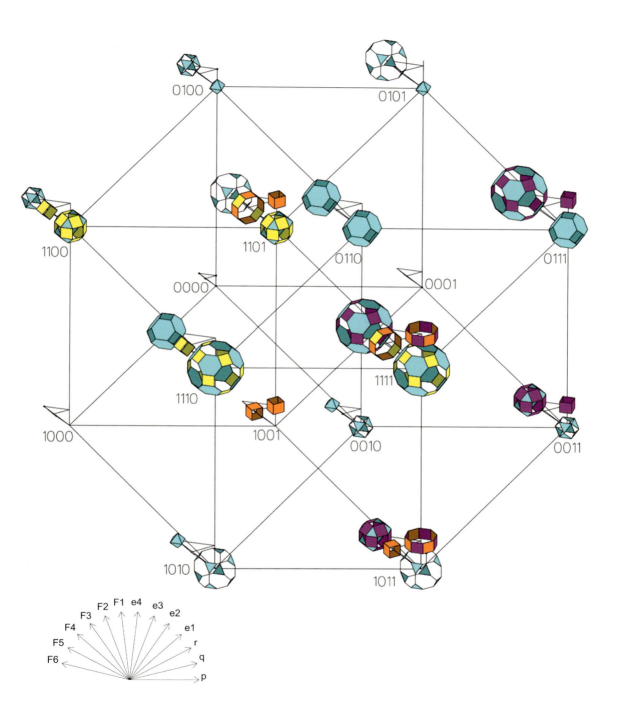

19
Fundamental region of (4,3,4) family shown with red and green faces removed from its 6 colored faces. These lead to infinite polyhedra, the faceted versions of curved minimal surfaces.

298　Coding, Shaping, Making

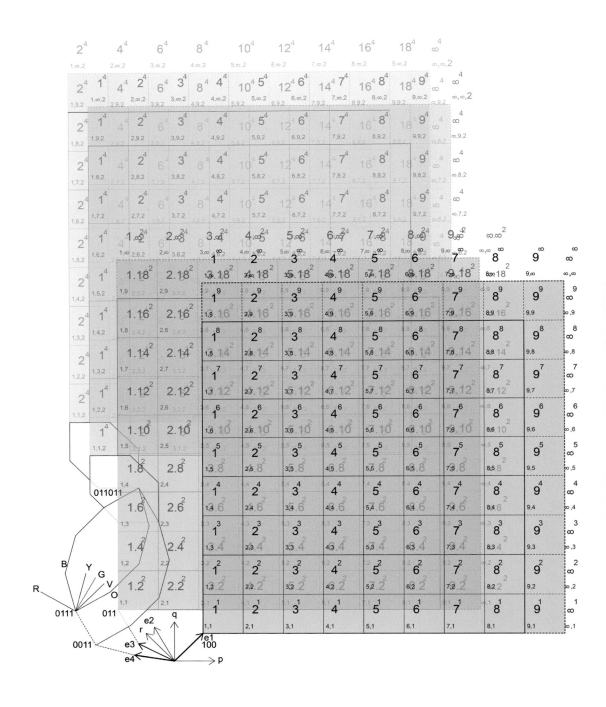

20
2D lattices of vertex symbols of families derived from (p,q,r) structures having red and green faces removed and organized in 13D space.

21 (next page)
The face-removal technique applied to the family of (6,3) multi-layered honeycomb of hexagonal prisms with its fundamental regions shown in a close-up from 9D space.

Face-Removal

Each (p,q,r) structure has 6 different faces colored in 3 complementary pairs of primary and secondary colors. Each complementary pair has faces touching point-to-point in dual locations. Removing any complementary pair leads to 3 classes of *infinite polyhedra*, one for each pair, each existing in 16 different states for any (p,q,r). Infinite polyhedra [22] are continuous plane-faced surfaces that divide space into two parts, inside and outside. These surfaces have identical vertices and, like any convex polyhedron, have only two faces meeting at each edge. An example is shown here (**fig.19**) with the complementary face-removal of red and green faces from the (4,3,4) family based on the packing of cubes and its derivatives. The six colors add 6D to the 9D lattice obtained earlier, extending it to the 15D microverse indicated by the region D in fig.10. It has many structures including the afore-mentioned infinite polyhedra which are faceted versions (low resolution) of triply periodic curved minimal surfaces (high resolution).

Lattices of Vertex Symbols

Infinite 2D lattices of vertex symbols of families of structures for any (p,q,r) derived by removing red and green faces (**fig.20**). Each symbol represents a distinct structure. The 2D lattice planes exist in 13D space, the same as the regions A, C and D in the microverse in fig.10. Similar lattices exist for the other two pairs of complementary colors.

Dual Edge-Removal

The face-removal technique is here applied to the multi-layered honeycomb of hexagonal prisms. Its fundamental region, a 1/24th portion of a hexagonal prism (a right-angled 30°-60° triangular prism), is shown and can be mirrored, rotated and arrayed to fill all space. Each fundamental region of the prism has 9 interior faces and 9 exterior edges, each edges perpendicular to corresponding faces (called a dual). The dual edges of faceted structures define networks (grids, lattices, space frames) and are composed only of edges. This displays a pair of complementary transformations where edge-removal is equivalent to face-addition, and vice versa. Each structure in **fig.21** shows the faces and a superimposed dual network in each fundamental region. Removed faces are replaced by added edges. All faceted structures, and their dual edge networks, are organized in a 9D cube indicated by the 9D coordinates below each fundamental region. The diagram is a close-up of a 6D cube portion of the 9D cube.

Curved Structures (Figs.22-31)

The straight-edged and plane-faced structures in **figs.10-21** can be continuously morphed to their curved counterparts through various methods. The examples that follow show three different methods that enable a continuum between

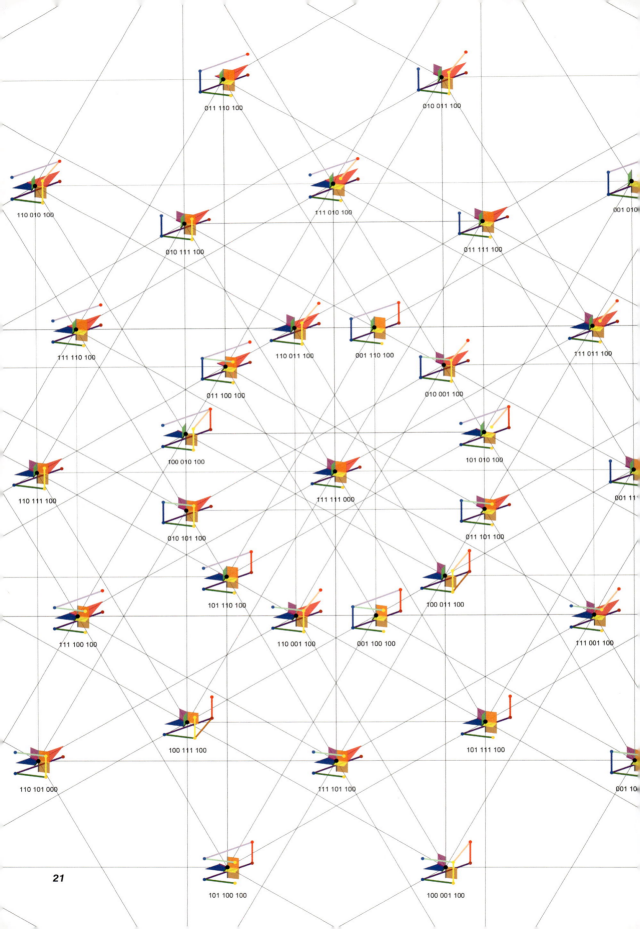

straight and curved structures, and also in families of curved structures. These examples demonstrate a bridge between the worlds of straight lines and curves which are generally considered separate. More interestingly, straight-line and flat-faced structures are a minority where curved-edges and curved-faces dominate. All curved structures fall within the 27 possibilities from combinations of three topological elements, edges, faces or cells, each existing in 3 states, concave (-1), flat (0) or convex (1) as mentioned earlier (p.241). The cells are either individual units or entire structures. The inventory of curved surfaces can be mapped on the vertices of a cube. The 3D taxonomy of curved surfaces presented provides a way to include saddle and plane surfaces as part of the same family which also includes some tensile-type forms with singularities. This leads to inter-conversion between morphologies of planar systems, minimal surfaces and tension membrane forms. The methods of curving are part of the morph tool kit.

Curved Polygons (Fig.22-23)

Polygons with straight and curved sides can be mapped in a 3D periodic table from which 2 parallel planes, $n=4$ and 5, are shown in figs.22,23. These planes extend to other values of n, and beyond to 3D.[16] A continuous morphing between the polygons within each plane suggests a continuum within n; a continuum between different values of n follows from **fig.13**.

The 2D periodic table of 49 states of octagons (**fig.22**) is based on pairs of angle numbers 1,2,3,4 ($n=4$ case, 1=45°) and shows a mix of straight- and curve-sided polygons. The * indicates the pairs are repeated 4 times so that 33* is within the 1/4th portion of the octagon and represents the full sequence 33333333 for the regular octagon. Similarly, 42* represents the square 42424242, and so on. Only seven of the 49 cases are polygons with straight sides, the rest are curved and fall on either side of the left-tilted diagonal. Polygons with concave sides on bottom left are mirrored to the ones with convex edges on top right. The angle numbers force the edges to curve. **Fig.23** shows a similar table of 81 symmetrical decagons ($n=5$ case) based on pairs of angle numbers in one-fifth portion of each (**video 27**). The * indicates the pairs are repeated 5 times, 43* represents the sequence 4343434343 for that decagon, 4444444444 is the regular decagon, 5353535353 the regular pentagon, and so on.

Curved Polyhedra (Fig.24)

Here, just as in the 2D singularity structures in **fig.2**, the topology forces a curvature on faces and edges leading to a class of curved polyhedra. A 2D lattice of finite and infinite curved polyhedra $(2p)^4$ corresponding to $(p,q,2)$ is shown (**fig.24**). Each structure (***top***) has red and green faces removed from one of the family of 16 structures analogous to the structure (0,1,1,0) in fig.18. The cross-sectional polygons are 2-sided (digons) which force the curvature on the faces. Not shown in the lattice are curved tubular versions of plane and hyperbolic tessellations. *Bottom left* figure shows a structure with vertex

---

[16] In 3D, the angle numbers of polygons are equivalent to the dihedral angles of polyhedra. Figs.22 and 23 represent cross-sections though polyhedra having curved faces with straight or curved edges.

Coding, Shaping, Making

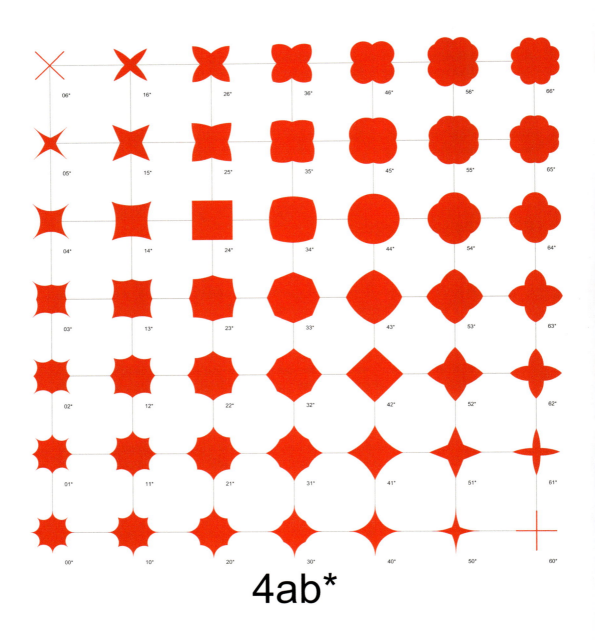

22
2D periodic table of 49 discrete states of octagons based on pairs of angle numbers ($n=4$) in one-fourth portion of the polygon. The * indicates the pairs are repeated 4 times so that 33* represents the regular octagon 33333333.

Morphoverse

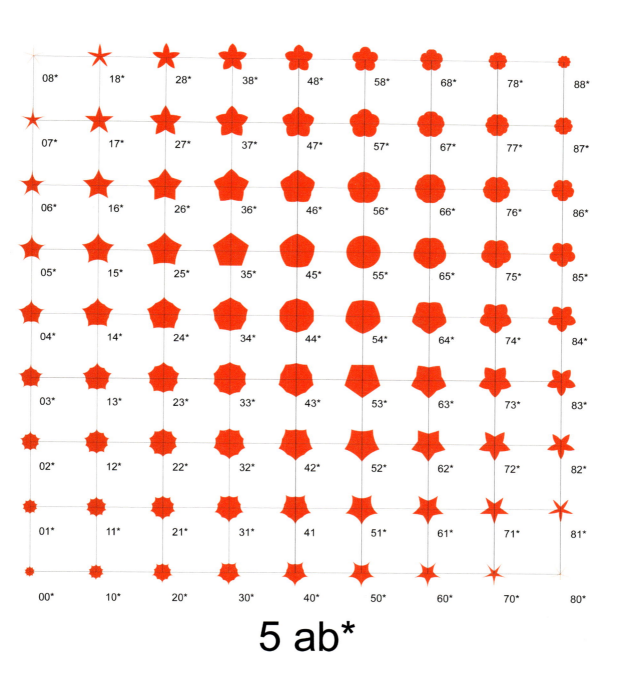

5 ab*

23
A 2D periodic table of 81 symmetrical decagons ($n=5$) based on pairs of angle numbers in one-fifth portion of each. The * indicates the pairs are repeated 5 times so that 43* represents 4343434343, a decagon with concave sides (**video 27**).

video 27

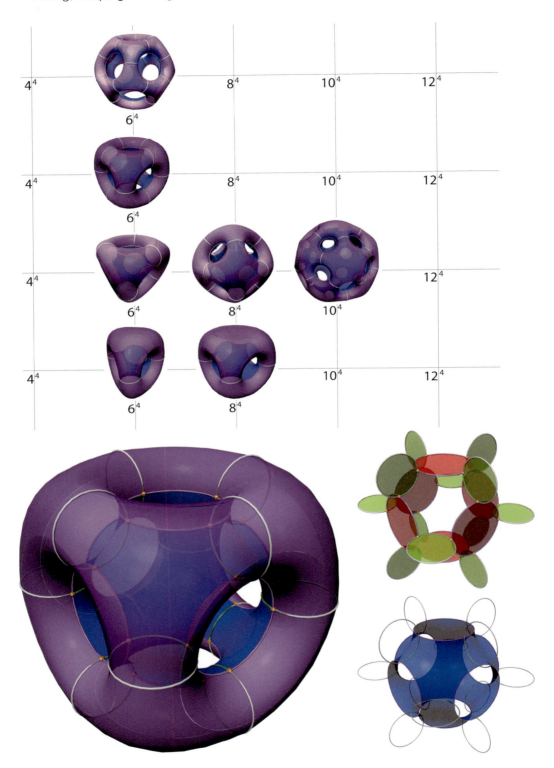

24
A 2D lattice of infinite polyhedra $(2p)^4$ derived from one structure within the $(p,q,2)$ family (*top*) shown here with red and green faces removed.

symbol 64 from the family (3,4,2) where *p*=3 and *q*=4. *Bottom right* shows two related face combinations embedded in the figure on left. A similar infinite 2D lattice exists for curved $(2q)^4$ structures.

Saddle Polygons, Saddle Polyhedra (Figs.25-31)

Saddle polygons and saddle polyhedral were discovered independently by Burt [23] and Pearce [24]. These can be derived from plane-faced structures like the ones in figs.18, 20, 21 through an iterative process of a special type of faceting termed *explosion* described in [25]. This adds new facets at each iteration to make the surface smoother as the facets become relatively smaller in relation to the size of the faces. The process bridges the divide between *discrete* and *continuous* through a graded recursive process of transformation, whereby the plane-faced versions lie on one extreme of a continuum and have the lowest resolution and smooth minimal surfaces lie on the other end with the highest resolution. This is similar to introducing the decimals between 0 and 1 described earlier as the binary continuum (p.116, 156).

Fig.25. A 4D periodic table of families of saddle polygons (4*p*) belonging to (*p*,1) group shown for *p*=3 thru 8 (*top*) and corresponding saddle prisms belonging to (*p*,2) group obtained by mirroring them (*below*). The largest prisms (in central location) within each family comprise only two 4*p*-sided saddle polygons with vertex symbol $(4p)^2$ and have two complementary pairs of *p* holes.

Fig.26. Three of the six families of half-saddle rings and saddle prisms for any (p,1) along with their intermediates corresponding to the axis code 101100000 (*top left*), 100100010 (*top right*), and 101100001 (*bottom*). Example is shown for the hexagonal family, *p*=6.

Fig.27. One of the six (6,3) families of saddle rings along with their intermediates corresponding to the axis code 101100001 obtained by mirroring the half saddle ring in the top left of fig.26.

Fig.28. Second example from the six (6,3) families of saddle rings along with their intermediates corresponding to the axis code 110100010 obtained by mirroring the half saddle ring in the top right of fig.26.

Fig.29. Third example from the six (6,3) families of saddle prisms along with their intermediates corresponding to the axis code 011010001 by mirroring the half-saddle prisms in the bottom of fig.26.

Fig.30. Miscellaneous examples of minimal saddle surfaces from the hexagonal family (6,3) with their codes. Corresponding structures exist for all (*p*,*q*) pairs in fig.11.

Fig.31. An assortment of examples of minimal surface saddle prisms with minimal surfaces from the family (*p*, 2), emanating from the infinite (*p*,*q*) pairs in fig.11.

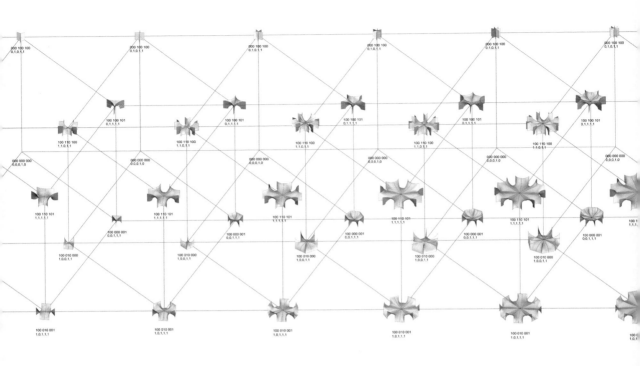

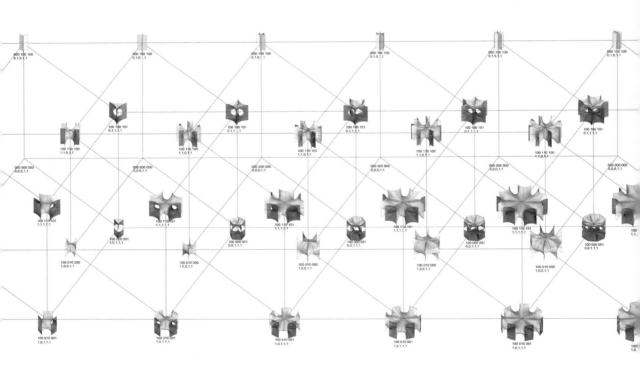

25
A 4D periodic table of families of saddle polygons (4*p*) (*top*) and saddle prisms (4*p*)² (*bottom*) obtained by mirroring them.

Morphoverse 307

26
Three of the six families of half-saddle rings and saddle polygons for any (*p*,1) along with their intermediates corresponding to the axis code 101100000 (*top left*), 100100010 (*top right*), and 101100001 (*bottom*). Example is shown for the hexagonal family, $p=6$.

308   Coding, Shaping, Making

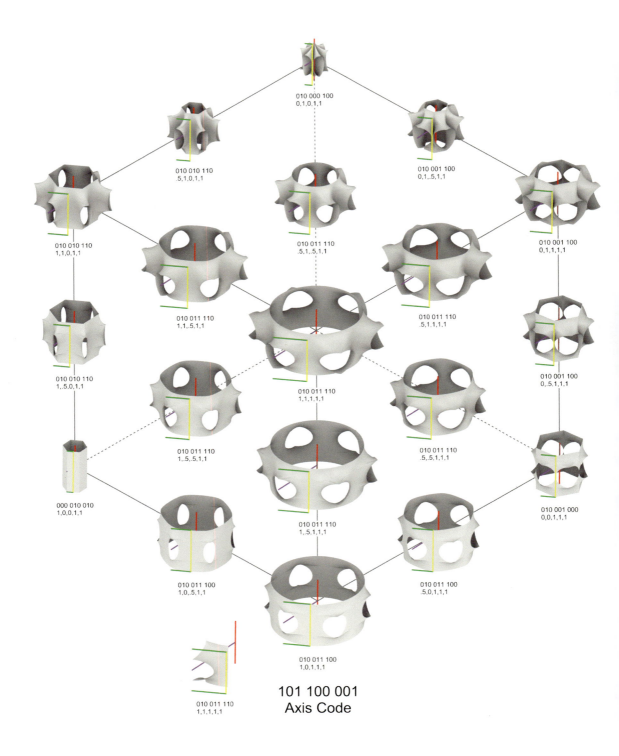

27
One of the six families of saddle rings along with their intermediates corresponding to the axis code 101100001 obtained by mirroring the half saddle ring in the top left of fig.26.

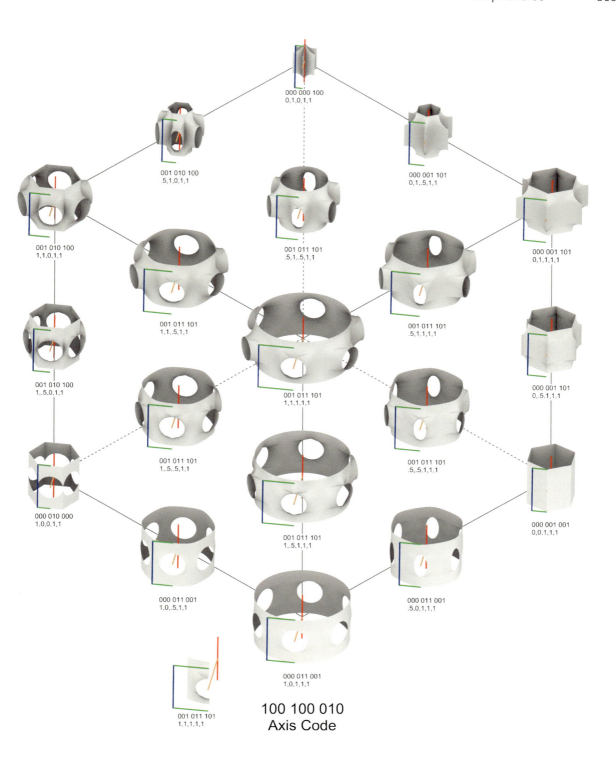

28
Second example from the six families of saddle rings along with their intermediates corresponding to the axis code 110100010 obtained by mirroring the half saddle ring in the top right of fig.26.

310 Coding, Shaping, Making

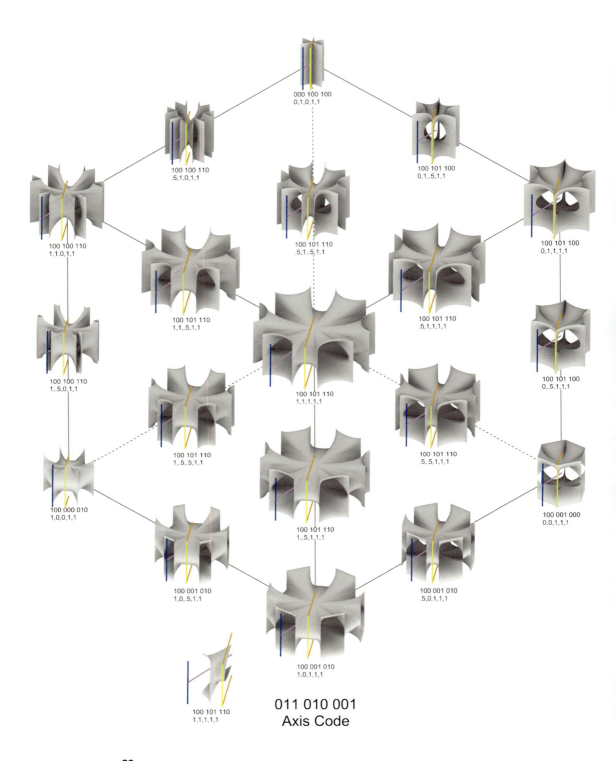

29
Third example from the six families of saddle prisms along with their intermediates corresponding to the axis code 011010001 by mirroring the half-saddle prisms in the bottom of fig.26.

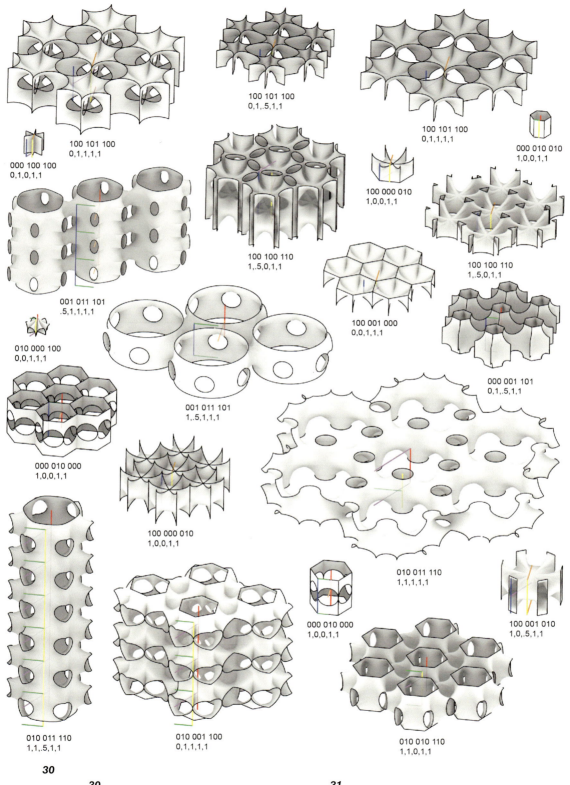

30
Miscellaneous examples of saddle surfaces from the hexagonal family (6,3) with their codes.

31
Miscellaneous examples of saddle surfaces from the families of *p*-sided prisms (*p*,2).

Coding, Shaping, Making

Variable Geometries

A variety of methods can be used to reduce the symmetry and regularity of the structures described so far. The deal with changes in geometry and are, in most part, *topology-preserving* transformations. All are devices for symmetry-breaking needed to achieve complexity in design. These geometric transformations include changes in length, angle, curvature, position, orientation, among the simple ones, and are part of the morph tool kit. Recursive and iterative procedures expand the kit of morphological tools further. A sampling of examples follows. Individual transformations can be combined with each other in any sequence and extent, the way artists combine three primary colors, to generate more complex forms.

Variable Stretching, Tilting, Tapering and Bending (Figs.32-38)

Albrecht Dürer's studies [26] on human proportions were the first example of grid transformations and influenced D'Arcy Thompson's celebrated chapter on changes in biological form.[17] Dürer's work marks the origin of morphing. His drawings indicate he was thinking of proportions as a 3D grid transformation system; this is clear from his front, side and top views of a human face in one drawing. He introduced transformations of scaling (stretching), shear (tilting), perspective (tapering), and hinted at other ways using optics and drawing devices which included curving. Dürer's implied 3D grid also anticipates Le Corbusier's *Modulor* [27] and Stanley Wysocki's 3D "Form Solid".[18] All these methods provide a way to break symmetry directly or through iterative use of same or a mix of transformational methods. Higher dimensions provide a way to extend these at a foundational level. Three illustrative examples are shown with geometric changes — variable lengths, angles, position and orientation along different directions of space. Variable curvatures of elements or entire space extend these tools further; this was shown with a few examples in Chapter 7 (p.241). Variable topologies in all earlier figures (**figs.2-31**) added to these geometric methods extend the transformational tool kit further. Iterative procedures on these transformations, especially on multi-cellular fundamental regions, lead to emergent morphologies.

4D Form Solid

The simplest symmetry-breaking device is by changing lengths, the same way a square can be variably stretched (scaled along two directions) into infinitely different rectangles. In 3D, this leads to Wysocki's "Form Solid" where a cube can be variably scaled along 3 directions (x-y-z) to generate infinitely different 3D block shapes, the shapes of the rooms most of us live in. These are also the bounding boxes of any form in 3D space, a method Katavolos used in his furniture variations. The same idea applies to a 4D cube which can morph into infinitely many states from combinations of 4-directional scaling (**fig.32**). Its "shadows" in 2D are variable octagons (zonogons) which we encountered in Chapter 7 (p.217).

---

[17] The last chapter "On the Theory of Transformations, or the Compariosn of Related Forms" [9].
[18] Personal communication, 1975. Wysocki's 3D Solid organizes an important class of affine transformations.

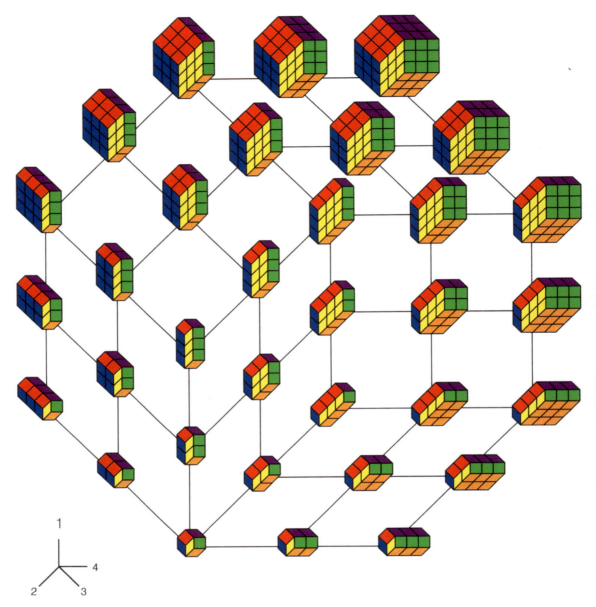

(4,3,3)(1,2,3,4)(3,3,3,3)

32
The outer shells of stretched hypercubes exemplify 4-directional (4D) scaling as an example of symmetrybreaking by changing lengths variably along different spatial directions.

Morphoverse 315

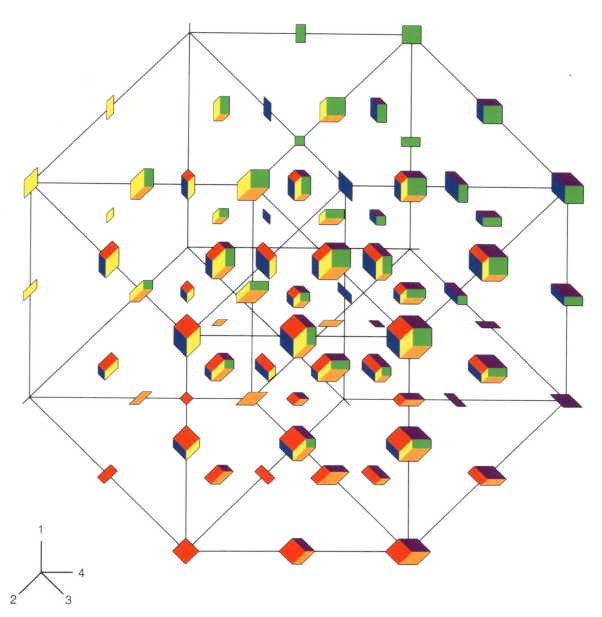

(4,3,3)(1,2,3,4)(1,1,1,1)(2)

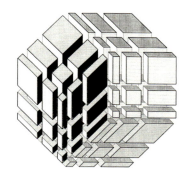

33
Continuous 4-directional scaling of a 4D cube along 4 axes morphs it to all its lower-dimensional states in a continuum (**video 32**). Bottom left shows a graded subdivision of cells of a 4D cube into stretched and tilted 3D blocks, combining tilting and scaling in one model.

video 32

The logic of 4D-scaling is captured in **fig.33** which shows a continuum of all possible individual states of a 4D cube in one diagram (**video 32**). This is interesting for several reasons. All combinations of scaling along the 4 directions lead to a continuum of changes of all its component elements – point (0D), line (1D), plane (2D), cell (3D), hyper-cell (4D). In a more general sense, this suggests that elements (building blocks) of form and space are a continuum. In addition, a 4D cube and its parts are mapped on a 4D cube and demonstrate a self-similarity between a structure and its meta-structure mentioned earlier as morphology-of-morphology. Further, it leads to a *4D Form Solid* (*bottom left*) comprising all cells in stretched as well as tilted states. *Stretching* (scaling) and *tilting* (shearing), two separate symmetry-reducing phenomena (called affine transformations), are integrated in one model. Adding *tapering* (perspective transformation like a cylinder becoming a cone) leads to a *5D Form Solid*, add *bending* (curved transformations) leads to a *6D Form Solid*. Adding multi-point perspectives and various ways of curving extend this tool kit further, and so on.

Variable Positions

This is a large class of transformations and an expedient device for symmetry breaking that can be applied to any topological element (vertex, edge, face or cell). The example shown here deals with changing the position of faces relative to a fixed center. It is in the context of asymmetric crystal growth with a higher dimensional coordinate system. It uses a 26D model (**fig.34**, *top*) showing 26 coordinate axes, numbered 1 thru 26, emanating from the center of a cube to its vertices (red, blue), edges (violet) and faces (yellow), where each point location along each axis represents a potential position of a plane normal to that axis. Examples are shown (**fig.34**, *bottom*) where the number codes comprise sequences of numbers within brackets, each bracket representing a group of axes of the same color, and each single number within the bracket represents the location of a plane perpendicular to an axis within that color group. Irregular number sequences lead to asymmetric crystal forms. The model provides a tool for symmetry-breaking in any 3D faceted form having axes that guide spatial movement and growth. It has applications to architectural morphologies, mass customized housing, synthetic crystals, gem-cutting, design of solar envelopes for buildings, to name a few.

Fig.35 shows the application of the 26D model to symmetry-reduction in one group of crystal forms related to tetrahedral and cubic (octahedral) symmetries. Additional axes, hence colors, can be added to generate more complex morphologies. The dual of this process leads to variable star configurations, for example, asymmetric radial branched forms. An interactive digital model is currently in development with Che-Wei Wang.

Variable Rotations

Rotations in 2D and 3D space can be applied to any elements of a structure or its components. For example, the edges can rotate (as in kinetic structures

Morphoverse

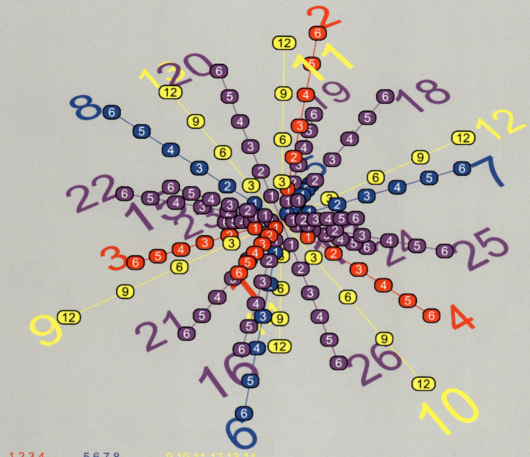

1 2 3 4   5 6 7 8   9,10,11,12,13,14
(v1,v2,v3,v4) (v5,v6,v7,v8) (v9,v10,v11,v12,v13,v14)

15,16,17,18,19,20,21,22,23,24,25,26
(v15,v16,v17,v18,v19,v20,v21,v22,v23,v24,v25,v26)

318   Coding, Shaping, Making

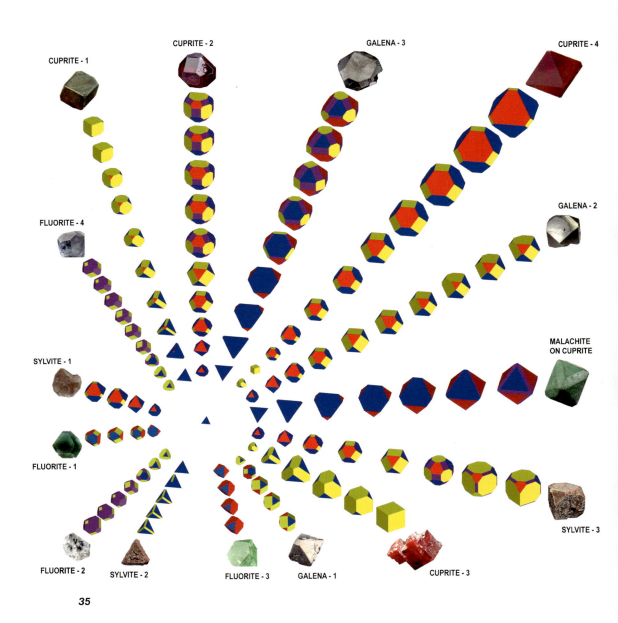

35

34 (previous page)
A 26D model (*top*) showing 26 coordinate axes emanating from the center of a cube to its vertices (red, blue), edges (violet) and faces (yellow), with point locations along each axis for a potential locations of a normal plane. Asymmetric locations on axes produce asymmetry as shown with 6 examples (*bottom*).

35
Application of the 26D model in Fig.34 to one group of crystal forms with symmetry-reduction related to tetrahedral and cubic (octahedral) symmetries (with Devanshi Agarwal, Jija Jadhav, Evan Hryshko).

from linear elements), faces can rotate as in "jitterbug"[19]-type or other rotary transformations which are not normal to the axis of the face, or cells can rotate in 2D or 3D. When entire structures rotate, it belongs to a related class of *"variable orientations"*.

Variable Angles

This example uses changing angle numbers (pp.215-220) as an expedient way to generate asymmetry. Angle numbers based on n dimension were introduced by projection in 2D and defined by number sequences for shape-generation.

The simplest case of triangles (**fig.36**), requiring only three numbers in a continuum (**video 25**), are shown in a 3D periodic table viewed corner-first with back faces not shown. Sequences with repeat numbers force symmetry on the shape, e.g. 111, 999, ... are equilateral triangles; 211, 311, 411, ... are isosceles triangles; three different numbers 123, 425, ... are scalene triangles. Bulk of the space generates asymmetry while symmetry is an exception. The illustration also shows a continuum of triangles. In an animation, symmetry appears fleetingly within a continuous flow of asymmetry, the same as the case of *Morphing Platters* (p.109). Similar tables for all polygons (and linear paths) can be mapped in corresponding spaces, 4D for 4-sided polygons, 5D for 5-sided, and, in general, pD for p-sided polygons.

Numeric Blobs, Squiggles, Amoebas

The example in **fig.37** (*top left*) shows an asymmetric number sequence of a non-convex polygon shown earlier (p.217) and based on $n=7$ case. Its number code has two parts, dimension $n=7$, and angle number sequence (6666...8). It can easily be converted into corresponding asymmetric amoeboid shape (*top right*) by rounding off the corners using an iterative procedure. The third number in the number code represents the level of iteration (6, in this instance) needed to reach a smoother corner by a repeated geometric operation like truncation in this instance. This is a 2D case of the iterative 3D faceting mentioned earlier to obtain saddle forms from planar versions (p.299). **Fig.38** shows a collection of 2D amoeboid shapes produced by the same method and applied to the non-convex polygons shown earlier (p.220). The 3D versions are hypersurface blobs mentioned in the last chapter (p.256).

The 2D amoeboid shapes are biomorphic polygons with rounded vertices and a continuously curved boundary composed of curved segments. In the history of art, design and architecture, the two strands, straight and curved morphologies, have generally been considered separate. The example of biomorphic polygons is one case where the discrete and the continuous lie on two ends of a line representing their continuum. This continuum integrates organic curved forms (non-Euclidean) with crystalline geometries (Euclidean) having straight lines and flat faces as states of each other. In Chapter 6 (p.193), a similar continuum between convex, flat, and concave surfaces was discovered in the *X-STRUCTURES* series as examples of physical emergence.

---

[19] "Jitterbug" is Bucky Fuller's term for rotating the faces of a hinged structure.

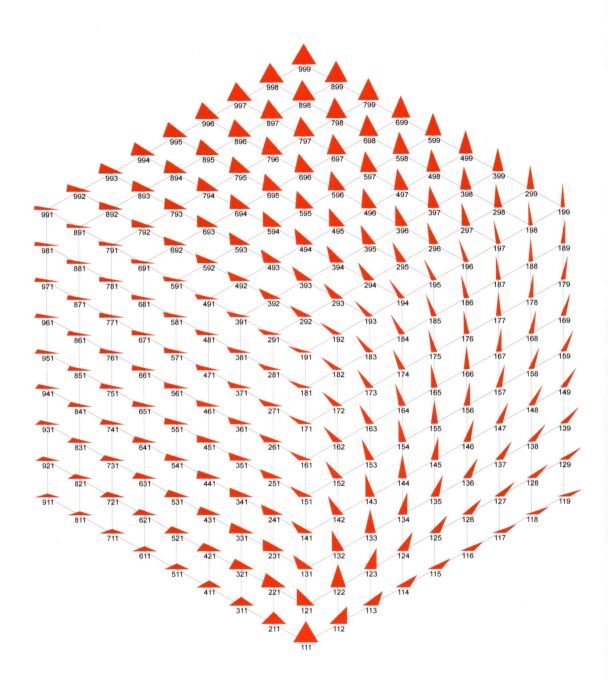

video 25

36
Symmetry-breaking by changing angle numbers. Asymmetric numbers lead to asymmetric triangles (**video 25**).

Morphoverse

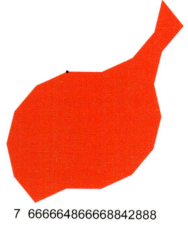
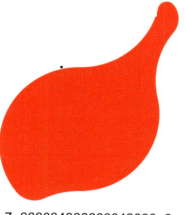

37

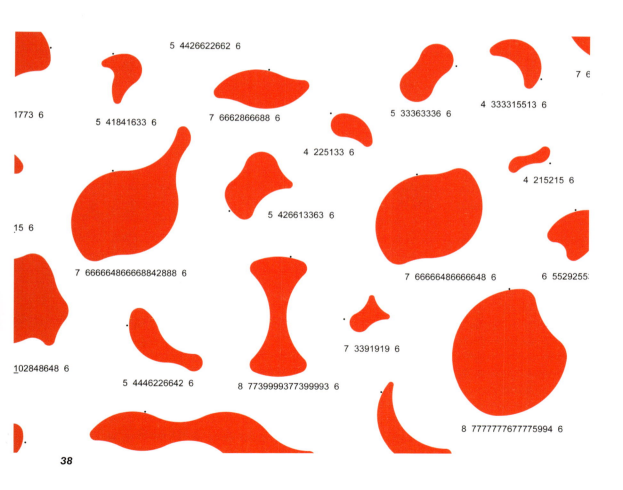

38

37
Angle number sequence for a non-convex tiles by fusing together two non-convex polygons ($n=7$) (*left*) and its "biomorphic" counterpart (*right*).

38
Asymmetric angle number sequences for non-convex polygons in their biomorphic amoeboid states.

Fruit Forms (Figs.39-41)

The curved polygons provide a starting point for the morphology of outer (and inner) shapes of fruits, seedpods, sections of cacti, and similar floral polygons in nature and human-made designs. For illustrative purposes, examples are shown with the 49 discrete states of $n=4$ polygons in **fig.34** which here define the transverse sections of fruits. These sections are combined with the only 9 possible vertical sections of fruits, seeds and seedpods, each with a vertical axis of a sphere and two poles at top and bottom of the axis. The two pole locations can each exist in only 3 states, each pole is either pulled up (1), pushed down (1) or remain unchanged (0). The combination of two poles, each in 3 states, leads to only 9 possibilities. The 9 classes of fruit forms derived this way are shown in (**fig.39**, *top right*) with the cross-section of a concave octagon on *top left*. Both sections are a continuum for any n. For another cross-section on *bottom left*, a continuum in the cross-section is shown with one intermediate on *bottom right*. **Figs.40** and **41** show two sets of the 49 states for two of the nine cross-sections.

    The model presented provides the base-level taxonomy of fruit forms (and seedpods) and deals only with the outer form (**video 33**). The interior fruit has a similar topology and provide an interesting example where the interior and exterior morphology are self-similar (**video 34**). The base model is being extended to more complex fruit morphologies with possible implications for future seedpod designs. Symmetry-breaking transformations described earlier expand the design space further. These sections apply to fruit forms, but also to architectural structures like tensile and pneumatic morphologies which comprise the broader space-time-mass morphoverse. The one-axis examples of fruits easily extend to 3D multi-axes structures.

 video 33   video 34

Sea-Shells

The infinite row of polygons in **fig.13** can be converted into corresponding helices through a continuous transformation by introducing a "rise angle", the helix can transform to a spiral by adding a "departure angle" for the deviation from a circle, and the spiral line can be dilated into a spiraling cone by introducing a "cone angle". These three combined generate a class of sea-shell forms organized in a 3D lattice. Adding the simple pigmentation scheme we saw in Chapter 4 (**fig.4**) extends it to 9D. The variable cross-section adds at least 2 more dimensions leading to a 11D model. Adding other features like ribs, bumps, undulations and more complex pigmentation patterns will increase the shell-form space further. **Fig.42** shows a selection of models from the 11D space and suggests the vast scope of simple shell morphologies that can be obtained this way.

    This model (jointly with Peter van Hage) used a 3D version of D'Arcy Thompson's method of a self-similar growth emanating from a fixed point of an asymmetric (scalene) triangle and adding a similar but larger triangle (gnomon) rotationally surrounding the point. This is a constructive procedure and different from David Raup's method of generating shells using a mathematical equation[28].

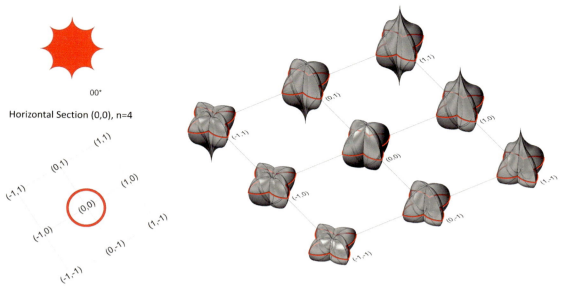

9 Classes of Fruit Forms based on 9 possible Vertical Sections shown for (0,0), n=4 case

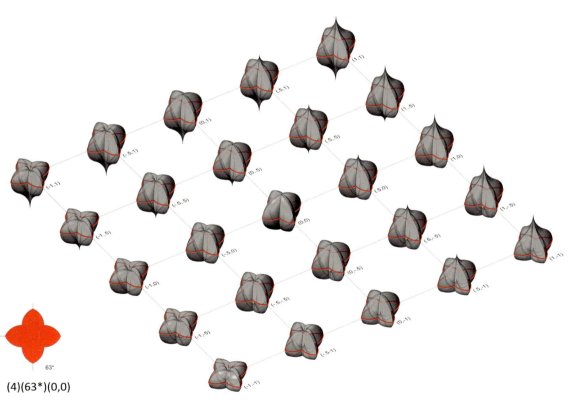

Form Continuum with 1 intermediate between 9 discrete states (n=4)

39
A 2D table of 9 classes of fruit sections (*top right*) extended with one intermediate in to 25 states (*bottom right*). Their cross-sections (red) are shown on the left for n=4 family.

324 Coding, Shaping, Making

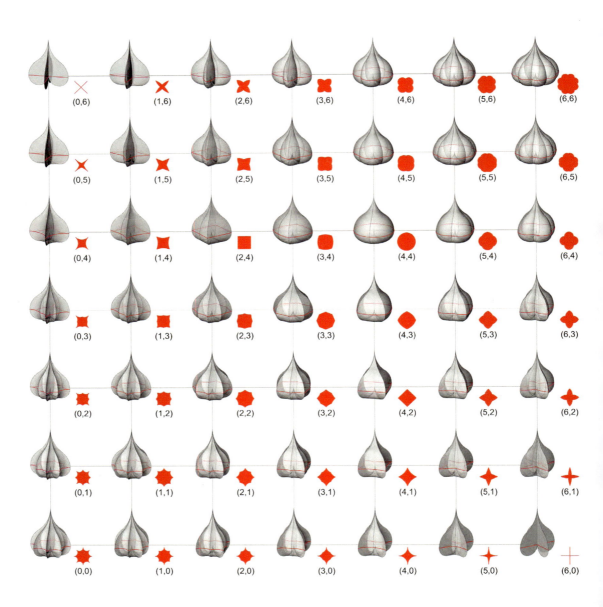

40

40
The 2D periodic table of 81 symmetrical octagons ($n=4$) in fig.34 is applied to one of the 9 classes of fruits (1,-1) in fig.36.

41
Close up of the 2D periodic table of 81 symmetrical octagons ($n=4$) in fig.34 is applied to another one of the 9 classes of fruits (-1,-1) in fig.36.

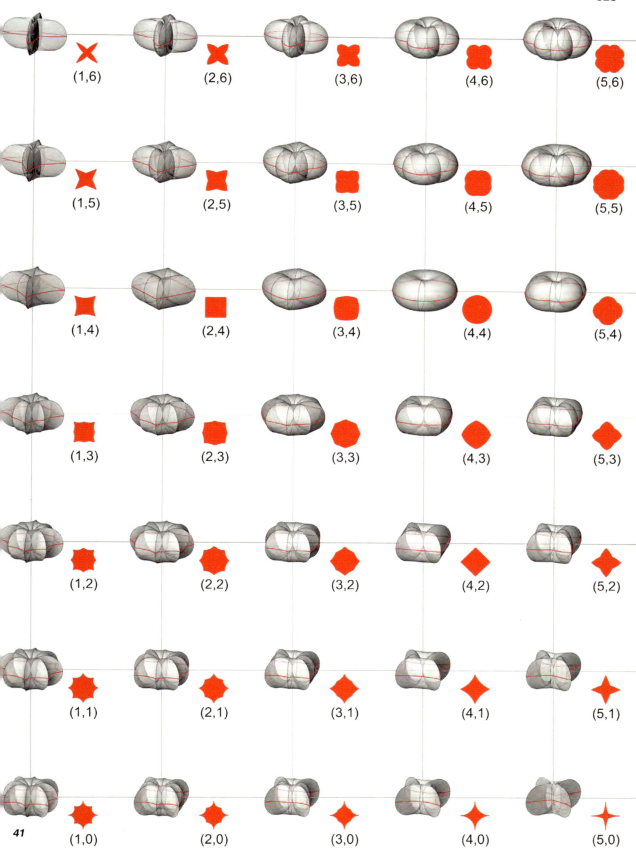

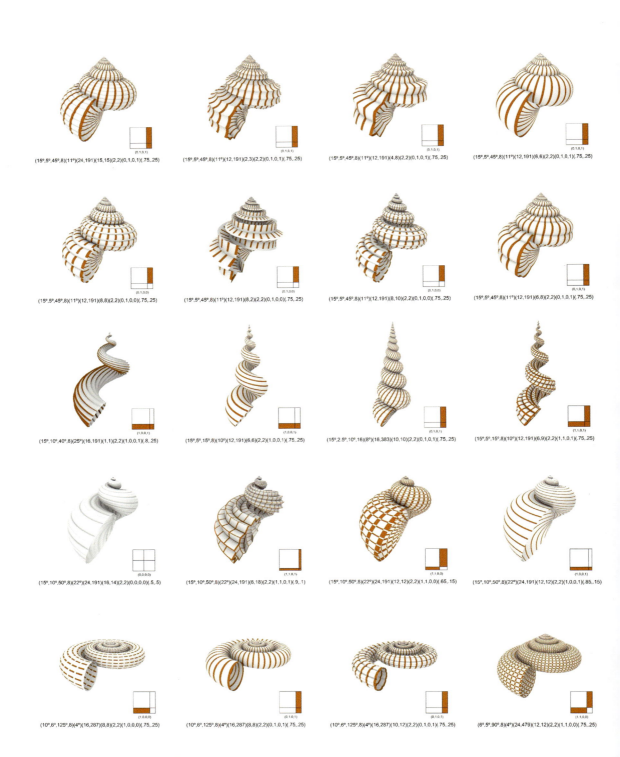

42
Various examples of sea-shell forms based on a 12D model by varying the spiral, cross-section and pigmentation pattern.

The two methods produced identical forms, one with a background structure provided by a radial coordinate system of space (Raup), the other a background-free constructive algorithm to build in 3D space from self-similar component parts. For designers, a constructive method of building is a natural preference.

Gaudi (Figs.43-46)

The twin processes of symmetry-breaking and symmetry-making are used for decoding Gaudi's work and are shown with a few examples [29]. **Fig.43** shows form development of 4 small portions of Gaudi's structures derived from simple source geometries (*left*, indicated by a binary code) to more complex target geometries (*right*) using step-by-step transformations, a single change at a time. The same 4 examples are part of a continuum of a family of 16 binary-coded structures in **fig.44** mapped in 4D-cubic space. This diagram suggests alternative pathways for form development as well as other possible source geometries from one family of structures related to the tessellation (4,4). Other (p,q) spaces open up more alternatives.

Stills from an animation sequence of Gaudi's *Portico A* (**fig.45**, *top*) show the morphogenesis from a simple high-symmetry source geometry (*top left* in the animation) to a reduced symmetry (*bottom right* frame) (**video 36**). It follows the sequence shown on the top in **fig.43**. 4 close-ups from that sequence are shown in the images below (**fig.45**, *bottom*).

A similar analysis of a portion of Casa Mila façade is shown in **fig.46**. Two close-ups of Casa Mila (*bottom left*) are shown with corresponding regularized topologies (*bottom right*) derived by imagining the entire façade as a tinker-toy system composed of nodes (blue/red) joined by struts (pink). 15 different nodes are used in the selected portions of the façade and are part of 64 minimal surface nodes (*top* image) organized in a 6D cube network. The nodes are derived from the Schwarz surface cell[20] [30] (located in the center) by adding or removing x-y-z directions to reduce symmetry. The source geometry is derived from (4,3,4). Other (p,q,r) values provide alternative starting points.

Emergence (Figs.47,48)

Examples of generative emergent designs based on a geometric automaton were shown in Chapter 4. Rule-based systems provide examples of *digital emergence* in contrast with *physical emergence* described in Chapter 5 and 6. Well-known examples use iterative rules applied over successive generations. In such emergent systems, e.g. L-systems, fractals and cellular automata (CA) and others, transformation rules are applied to the initial geometry and topology ("initial conditions") and at each step in successive generations of the form development

---

[20] A 3D cell of one of the first triply-periodic minimal surfaces discovered by Hermann Schwarz in the 19th century. The near-similarity of the main balcony space of Casa Mila (5 holes) with a Schwarz surface cell (6 holes), in a way, anticipates topological architecture. This proto-example of topology in architectural form-making also has the column nodes with beams and slabs as portions of Schwarz cells, albeit "solid" ones, hinting at the idea of a "minimal solid" (as an analog of "minimal surface") in the vocabulary of form.

1011

Parc Guell, Portico A

1101

Parc Guell, Portico B

0011

Sagrada Familia
(facade close-ups)

0011

Casa Mila
(facade close-up)

43
4 examples of portions of Gaudi's structures derived from 4 regular structures (*left column*) by systematic transformations (*top to bottom*): John Gulliford, Christian Lopez, Brian Hopkins, Joon Bae Park at Pratt Institute.

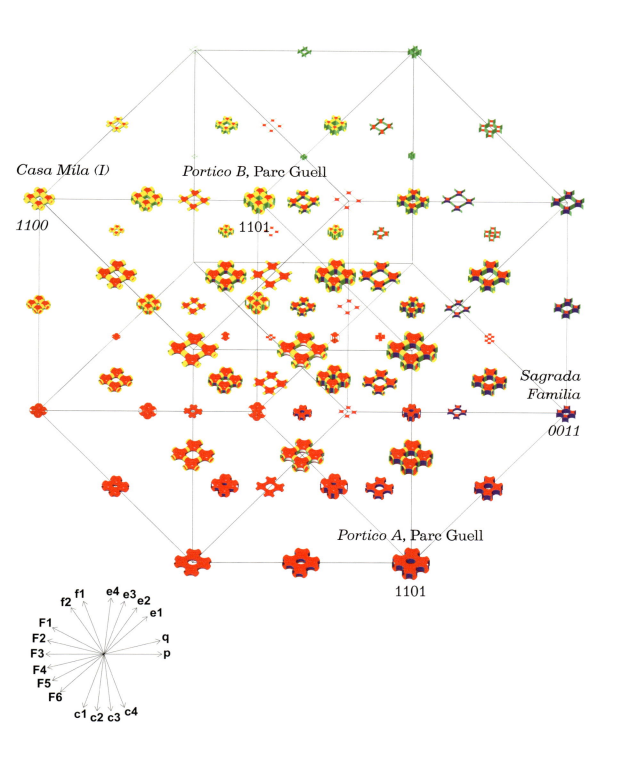

44
4 examples of Gaudi in fig.17.2 as part of a continuum of a family of structures (with Reza Schricker, see **video 35**).

video 35

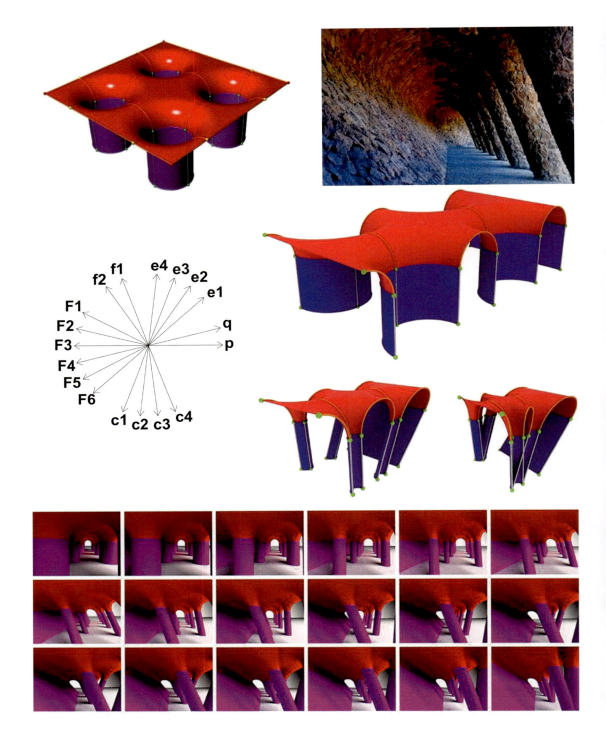

video 36

45
Morphogenesis of Gaudi's Portico A (*top right*) by sequential transformations (*middle*) of a regular minimal surface (*top left*), and stills from an animation sequence (*bottom*, **video 36**); with John Gulliford.

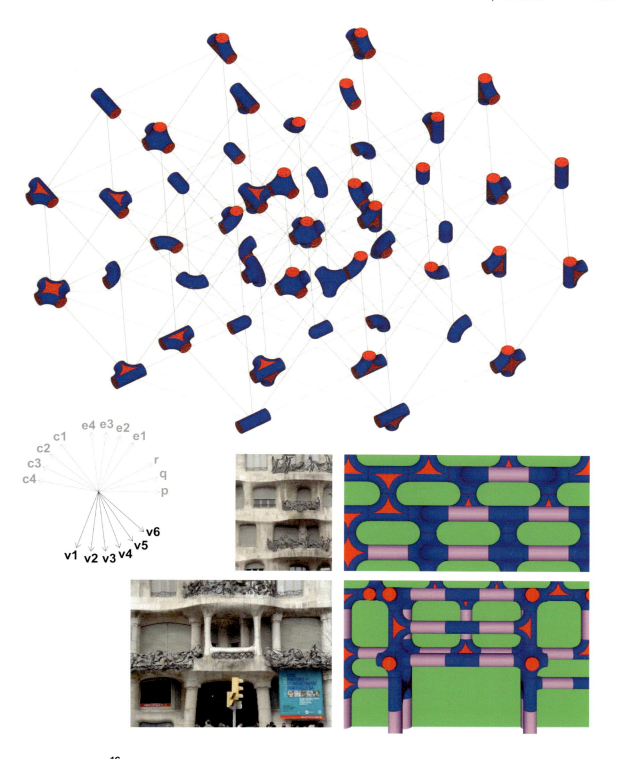

46
64 minimal surfaces, arranged on a 6D cube (*top*), as regularized cells which define the building blocks of the façade of Gaudi's Casa Mila (*left, bottom*) and their equivalent regularized topologies (*right, bottom*).

process. Due to space restriction, only one example of CA based on Wolfram's rules [31] is described here briefly. It builds on the earlier application of Wolfram's Rule 30 (p.74) which is revisited in the context of the morphoverse.

The mapping of Wolfram's 256 rules of 1D cellular automata on the vertices of an 8D cube was an application of a higher dimensional organization scheme (see note 15, p.79). It used the 8 generators in **fig.47** (*top*) which include Wolfram's rules 1, 2, 4, 8, 16, 32, 64, 128. All others were derived as combinations of these 8. One subset, a 4D cube of the 16 similar 4D cubes arranged on the vertices of a larger (self-similar) 4D cube to build the 8D cube, is shown in **fig.48**.[21] It shows a subset of 16 rules, rules 16-31 as indicated. Rule 16 is the local "origin" from which rules 17, 18, 20 and 24 emanate along four directions. These add 1, 2, 4 and 8 to the rule numbers along the four axes of the local 4D cube. Conservation of complementary sums, an important hallmark of the higher dimensional representations, is captured in the rule numbers. All rule numbers located diametrically across the center of the 4D cube add up to 47 in this example.

CA Continuum (Fig.48, *bottom*)

In Chapter 3 (p.74), a comment was also made that the pattern continuum displayed in much of the work was absent in CA patterns. That comment is now corrected with the observation that though the CA patterns don't appear to morph continuously from one to another, their rules do. This is indicated in the diagram in **fig.47** (*bottom*), where real numbers are introduced in the original binary rule description. For convenience, only 4 Wolfram rules on the vertices of one square of the 8D cube are shown. In addition, one intermediate on the mid-point of each edge, and one in the mid-point of the face, are shown with a gray cell, mid-way between white and black cells. The intermediate gray cells are indicated with .5. Introducing other shades of gray in between makes this a continuum within the rule set. The *rule continuum* extends over the entire 8D space and provides an example of *continuous cellular automata*. A simpler example is a continuum in the 16 rules of the 2-cell 1D cellular automata (p.117, **figs.13A,B**).

Rule-based geometric-topologic emergent systems expand the morphoverse considerably. However, they rely on the initial conditions provided by non-emergent geometries-topologies presented in most of this chapter. In cellular automata, fractals, L-systems and others, the geometry-topology of the initial conditions are excerpted from subsets of ordered non-emergent structures (like the ones shown in most of this chapter) and provide the geometric seed on which rules of emergence are applied. The fundamental regions of all single-cell and multi-cell orderly structures described earlier are potential "initial conditions" for a vast class of emergent structures. The vertices of the higher dimensional morphoverse of ordered non-emergent structures branch out from each vertex to accommodate emergent morphologies.

---

[21] This work (with CA modeling by Neil Katz) has remained unpublished since 2004 except for the diagram shown here which was first published recently in [15, p.120].

Morphoverse 333

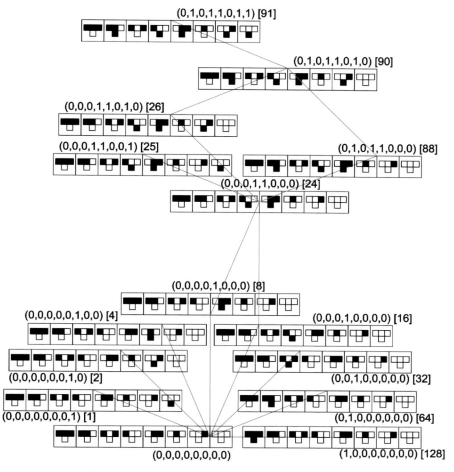

(Cartesian Co-ordinates) [Wolfram Rules]

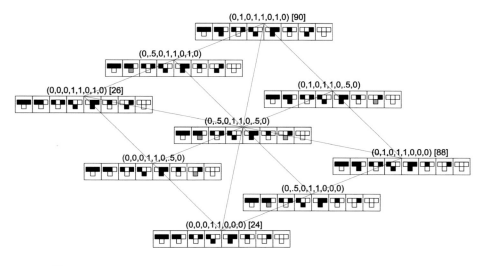

47
Wolfram's 256 1D cellular automata rules with its 8 generators for its 8D mapping (*top*). A rule continuum for 4 rules indicated with one intermediate with grey cells (*bottom*).

334 Coding, Shaping, Making

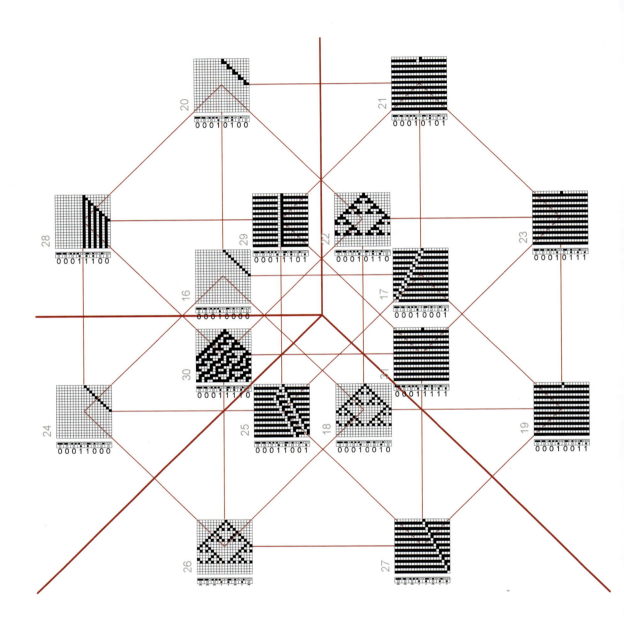

48
One of the 4D cubes from the set of 16 within the 8D cube mapping of Wolfram's 1D CA rules.

Physical Form

For *physical form*, the morphoverse needs to be connected with similar universes of physical processes and materials that exist in parallel. The combined universe of the Form-Process-Material triad provides the fundamental palette for physical making. In physical nature, form, process and material are one. Form, within the morphoverse as presented here, is a *space-time* phenomenon. Physical process and material add mass, making physical form a *space-time-mass* phenomenon, a larger and more complex space. Since process and material also exist in space-time, these two phenomena must be guided by similar unifying principles as the space-time aspects of form. This makes mapping the *process universe* and the *materials universe* an offshoot of the *form universe* which is part of the larger universe of physical form. Physical form requires material (matter) to build with. This segways into elementary particles, chemical elements and molecules as the (chemical) building blocks of materials that enable physical human designs in the world which leads us to the molecular basis of life captured in DNA described in the next chapter.

Evo Devo, Morph Tool Kit

The rapid developments in Evo Devo are providing an understanding of how biological form, as it grows (development) and transforms over generations (evolution), and genetics, are unified by a universal *gene tool kit* across organisms. Leaving aside the complex chemical, physical and biological aspects of morphogenesis in biological form, and restricting only to spatial-temporal aspects, a few commonalities in organisms raise the question: Is any portion of the morph tool kit embedded in DNA? If yes, is it embedded within the gene tool kit or does it exist in parallel? Two examples are presented here selected from many others. These provide hints of a connection between the two.

Spatial Building Blocks

The fundamental spatial building blocks of all 3D structures – living, non-living, human-mad - are dimensional elements: points (V), lines (E), planes (F) and cells (C) described as "solids", "volumes" or "blocks" in the design arts as mentioned earlier. These 4 can transform to each other in 16 different ways (Chapter 9, p.383). These 16 possibilities describe all possible transformations between these four spatial elements at any scale independent of the material or process they are built with. Biological forms, non-living and human-made ones, are restricted to these 16 classes of changes in the (dimensions of) shapes of component parts. In addition, in the next chapter, these transformations are shown to be isomorphic with the only 16 possible "snips" or SNPs, single nucleotide polymorphisms,[22] in genetics. For this reason, these 16 transformations are described as the *snips of form* (Chapter 9, p.376; **video 37**; see also p.315, **video 32**).

---

[22] See, for example, https://en.wikipedia.org/wiki/Single-nucleotide_polymorphism

9 Fruit Sections

The 9 classes of fruit sections mentioned earlier (fig.36, *top right*) are the result of a sphere with a vertical axis (defined by the axis of the fruit stem) and two pole positions located at the top and bottom of the axis. The point of attachment at each polar point can exist in only 3 states, either pulled outwards (1) away from the center of the fruit, pushed inwards (-1) towards the center, or neither (0). Only 9 combinations of the top and bottom polar positions are possible giving rise to only 9 classes of fruit forms. Disregarding the exact shapes of fruits and considering them as schematic spheres, the pear is (1,-1) where the top is pulled outwards and the bottom is pushed inwards, the apple or the gourd is (-1,-1) with the top and bottom pushed inwards, the banana is (1,1) with the top and bottom pulled outwards, some figs are nearly (1,0) where the top is pulled outwards and the bottom is either flat or slightly curved. And so on. The continuum of the 9 states introduces decimal places between 0 and 1, leading to a continuum of fruit sections. These 9 are the only combinatorial classes of idealized morphologies of fruit forms. All fruit forms must come from these options. Needless to say, the exact shapes of fruits will take us into morphometrics, fruit development into genetics of growth and fruit evolution into relationships across the tree of fruit morphologies, but these changes will be restricted within the bounds of the 9 possible classes of fruit sections.

Two simple illustrative examples, the snips of form and fruit sections, have been presented as potential candidate applications of parts of a morph tool kit to biological form. Other examples can be extracted from the *Morphoverse*. In biological form, the transition between bilateral and radial is another candidate area of application of the morph tool kit. How these are embedded within the genetic code of living systems is a question for Evo Devo scientists where early evidence is emerging. The x-y-z scaling mentioned earlier (p.313), and found in the shapes of beaks in Darwin's finches (genus *Geospiza*), is regulated by the expression of growth proteins Bmp4 and CaM[23] [32,33] which controls growth along different spatial axes. The handedness (chirality) of the spiral in the snail (genus *Lymnaea*) is controlled by one gene, nodal-Pitx gene[24] [34]. These two examples, scaling and handedness, are space-time phenomena independent of biology and common to non-living and human-made systems and are part of the morph tool kit.

Growth and Form Continuum

Form continuum within higher dimensional networks offers a natural opportunity to model growth of form. Any form, or an element of form, located at any vertex of the network, can grow along any line emanating from this vertex to another vertex, then another, in a continuous sequence. The meandering zig-zag line connecting a linear sequence of points within the network defines the process of form development.

---

[23] Bone Morphogenetic Protein 4 (Bmp4) and Calmodulin (CaM); elongated beaks of finches that probe cactus flowers/fruit (species G. *conirostris*) have high CaM, the ground finches with short beaks and large depth/width for crushing hard/large seeds have high Bmp4 [33].
[24] This was demonstrated in the species *L. stagnalis*, a freshwater snail [34].

The zig-zag line is a morphogenetic pathway and defines a map of morphogenesis. The line can branch (bifurcate, trifurcate, etc.) at any point within the process. The direction of time is the direction of the zig-zag line or zig-zag branching, and the selection of points in the network is based on design factors at each point location. The morphogenetic pathway can meander within a hypercube or hyper-cubic lattice (within the same morph gene) or between different hypercubes (between different morph genes) to guide form development. This permits gene-mixing of form and generates hybrids. Complexity of form results by adding more morph genes and more zig-zags. Irregular growth involves compounding the transformations from the morph toolkit in any combination and applying them to the entire form and/or its components in parallel or in sequence.

For physical growth of 3D structures using DNA technologies, the most promising candidates for morph genes are likely to be the ones that are 4-based to match the 4-base structure of DNA sequences. This is addressed further in Chapter 9.

Mapping the Morphoverse

Mapping is the 21st century frontier. Individual genomes, genealogies and all-known species are being mapped in the tree of life. Stars and galaxies are being mapped in a 3D map of the universe. Massive amount of data is being visualized, dictionaries are becoming visual, as are the encyclopedias. This work on the morphoverse marks the beginnings of mapping the universe of form into a unified framework. It crosses the boundaries of art and science, transcending their separations and uses visual-mathematical tools for mapping. Mathematics-based visual language is universal, hence pan-cultural, and bypasses the limitations of spoken languages.

The tiny fraction of the morphoverse presented here sets the stage for further extensions into its many branches. The choices presented here have been limited by available space and the advancement of the work has been slow due to lack of funding. In addition, topology provides the kernel of the morphoverse though geometry pervades by providing infinite variations of a single topology. In addition, topology is quantized and is based on integers (natural numbers), making it easier to equate form and number. In [15, pp.16-17], I mentioned the isomorphism between number and form, suggesting the equivalence of the *universe of number* with the *universe of form*. However, as I also mentioned, this comes with a caveat. Since a number can represent the topology of form in more than one way, for example, 3 can represent the number of sides of a triangle, or the number of lines radiating from a point, or three points in a row – all examples of different graphs or networks — it follows that the universe of form is larger than the universe of number. This makes mapping the morphoverse (topoverse) more challenging.

Nonetheless, the morphoverse provides designers a rich source which can be mined continually to explore and discover new formal possibilities at scales ranging from the nano- to macro- and beyond. At the human scale, it can tie-in with industrial production and manufacturing to invent new products, new architectural morphologies and new art forms. The application of the method to other disciplines provides a tool for discovering the deep structure of things and phenomena, both physical and conceptual. The scope is unlimited.

DIY Microverses

As mentioned in the opening paragraph of this chapter, the "infinite infinities" in the morphoverse make it impossible to map since it is like counting to infinity, an impossibility. In its vastness, the morphoverse provides infinite pathways to navigate and mine them within the ever expanding morphoverse. It thus provides an inexhaustible source for design, as nature does for us, and ensures the longevity of design arts. It also permits the emergence of personalized DIY micro-morphoverses, microverses, to explore options by those in the design arts and other fields driven by mathematical and computational tools and require visualization for exploring systematic design possibilities. The morphoverse will continue to unfold as long as we as a species continue and have the means to visualize. However, the basic underlying principles captured in the underlying higher dimensional structure of the morphoverse, and the proposed universal morph tool kit when fully established, will persist along with other landmarks specified by plateaus of topology and geometry that provide it deep structure within the morphoverse.

REFERENCES

1. Lalvani, Haresh (1981) *Multi-dimensional Periodic Arrangements of Transforming Space Structures*, PhD Dissertation, University of Pennsylvania, Published by University Microfilms, Ann Arbor, 1981.

2. Lalvani, Haresh (1982) *Structures on Hyper-Structures*, (self-published) Lalvani, New York, 1982; a slightly revised version of [1].

3. Lalvani, Haresh (1982), Patterns in Hyper-Space, (self-published) Lalvani, New York, in connection with the exhibition at the St. Boniface Chapel, Cathedral of St. John the Divine, New York in the same year.

4. Lalvani, Haresh (1990) Continuous Transformations of Subdivided Surfaces, In: Geodesic Structures, Guest Editor Tibor Tarnai, *International Journal of Space Structures*, Vol.5 Nos.3 and 4, Multi-Science Publishing, U.K.

5. Lalvani, Haresh (1994) Periodic Tables of Buckminsterfullerenes and Related Structures, Proceedings, *Katachi U Symmetry*, pp.118-122, Tsukuba University, Japan.

6. Lalvani, Haresh (1996) Higher dimensional Periodic Table of Regular and Semi-regular Polytopes, In: Morphology and Architecture, Guest Editor, H. Lalvani, *International Journal of Space Structures*, Vol.11, Nos.1 & 2, pp.155-171, Multi-Science Publishing, UK.

7. Lalvani, Haresh (1998) Visual Morphology of Space Labyrinths: A Source for Architecture and Design, In: *Beyond the Cube, The Architecture of Space Frames* by Francois Gabriel, pp.406-26 and color plates (6-14), John Wiley.

8. Carroll, Sean B. (2005) Endless Forms, Most Beautiful, The New Science of Evo Devo, W.W. Norton.

9. Thompson, D'Arcy, W. (1942) On Growth and Form, Cambridge University Press.

10. Marks, Robert W. (1973) *The Dymaxion World of Buckminster Fuller*, Doubleday Anchor Books.

11. Le Ricolais, Robert (1973) In: VIA 2, *Structures, Implicit and Explicit*, Eds. Bryan, J. and Sauer, R., University of Pennsylvania, PA.

12 Otto, Frei, Ed. (1973) *Tensile Structures*, Vols. 1 and 2, MIT Press.

13 Tyng, Anne (1969), Geometric Extensions of Man's Consciousness, *Zodiac* 19, pp.130-62.

14 Lalvani, Haresh (2010) Morphological Genome for Design Applications, *US Patent 7805387* B2, Sep 28, 2010, filed Feb 15, 2006.

15 Lalvani, Haresh (2018) Morphological Universe: Genetics and Epigenetics in Form-Making, *Symmetry: Science and Culture*, Vol. 29, No. 1, 240pp, January.

16 Lalvani, Haresh (1989) Structures and Meta-Structures, In: *Abstracts, Symmetry of Structure*, Budapest, Hungary, August.

17 Lalvani, Haresh (1993) Metamorphic Tiling Patterns Based on Zonohedra, *U.S. Patent 5,211,692*, May 18.

18 Lalvani, Haresh (2020) 4D-Cubic Lattice of Elements, *Foundations of Chemistry* 22:147-194 (published online, November 14 2019), Springer.

19 Lalvani, Zaran (2017) Zero-Cyclic Sum Property in Graphs, Supplement E in [18].

20 Coxeter, H.S.M (1973) *Regular Polytopes*, Courier Corporation.

21 Fuller, R. Buckminster (1975) *Synergetics: Explorations in the Geometry of Thinking*, MacMillan.

22 Burt Michael, M. Kleinman, and A. Wachman (1974) *Infinite Polyhedra*, Technion, Haifa, Israel.

23 Burt, Michael (1973) *Saddle polyhedra, Zodiac 21,* Olivetti Co., Italy.

24 Pearce, Peter (1978) *Structure in Nature is a Strategy for Design*, MIT Press, Cambridge, MA.

25 Lalvani, Haresh (1977) Transpolyhedra, Dual Transformations by Explosion-Implosion, Lalvani (self-published), New York.

26 Durer, Albrecht (1528) *Four Books of Human Proportions*, Nuremberg. Original online: https://archive.org/details/hierinnsindbegri00dure/page/n3/mode/2up?view=theater

27 Le Corbusier (2004) (First published in two volumes in 1954 and 1958) *The Modulor: A Harmonious Measure to the Human Scale. Basel & Boston: Birkhäuser.*

28 Raup, David (1966), Geometric Analysis of Shell Coiling: General Problems, *Journal of Paleontology*, Vol. 40, No. 5 (September), pp. 1178-1190.

29 Lalvani, Haresh (2004) Metamorphology and Gaudi, In: Shell and Spatial Structures from Models to Realization, *Extended Abstracts, IASS 2004 Symposium*, Montpellier, France, Ed. Rene Motro, pp.40-41.

30 Schwarz, Hermann A. (1890) *Gesammelte Mathematische Abhandlungen*, Vol.1, Julius Springer, Berlin.

31 Wolfram, Stephen (2002) *A New Kind of Science*, Wolfram Media, Champaign, IL.

32 Abzhanov, Arhat, Winston P. Kuo, Christine Hartmann, B. Rosemary Grant, Peter R. Grant & Clifford J. Tabin (2004), Bmp4 and Morphological Variation of Beaks in Darwin's Finches, Science, New Series, Vol. 305, No. 5689 (Sep. 3, 2004), pp. 1462-1465.

33 Abzhanov, Arhat, Meredith Protas, B. Rosemary Grant, Peter R. Grant and Clifford J. Tabin, Science, (2006) The calmodulin pathway and evolution of elongated beak morphology in Darwin's finches, *Nature*, Vol. 442/3, August 2006, pp.563-67.

34 Kuroda, Reiko (2010), How a Single Gene Twists a Snail, *Integrative and Comparative Biology*, Vol. 54, No. 4, pp. 677–687.

9

ABIOGENESIS AND THE FUTURE OF ARCHITECTURE

"We are made of star-stuff. We are a way for the cosmos to know itself."
Carl Sagan[a] (1980)

"I had spent the previous afternoon making cardboard cutouts of these various components [A,T,C,G], and now,...I could shuffle around the pieces of the 3-D jigsaw puzzle. How did they all fit together? Soon I realized that a simple pairing worked exquisitely well: A neatly fitted with T, and G with C. Was this it?... It was so simple, and so elegant, that it almost had to be right."
James D. Watson[b] (2017)
(recalling his discovery in 1953)

"[In future] Designing genomes will be a personal thing, a new art form as creative as painting or sculpture."
Freeman Dyson[c] (2007)

"Chemistry is basic structure, ergo architecture."
R. Buckminster Fuller[d] (1959)

The classic film, *Powers of Ten* by Eames and Morrison, visually captures the link between the smallest known building blocks, the elementary particles, and builds up to the atom, molecules, macro-molecules, cells, organs, organisms, and so on, by scaling up in powers of 10 to reach astronomical sizes like the solar system, galaxies,

---

[a] Carl Sagan, *Cosmos, a Personal Journey* (TV Series, 1980), Episode 1: The Shores of the Cosmic Ocean.
[b] James Watson, February 28 1953, Cavendish Laboratory, University of Cambridge, on discovering the Watson-Crick base pairing. In: DNA, *The Story of the Genetic Revolution* (2017), James D. Watson, Revised edition, Alfred A. Knopf, Introduction, p.xi.
[c] Dyson, Freeman (2007) Our Biotech Future, July 19, *The New York Review of Books*.
[d] Fuller, Buckminster R. *Ideas and Integrities*, 1959, p.75. Quote shared by Thomas Zung (personal communication, Oct 30 2021).

galaxies clusters and the universe. Powers of ten is a convenient exponential scale for nature, a descriptive scale and not one that emanates from the physical structure of matter. In the plaque for Pioneer 10 spacecraft, Carl Sagan and Frank Drake [2] proposed a universal scale based on a physical property of hydrogen,[1] the smallest atom and the most abundant element in the universe, as a measure for universal length and time. In the plaque, this was graphically notated in binary numbers. This is a jump in scale from the Planck level in particle physics which defines the metrics of the fundamental phenomena of space, time, mass, current and temperature in Planck units[2] [3]. These five, with two more, determine the morphological space of physical phenomena (**video 38**). Leaving aside the metrics of scale, it is the structure of the atom and its parts, and how these parts are arranged in 3D space that determine different chemical elements which define materials we use in the design arts. These elements have been neatly organized in a periodic table of elements based on their inner structure.

Periodic Table Of Chemical Elements

The concept of periodic tables of form in the last chapter, and parts of previous chapters, traces back to the periodic table of chemical elements proposed by Mendeleev more than 150 years ago [4,5]. This, in its present form [6], is taught to high school students worldwide. For the design arts, it provides a foundation, a starting point of all building materials we use and future materials. The periodic table in chemistry includes the set of 120 different atoms that define the chemical elements which combine to self-build physical structures at all scales from living to nonliving to human-made. Architecture, all physical arts, and all sciences and technologies of physical making, are ultimately tied to the structure of physical matter which is determined by its chemical and physical building blocks, the atoms and the elementary particles.

Scerri points out that *"more than 1000 versions of the periodic table have been published so far"* [7]. These have appeared in the literature in different books [8,9] and a collection is available at the website curated by Leach [10]. This raises the question: if nature is one, shouldn't there be one periodic table of chemical elements? The opinion is divided with supporters on each side, one-table vs many-tables. The many-tables argument is captured in the chemist Martyn Poliakoff's words: *"I regard PT as a tool like a hammer and, just like tools, you have different forms for different purposes (e.g. a claw-hammer and a mallet). There just isn't a "right" or "wrong" form"* [11]. Like Scerri, who has suggested there is a *"best form"* of the periodic table [12], I hold the one-table view since nature is one and a set of universal building blocks underpinned by a unified organizational structure makes design sense.

---

[1] The "hyperfine transition" of the hydrogen atom (as it flips its spin) was used as the basic unit of space (length) and time. It was represented by the binary digit 1. It equals 21 cm and .7 nano-seconds.
[2] In his 1899 paper [6], Planck stated *"These necessarily retain their meaning for all times and for all civilizations, even extraterrestrial and non-human ones, and can therefore be designated as 'natural units'"*. The international S.I. units of mass, length, time, temperature and current have switched to physical constants as the basis for the metric of physical phenomena. This is in the same spirit of Planck who used physical constants to designate his 'natural units'.

video 38

4D Periodic Table

Mendeleev's periodic table and its sequel developments underwent an important change with the introduction of the four quantum numbers of the atom the principal quantum number n, the azimuthal quantum number l, the magnetic quantum number m, and the spin quantum number s.[3] Using these four quantum numbers as generators (vectors), the author mapped the permissible values of these quantum numbers in 4D Cartesian coordinates (n,l,m,s) which led to a 4D periodic table of elements shown in **fig.1** [13]. It occupies a finite portion of a 4D-cubic lattice determined by the known range of values of the four quantum numbers. The animations of this table were first presented at *MD150* [14] (**video 39**), a conference commemorating the 150 years of the publication of Mendeleev's original paper of 1869, at St. Petersburg where Mendeleev developed his celebrated work. As we saw in Chapter 7, the 4D-cubic periodic table (4D PT) can be built in 3D space after projection. In the examples shown, it uses the 3 directions of the cubic node representing an element and adds one more spatial direction along its cell diagonal as a preferred fourth dimension projected in 3D space though any other angle will do. In the example shown, s permits the 4D PT to be split into two 3D cubic PTs joined by s, the 4th dimension. This is similar to joining two cubes with a fourth dimension to make a 4D cube (hypercube, tesseract). Alternatively, a 4D PT can be built with any one of the quantum numbers as the 4th dimension. In the representation shown, n and m have the same directions as in Mendeleev with the difference that l and s have been separated as independent dimensions in (n,l,m,s) space.

A derivative extended 4D PT of chemical elements is shown in **fig.2** where the vertical axis n is calibrated so all elements are visible in the "front" view in increasing atomic number sequences. The idea of reading the PT so all elements are visible in a continuous numeric sequence has been a design requirement for PTs influenced by the print media, and a challenge for 3D PTs. This is an extended 4D PT in 3D shown in two different views. Two additional variants of the extended 4D PT are shown, one in 3D (**fig.3**) and the other in 2D (**fig.4**). The numbers zig-zag in 4 directions (*right*). These are some of the many variants that the author has developed. All are topologically equivalent, i.e. they preserve their connectivity with their neighboring elements in the 4D structure governed by an Euler-Schlafli type equation[4] that relates the number of topological elements (vertices, edges, faces, cells) in the PT. This fixes the topology of the 4D PT.

An interesting result of the 4D organization is that the elements are organized neatly into the 4 known electronic blocks (s-block, p-block, d-block and f-block), each block having elements arranged in linear sequences of

---

[3] Scerri [12] provides a history of the sequence in which the quantum numbers were introduced into the periodic table by physicists: n by Bohr, l by Sommerfeld, m by Stoner and s by Pauli (Chapter 7, pp.192-203).

[4] $N0 - N1 + N2 - N3 + N4 = 1$, where $N0, N1, N2, N3, N4$ are the number of vertices, edges, faces, 3D cells and 4D cells, respectively, in the 4D PT.

video 39

Abiogenesis and the Future of Architecture 343

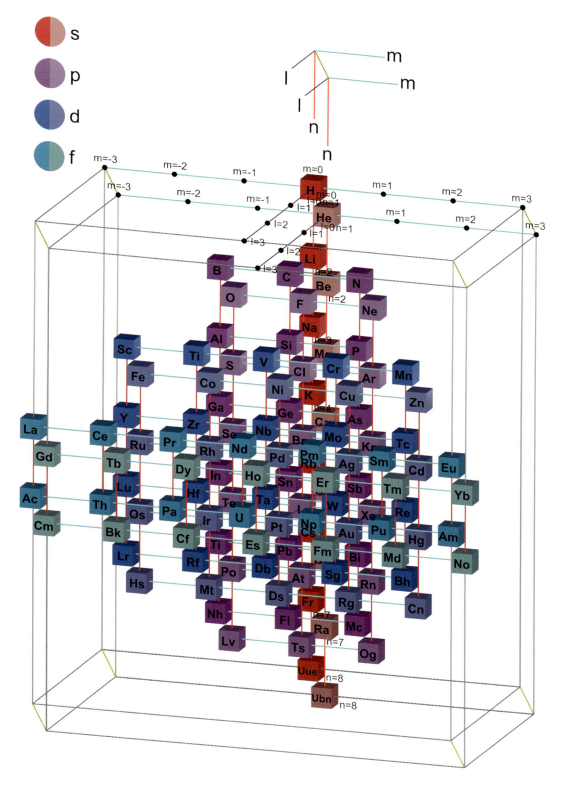

1
A 4D periodic table of chemical elements in (n,l,m,s) space, each specifying a different quantum number.

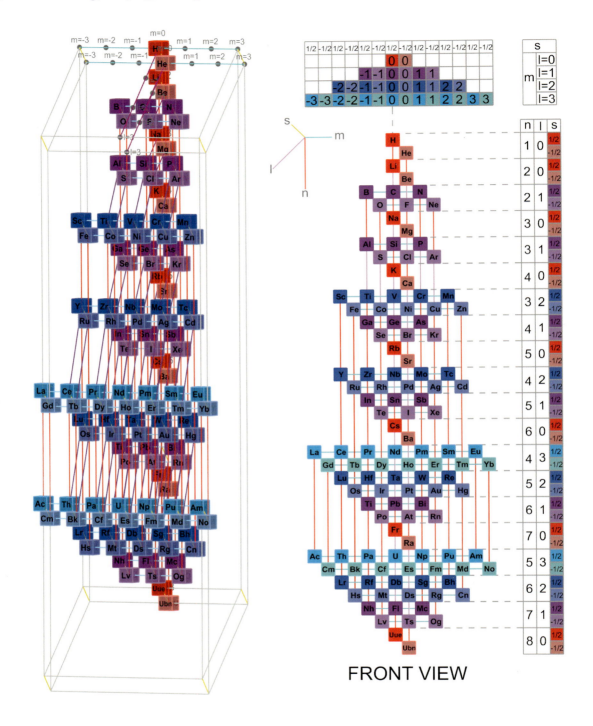

2
Extended 4D periodic table of chemical elements in (n,l,m,s) space with the vertical axis n calibrated so all elements are visible in the "front" view in increasing atomic number sequences.

Abiogenesis and the Future of Architecture   345

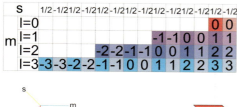

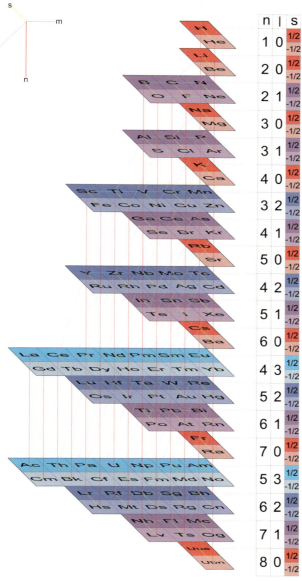

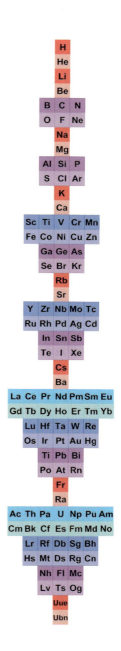

3
A variant of the extended 4D periodic table of chemical elements in fig.2 in 3D.

4
A variant of the extended 4D periodic table of chemical elements in fig.2 in 2D.

increasing atomic number. The blocks are indicated by 4 different colors in the PTs shown, with each color in two shades to indicate two sets, each set having a different spin quantum number representing its two states, s=1/2 and -1/2. One conservation rule in the 4D PT is complementarity whereby the sum of quantum numbers of any pair of elements located in opposite positions around the center of any block are the same.[5] Another conservation rule in the 4D PT is the zero-cyclic sum.[6] These conservation principles are not evident in existing periodic tables. In addition, several existing periodic tables can be derived from the 4D PT by various topology-preserving geometric transformations (rotations, elongations, etc.) and lead to a periodic table of periodic tables.

118D Periodic Table of Chemical Compounds

The 4D periodic table of elements can be extended systematically to include molecules and compounds. This extended 4D periodic table uses each of the 118 known elements (or 120 including the two that have not been found so far) in the 4D table as a new sub-vector. The vectors of the 118-vector-star are indexed v1, v2, v3, v4, v5, ... v118, where the suffixes represent the atomic number of corresponding elements, and each vector represents an independent direction in 118D space. The space of the compounds is a 118D cubic lattice which is embedded within a 4D cubic super-lattice defined by the 4 quantum number coordinates. This provides a point location for all combinations of all elements in one space and is the starting point of an inclusive generative taxonomy for compounds. Herein, each compound is indexed by specific coordinate locations and new compounds can be generated systematically.

As an example, eight of these 118 vectors, v1, v2, v5-v10, are shown in **fig.5**; the vector numbers correspond to the atomic number of each element, e.g. v1 for atomic number 1, v2 for atomic number 2, and so on. Two are from the s-block n=1 elements H (v1) and He (v2), and six from the p=block n=1 elements, namely, B(v5), C(v6), N(v7), O(v8), F(v9) and Ne(v10). Each vector axis comprises molecules or multiples of each element which act as generators for compounds. For example, vector v1, specifies H (1,0,0,1/2), H2 (2,0,0,1), H3 (3,0,0,3/2), ...; vector v8 specifies O (2,1,-1,-1/2), O2 (4,2,-2,-1), O3 (6,3,-3,-3/2), and so on. Classes of compounds corresponding to each compound of each element can be systematically and exhaustively generated by combining vector with others within subsets of the 118D space. A few examples of portions of sub-lattices are shown in **figs.6-9**.

An infinite 2D periodic table of hydrocarbons defined by v1, the Hydrogen axis, and v6, the Carbon axis is shown in **fig.6**. These two axes are defined by the 2-vector-star shown below the table. Some formulae for classes of hydrocarbons are indicated as diagonals within this 2D lattice. **fig.7** shows a 3D periodic table of compounds of Nitrogen (N, v7), Oxygen (O, v8) and Hydrogen (H, v1) in HNO space. The 4D quantum coordinates are shown for each molecule and compound,

---

[5] See figs.8-11, 31 and Table 3 [13]. Complementarity rule works for different properties of elements, e.g. numbers, which include quantum number, atomic number and mass number.

[6] The sum of the differences in values of properties of elements (e.g. atomic number, mass number, mass, mass excess, nuclear binding energy, etc.) located on the vertices of any closed cycle, reading either clockwise or counter-clockwise, equals zero.

Abiogenesis and the Future of Architecture

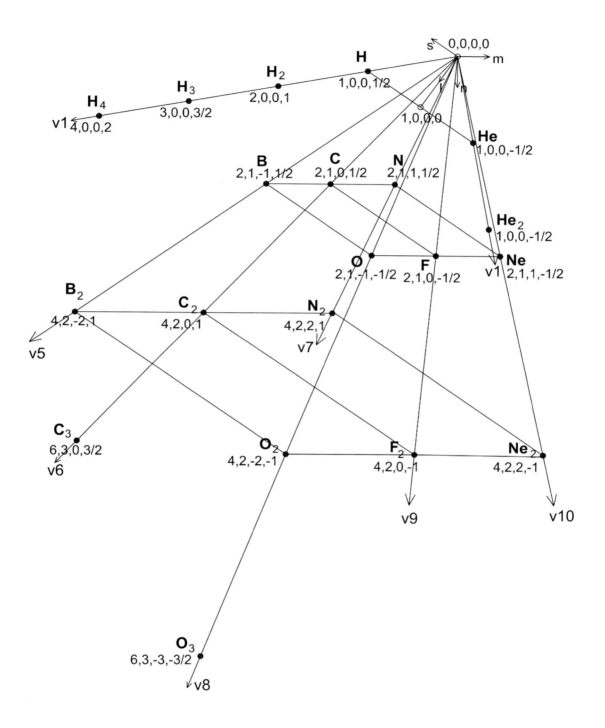

5
The 4D periodic table of chemical elements expanded to a 118D lattice of compounds and molecules. Only 8 vectors, which act as generators of the expanded 8D table and encompass the elements with atomic numbers 1, 2, 5-10 are shown.

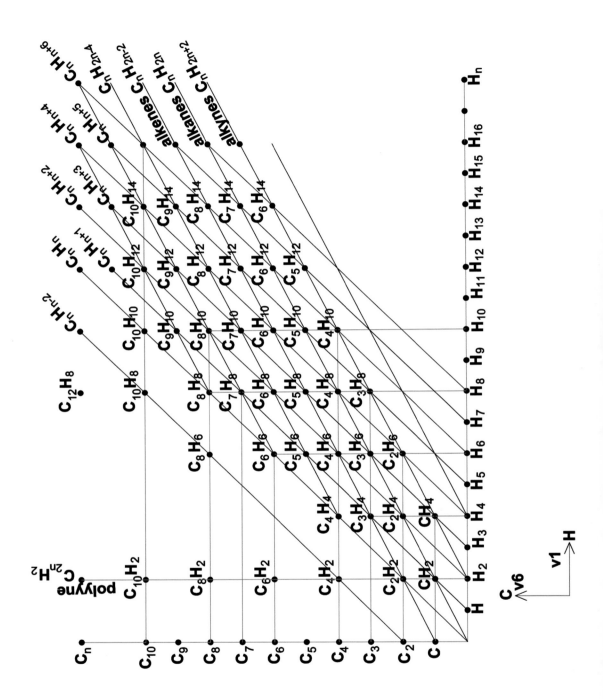

6
The infinite 2D periodic table of molecules and compounds of hydrogen (H) and carbon (C).

Abiogenesis and the Future of Architecture   349

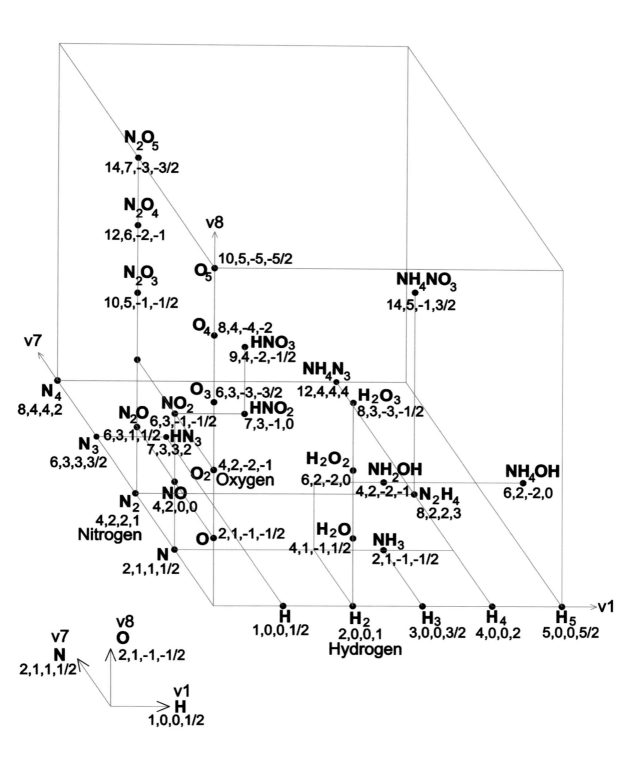

7
The 3D periodic table of molecules and compounds of hydrogen (H), nitrogen (N) and oxygen (O) in HNO space.

350 Coding, Shaping, Making

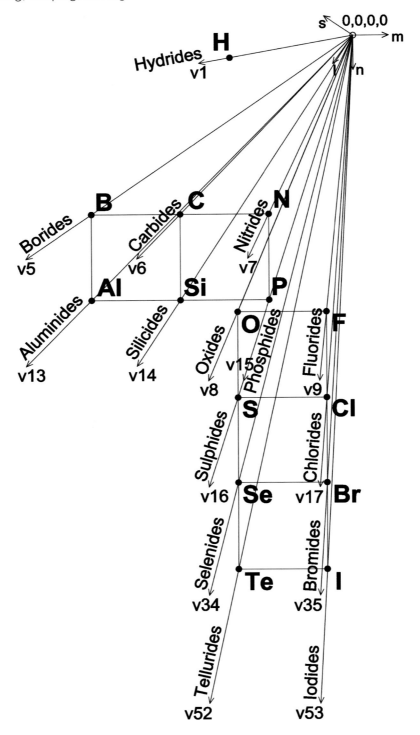

8
The vectors of 14D space of oxides of fourteen elements with atomic numbers 5-9, 13-17, 34, 35, 52, 53.

Abiogenesis and the Future of Architecture 351

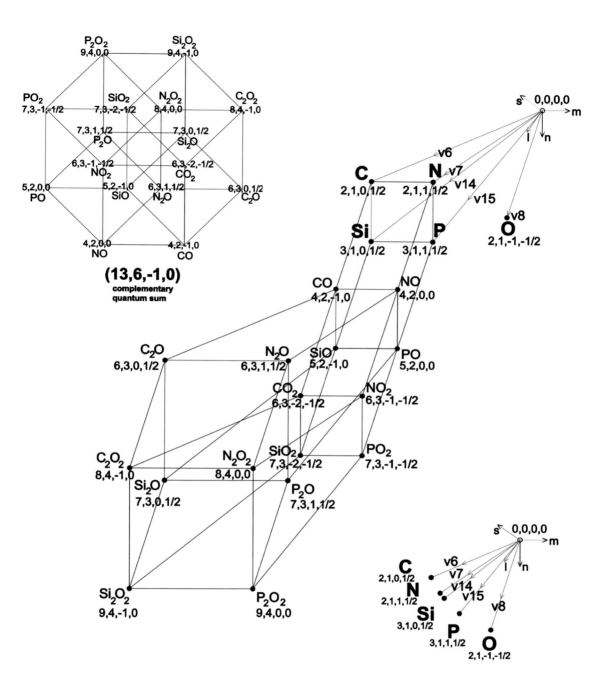

9
The 4D space of oxides of C, N, Si, P shown as the hypercube (*top left*) and as sub-space of the 118D space of compounds emanating from the origin (0,0,0,0).

the generating 3-vector-star is shown below. **Fig.8** shows the method for generating families of compounds of various elements generated by their corresponding vectors: v1 (H) generates hydrides, v5 (B) generates borides, v6 (C) carbides, v7 (N) nitrides, v8 (O) oxides, v9 (F) fluorides, and so on. Vectors of 14D space to generate 14 families of compounds are shown as an example. **Fig.9** shows 16 oxides of four elements, C (v6), N (v7), Si (v14) and P (v15). The vector-star is shown on the bottom right. This is a smaller subset of possible oxides of these compounds since the set shown is restricted to one or two atoms of each element. The 16 compounds are defined by a 4D cube in a "perspective" space since the generating vectors are radiating from a point at the origin (0,0,0,0).

The image on top left shows the same 4D cube in a more familiar orthographic view. In this centrally symmetric view, it is easy to see complementarity between compounds – the sum of quantum numbers of the 8 complementary pairs of compounds on opposite locations around the center of the 4D cube are equal and conserved at (13,6,-1,0).

4D Abiogenetic Space

The abiogenetic space (abiotic or prebiotic space) maps the possibilities of all molecules and compounds built from the five atoms that make up DNA, namely, H, C, N, O, P (in increasing atomic numbers 1,6,7,8,15, respectively). Since phosphorus (P) comes later in the atomic number sequence, the early abiogenetic space, the space of the chemical origins of life represented by DNA (or RNA), comprises the first four of these atoms which can be mapped in the 4D HCNO space. This space is determined by corresponding vectors (v1, v6, v7, v8) to encompass all compounds from H,C,N,O (**fig.10**). The early abiogenetic space emanates from the origin 0 (0,0,0,0) on bottom left and provides potential chemogenetic pathways from simpler compounds to the DNA-RNA bases A, T, C, G, U on the upper right of the diagram. This space provides a map for molecular evolution from simple to complex molecules. The important issue of isomers, molecules with the same 4D (H,C,N,O) coordinates but a different 3D spatial organization of atoms, requires extending this space from each vertex of the 4D lattice into additional branch/lattice type structures to complete the space of possibilities. The morphology of the isomer space, when established, will mark an important component of the morphogenesis of molecules from early elements to life molecules.

Figs.11 and **12** show a fuller mapping of a small portion of the earliest portion of HCNO space nearest the origin (0,0,0,0) at bottom left. All compounds with no more than 2 atoms of each element, H, C, N, O are shown. They are mapped exhaustively in a 2x2x2x2 (2^4) 4D lattice which has 81 point locations for all possible combinations of these four elements. The general case is m^n, where m is the number of atoms of the same element and n is the number of different atoms, 4 in this case, which also defines the dimension of the morphological space of possibilities. The isomers (marked with #) will add branching from each node as mentioned earlier so each isomer can have its own distinct point location in this space. The surprise in the mapping shown is that nearly all 81 possible combinations (excluding the

Abiogenesis and the Future of Architecture 353

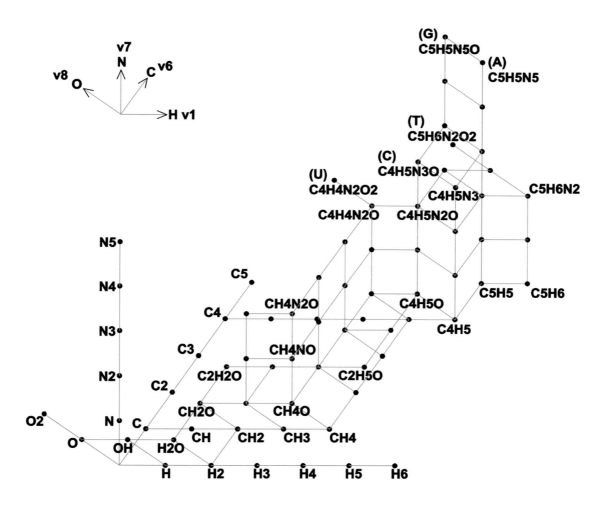

10
Early abiogenetic space is the extended 4D HCNO which includes the DNA and RNA bases on the upper right indicated by their symbols A, T, C, G, U.

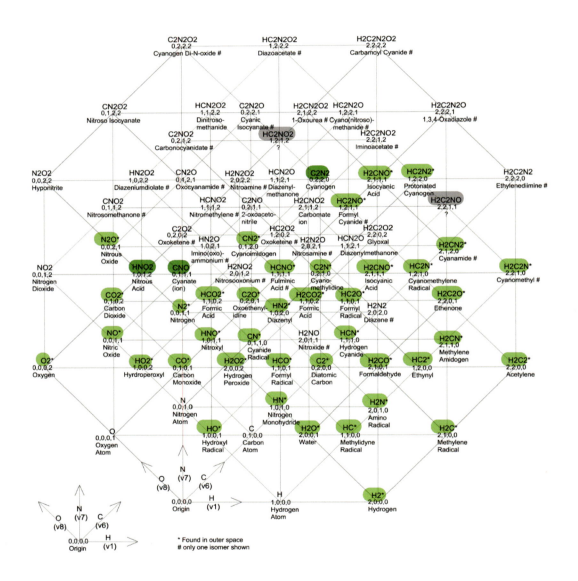

11
Early HCNO space for up to 2 atoms for each element in a 2x2x2x2 subdivided hypercube. Elements found in outer space are indicated by *, isomers by #.

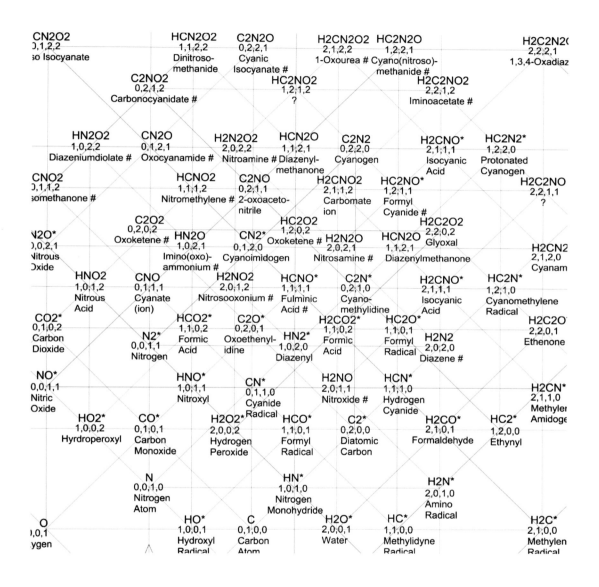

12
Close-up of the HCNO space in fig.15.

origin (0,0,0,0) and two other compounds) exist and have a name.[7] Some have been found in outer space (marked with * and highlighted in green) and lie closer to the origin, some others will be found in outer space in future (three examples in darker green were discovered since the author first began mapping this diagram over 5 years ago), some others exist naturally on earth or are synthetically produced in the laboratory.

The 4D lattice can be extended to include more HCNO compounds from the same 4 elements though, as the space increases, more unfilled point locations are found. Excluding examples that are not possible due to physical, chemical, or mathematical constraints, mapping the space provides an opportunity to discover or invent new design possibilities. This space defines the domain of design-artists (artists, designers, architects and makers) who share the same spirit with synthetic chemists and synthetic biologists, the spirit of exploring new design possibilities and thereby extending or filling the space of possible designs. The open-ended exploratory nature of this space acts as an interactive taxonomic device by default. My remark on the new geometry of *SEED54* addressed this spirit that shapes the space of designers in both art and science: "*SEED54 has a new geometry. We don't know yet if it exists in nature. Maybe someday it will, with or without our help*" [16].

Two DNA Technologies

DNA technology has two strands, biological and non-biological. Biological applications of DNA technology retain the features of living systems and have been used in the design of genetically modified foods, genetically engineered organisms, medicine and therapeutics, forensics and genetic fingerprinting, human genealogies and for advancing basic scientific understanding of how nature works [17]. Falling broadly under synthetic biology (recombinant DNA, genetic engineering, biotechnology), it has two schools of thought, one by modifying existing living systems and the other, a more recent one, aimed at creating new life. Both alter nature by introducing living structures absent in nature. The former, in its current state, includes gene-editing which has rapidly accelerated with the development of several methods including CRISPR-Cas9 [18], a very precise genetic cut-and-paste method for replacing and inserting genes into existing genomes at precise locations. This leads to designer (micro-) organisms that can address many societal needs besides therapeutics including planetary environmental issues of diminishing resources, new energy sources, new ways of purifying waste, removal of toxins, reducing carbon dioxide, and so on. Dyson foresees DIY genetics[8] [19] and Church envisions the limit case in therapeutics where all human diseases are eliminated [20]. There is considerable

---

[7] The primary database used by the author is PubChem, National Library of Medicine, National Center for Biotechnology Information, a NIH database, https://pubchem.ncbi.nlm.nih.gov/. For the isomers, only one name is shown in fig.15.

[8] "*There will be do-it-yourself kits for gardeners who will use genetic engineering to breed new varieties of roses and orchids. Also kits for lovers of pigeons and parrots and lizards and snakes to breed new varieties of pets. Breeders of dogs and cats will have their kits too*" [19].

interest within architecture and physical arts in incorporating synthetic biology into design leading to very exciting work[9] [21-25].

The second school, pioneered by Venter and his colleagues [26], deals with designing new synthetic organisms using synthetic DNA sequences designed in the computer and assembled in a wet lab environment. The striking example is the synthetic organism *Mycoplasma mycoides* JVCI-syn1.0, the first synthetic bacterium introduced by humans in 2010.[10] To address the concern about a synthetic genome inter-mixing with natural genomes, JVCI-syn1.0 had distinguishing "watermarks" with author's names and other identifying information as well as three quotations, one each from James Joyce, Robert Oppenheimer and Richard Feynman, in the synthetic DNA sequences.[11] This sets it apart from natural genomes. This was followed by JVCI-syn3.0, a smaller version with a minimum genome to sustain life, in 2016.[12] The difference between the two ways to go beyond nature, one by modification, the other by introducing new designs, is captured in Venter's words *"If you want to make a few changes, CRISPRs are a great tool .... But if you're really making something new and you're trying to design life, CRISPRs aren't going to get you there"* [27].

Both approaches raise ethical as well as bio-hazard and bio-terrorism concerns about altered or new designed life forms, a continuing topic of intense collective introspection, self-regulation, and bioethics guidlines over the past four decades [28,29]. The mitigation strategies range from a moratorium on genetic experiments on humans to the introduction of "suicide genes" and "kill switches" in synthetic life.[13] These techniques will extend to growing architecture, art and design.

Non-biological applications include DNA nano-structures pioneered by Seeman for nano-scale construction of various geometric skeletal 3D structures [30], Adelman's DNA-computing which solved the Königsberg 7 bridges problem[14] in a test tube [31], Rothemund's "DNA origami" to build raster-filled shapes [32], physical design of a DNA computer built from DNA tiles by Woods et al [33], a walking machine built with DNA by Jung et al [34], and so on. Architectural, structural or surface components of manufactured products at human scale and larger of these types of non-biological DNA nano-technologies are a promising direction which head towards the autonomously growing chemical architecture envisioned by Katavolos [35], or cymatics-based architecture by Giorgini [36], to name the early pioneering works. Synthetic information-carrying self-replicating molecules other than DNA are well-suited for such non-biological applications of DNA especially in the area of building nano-structures for various applications and scaling them up to macro-scale applications like architecture. Any one of the

---

[9] See, for example, the works of Joachim Mitchell, Neri Oxman, David Benjamin, Rachel Armstrong, Skylar Tibbits and others.
[10] This led to Venter's question and answer: *"Can we grow in the laboratory an organism that represents a brand new branch on the tree of life, a representative of what some like to call the Synthetic Kingdom. In theory, at least we can"*, Venter [21], p.131.
[11] Venter [21], pp.88-89, 124-125.
[12] Syn3.0 had 473 genes of which 149 genes had "unknown biological functions".
[13] Venter [21], p.157.
[14] To walk through the city of Konigsberg by crossing each of the seven bridges once. This is a problem devised by the mathematician Leonhard Euler and is considered as the beginning of graph theory. https://en.wikipedia.org/wiki/Seven_Bridges_of_K%C3%B6nigsberg

minimal DNA-type double helical molecules presented here, if proven to be viable, provides a candidate starting point.

On Minimal Structures

Conservation of available resources, planetary and beyond, requires us to look for an economy of means and materials. In view of our strained natural resources and, as we extend our wings beyond our home planet, this guiding principle applies to design at all scales. This, in principle, includes conserving the number of atoms (or elementary particles) in design at all scales from molecules to buildings. The idea of minimal structures is well-known in architecture from the pioneering works of Bucky Fuller with his design strategy of doing "more with less" and Frei Otto's widely influential work in the development of lightweight structures. There has been considerable work in this area over the last six decades and is even more relevant now in view of the global environmental crisis. Minimal structures, and minimal processes, are an essential cornerstone in how we address planetary issues of conservation of material resources as well as industrial fabrication and manufacturing processes.

This strategy has driven the works in earlier chapters related to industrial production (Chapters 1-3). It was the main takeaway in the invention of new methods of making expanded and adaptive structures in Chapters 5 and 6, where a fixed amount of material was being redistributed to make 3D structures from 2D while acquiring additional attributes like lightness, transparency, strength, curvature and increased surface area or volume from the same material at no additional cost besides forming. It was also the lesson from the *Morphing Platter* series in Chapter 4 where the same mass was re-distributed to produce unique variations of algorithm-driven emergent designs. It was extended to minimal surfaces (Chapter 7) and their taxonomies (Chapter 8). The minimal DNA-type molecular structures presented next display the same underlying thinking.

Minimal DNA-Type Molecules

Notwithstanding the complexity of the chemistry, physics and biology of the DNA molecule, the morphologic possibilities of DNA alternatives provide a challenging area for designers. A method of generating an inventory of designs of minimal DNA-type double helical molecules is presented.[15] The DNA molecule is whittled down to its bare minimum to retain the fundamental features of information encoding and potential replication using the two sets of complementary base pairs, each pair joined by hydrogen bonds. The minimal molecules constructed this way are speculative design possibilities, some may be possible, some may require different chemical environments than water, some may have a natural connection with pre-DNA molecules and may exist in outer space or may appear in future, and some others may inspire new developments in the design of information coding, and

---

[15] This aspect of the work has been in progress since 2015.

potentially self-replicating, molecules and DNA alternatives that will emerge in future.

Design Criteria

The minimal DNA-type molecules presented mimic the canonical Watson-Crick DNA structure [37,38] and are composed of only H, C, N, O atoms; phosphorus is removed from the helical backbone by restricting atoms from early abiogenetic space. The proposed molecules have the following design features adapted from DNA:

1. Two minimum "anti-parallel" helical molecules that support two pairs of complementary minimum bases excerpted or adapted from Adenine (A), Thymine (T), Guanine (G) and Cytosine (C). Anti-parallel feature requires the double helix to have an axis of 2-fold symmetry perpendicular to the helical axis.
2. The minimum bases are made from H, C, N, O atoms and are paired to make elongated hexagonal configurations Hex1 and Hex2 present in DNA (**fig.13**) and enabled by the hydrogen bonds (dotted lines). The two pairs are defined by two different triads of NCN and NCO (**fig.14**) with C located on the opposite ends of the horizontal axis in Hex1 and Hex2. A third triad, OCO, is added though absent in DNA.
3. The minimum base pairs are flat and enabled by the sp2 C atoms which have a 3-way connection meeting at 120 degrees.[16] This configuration fixes the inner C-C distance and is the same as in DNA.
4. The minimum base pairs occupy parallel rungs of the double-helical DNA-type ladder and their sequence along the axis of the helix determines the genetic code.
5. Each minimum triad pair is illustrated with an outer sp3 C (with a 4-way connection meeting at the central angle of a tetrahedron, 109.4712 degrees) on either sides in Hex1 and Hex2 and attached to the inner sp2 C atom. The outer sp3 C is part of the helical molecular chain. The distance between the two outer sp3 C atoms, same as in Hex1 in the G-C pair in DNA, fixes the width of this group of minimal helices. Alternative groups are produced by replacing the outer sp3 C with sp2 C or N.
6. The minimum helical molecule is linearly periodic. It has an asymmetric repeat unit which joins atoms between the rungs of the helix. This asymmetry permits the anti-parallel helices as in DNA.
7. The helical backbone is built from H, C, N, O atoms. It has a minimum of 3 bends in the repeat unit.
8. The distance between the rungs of the ladder along the helical axis is the same as in DNA and equals 3.4 angstroms. This, along with the number

---

[16] Carbon exists in 3 states which have different connectivity with neighboring atoms, namely, 4-way, 3-way and 2-way connections. The 4-way connection is the sp3 C which has 4 single-bonds (as in diamond), 3-way is the sp2 C and has one double-bond and two single bonds (as in fullerenes or graphene), and 2-way has one triple-bond and one single bond. In the DNA bases, the 3-way sp2 carbon is used, the helical backbone of DNA has some 4-way sp3 carbon atoms in its pentagonal rings. The flatness of Hex1 and Hex2 is enabled by sp2 C. Alternatively, nitrogen atoms can provide alternative flat base pairs.

of bends in the repeat unit of the helix, determines the angle of turn of the helix.

Minimal Triads

Fig.13 shows the two base pairs from DNA, Adenine-Thymine (A-T) (top) and Guanine-Cytosine (G-C) (***bottom***) joined by hydrogen bonds (dotted lines). The pairing defines the central configurations Hex1 and Hex 2 which have an elongated hexagonal geometry. The hydrogen bonds have 2-fold rotational symmetry in Hex1 and mirror-symmetry in Hex2 (**fig.14**). The paired molecules comprise NCN or NCO triads, with sp2 C (represented with a 3-way connection) positioned along the horizontal axis of the hexagon, with N or O bonds emanating from C along the two angles on the inner side of the hexagon, and the third bond with H on the outer side of C along the horizontal axis. The number of available triads are 4 for NCN and 5 for NCO. Two more triads are possible when OCO[17] is added. This makes a total of 11 triads with sp2 C as the central atom of the triad as in DNA (**fig.15**).

Precursor Molecules

The triads in **fig.15** are candidate precursor molecules for Hex1 and Hex2 portions of DNA. For example, formic acid and formamide have been found in outer space [39,40], and their extended versions, after replacing H with CH3, lead to acetic acid and acetamide which have also been found in outer space [41,42]. Minimal DNA structures constructed from molecules found in outer space would suggest their potential roles as precursors to DNA.

Triad Pairs

The combinations of 11 triads lead to 18 *triad pairs* joined by hydrogen bonds where each triad pair is equivalent to a complementary base pair in DNA. Of these, 14 are shown (**fig.16**),[18] 10 pairs in Hex1 configuration and 4 in Hex2. This comprises the initial palette of base pair molecules for constructing minimal-DNA options.

Connection to Helix

Connecting the triad to the helical backbone requires replacing the outer H on the horizontal axis in each triad pair with another atom, either C or N, leading to CH3, CH2 or NH2 as the only options. These are *connection atoms* which become part of the backbone. Five examples from the total of 14 are shown where the outer H in the triad is replaced with CH3 on either end (**fig.17**). The outer C-C distance (with CH3 added) establishes the width of minimal double-helical molecules and is the same as the Hex1 portion of DNA and slightly different from the Hex2 portion. Replacing H with CH2 leads to 8 additional

---

[17] These two molecules include one each that make Hex1 and Hex2 configurations.
[18] 4 paired pairs using triads HCNO2 (3) and Formic Acid (2) and comprising OCO triads are removed.

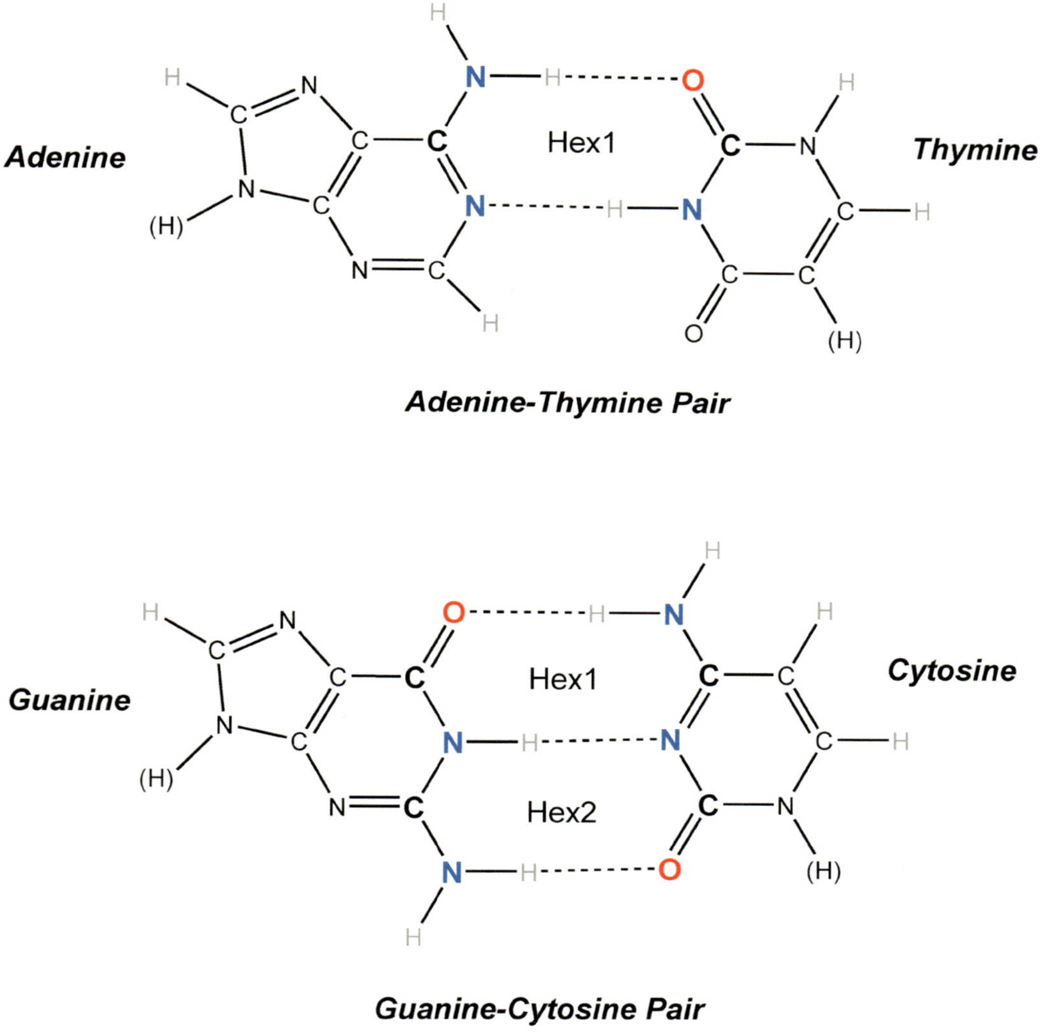

Adenine

(H)

Hex1

Thymine

(H)

Adenine-Thymine Pair

Guanine

(H)

Hex1

Hex2

Cytosine

(H)

Guanine-Cytosine Pair

13
The two complementary base pairs from DNA, Adenine-Thymine (A-T) (*top*) and Guanine-Cytosine (G-C) (*bottom*), make the central configurations of elongated hexagons Hex1 and Hex2.

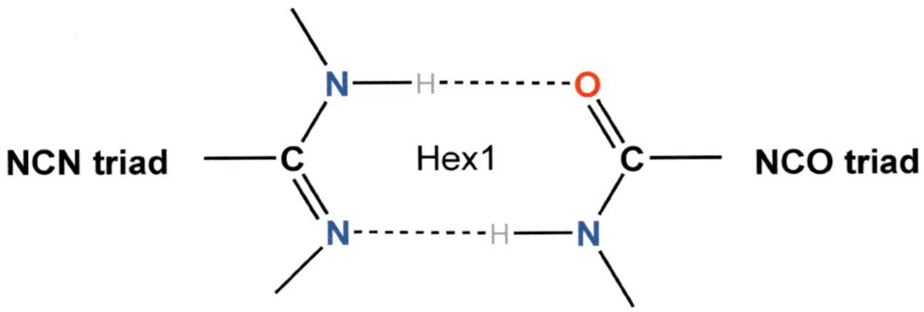

NCN triad —C〈Hex1〉C— **NCO triad**

From A-T Pair

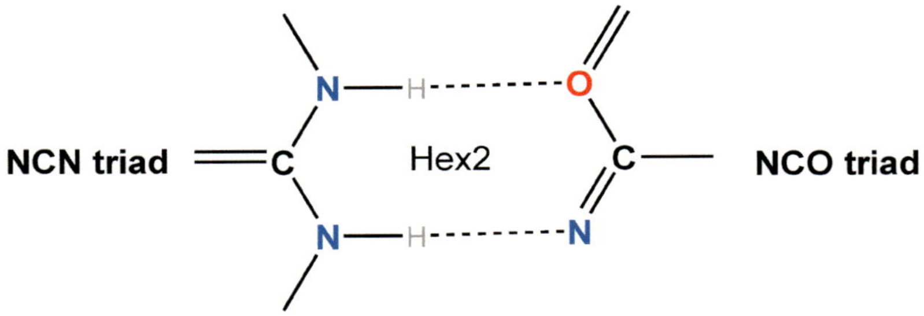

NCN triad =C〈Hex2〉C— **NCO triad**

From G-C Pair

14
Two ways to enable hydrogen bonds (dotted lines) by placing hydrogen atoms in rotational or mirror symmetry within Hex1 and Hex2. The hydrogen bonds join two triad molecules, NCN and NCO excerpted from the bases.

Abiogenesis and the Future of Architecture   363

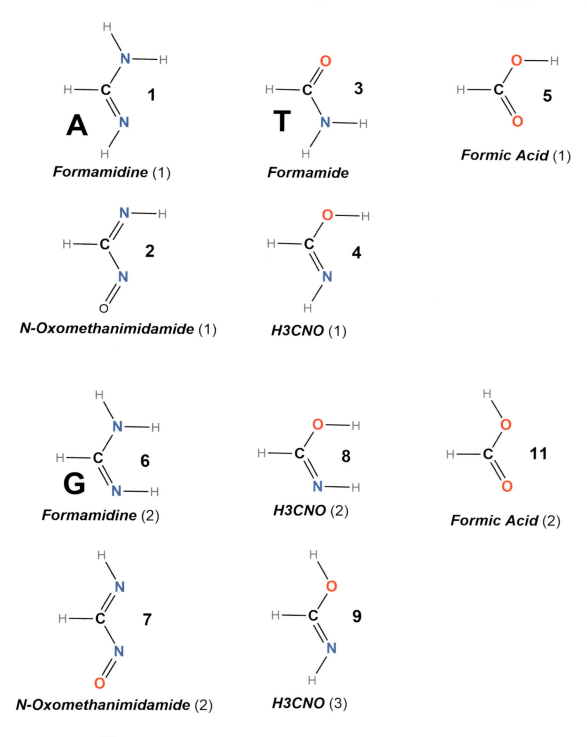

15
The 4D periodic table of chemical elements expanded to a 118D lattice of compounds and molecules. Only 8 vectors, which act as generators of the expanded 8D table and encompass the elements with atomic numbers 1, 2, 5-10 are shown.

16
14 triad pairs from the 9 simple triads selected from the 11 in fig.15 (excluding triads #9 and #11), 10 in Hex1 and 4 in Hex2 configurations.

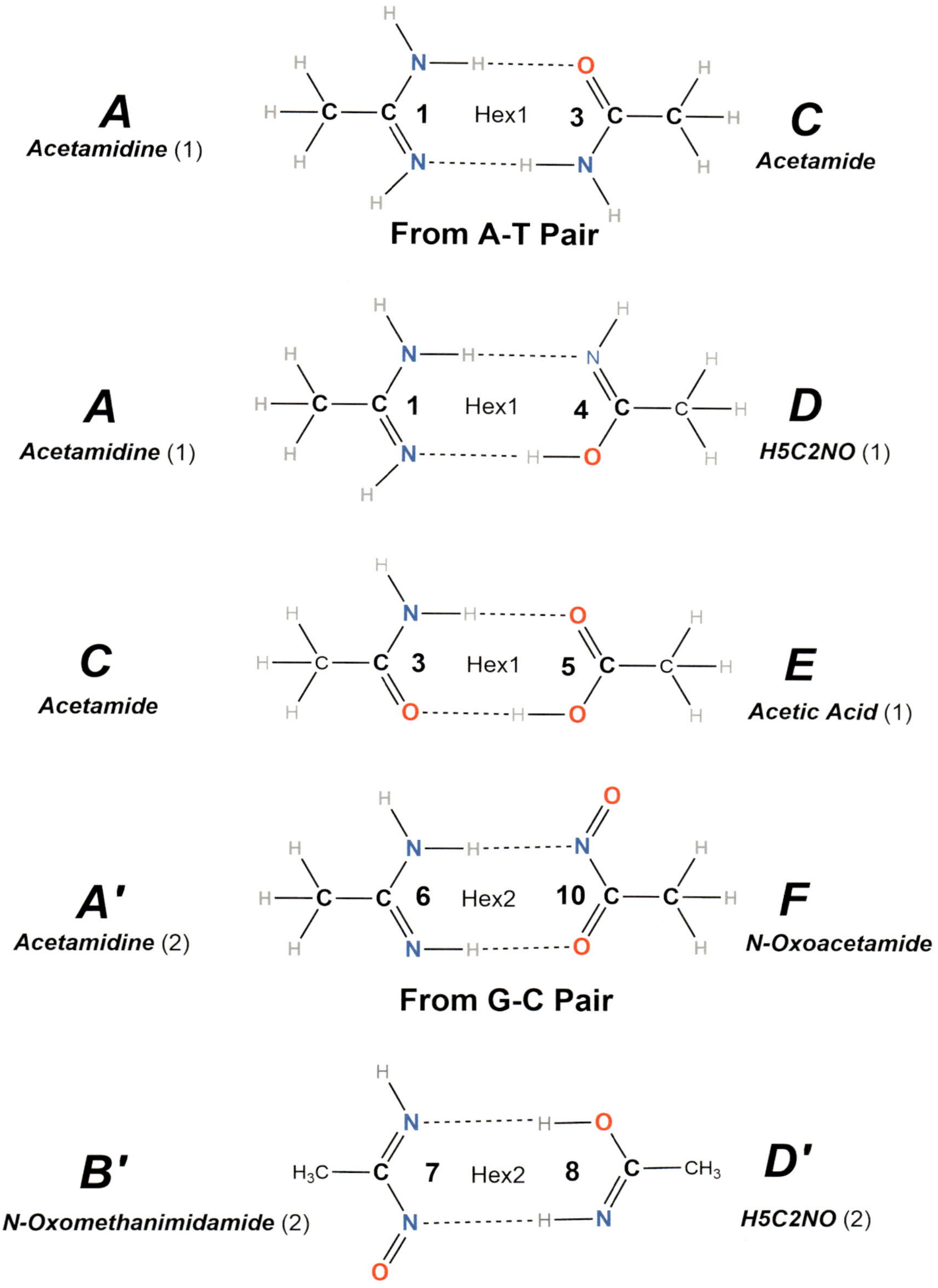

Abiogenesis and the Future of Architecture 365

A *Acetamidine* (1) — 1 Hex1 3 — **C** *Acetamide*

From A-T Pair

A *Acetamidine* (1) — 1 Hex1 4 — **D** *H5C2NO* (1)

C *Acetamide* — 3 Hex1 5 — **E** *Acetic Acid* (1)

A' *Acetamidine* (2) — 6 Hex2 10 — **F** *N-Oxoacetamide*

From G-C Pair

B' *N-Oxomethanimidamide* (2) — 7 Hex2 8 — **D'** *H5C2NO* (2)

17
Five of the 14 triad pairs in fig.16 in Hex1 and Hex2 configurations are shown with CH3 replacing H on the two outer ends. CH3 is part of the two helical molecules.

options while NH2 yields another 14. The total available options for base pairs equals 36.

Paired Pairs

The triad pairs can be combined to produce *paired pairs* equivalent to the two complementary base pairs in DNA. For triads extended with CH3 and NH2, a total of 91 *paired pairs* are possible for each; for CH2, this total equals 22. The total inventory of possible paired pairs equals 204. The linear sequences of any paired pairs from this set defines a genetic code built from minimal molecules. To complete the double-helical molecule, the paired pairs need to be joined to two complementary helical molecules, one pointing up and the other pointing down to satisfy the anti-parallel requirement.

Helical Molecules

The triads can be joined to a helical chain molecule to construct one strand of the minimal DNA molecule and the second strand can be obtained by using any paired pair. A few molecules found in outer space are used here to build the helical molecule. The minimal helices are periodic molecular chains with a *repeat unit* of translation bound by the connection atoms. The repeat unit can have 3 or 4 zig-zag bends, or more, where each bend represents a bond between successive connection atoms. To ensure complementarity, the repeat unit must have asymmetry. A 2-bend helix, with 2 bends between connection atoms will have only one atom in between, resulting in a mirror symmetry within the repeat unit. This will prohibit anti-parallel helices. A 3-bend helix, with 2 atoms within the repeat unit, will permit anti-symmetry with two different atoms and is the minimal helix for minimal DNA-type structures. This leads to 9 possible repeat units, hence 9 candidate double-helices. A 4-bend helix, with 4 bends and 3 atoms in the repeat unit, provides 30 design options for asymmetry. The three isomers — acetic acid, methyl formate or glycolaldehyde (**fig.18**, *top*) — each made from H, C, O atoms, and discovered in outer space [41,43,44], can be used to build a linear HCO molecular chain. Each of these can be used to build periodic molecular chains (*bottom*) which can be bent into a helix. A chain made from glycine, the first amino acid found in outer space [45], provides another alternative. Triads can be attached to the chain at the repeat units to build a range of minimal DNA-type molecules.

Double-Helical Molecules

The combined inventories of paired pairs and 4-bend helices produce 6,120 DNA-type double helical molecules. The same number of paired pairs combined with the 9 3-bend helices produces 1,836 double-helical molecules. Any one of these nearly 8,000 molecules, if shown to be viable, is a candidate for a prebiotic information coding (and potentially self-replicating) molecule as a precursor to DNA or RNA. The others, also if viable, would be candidates for abiotic technologies. Additional design features, for example, equivalent to the

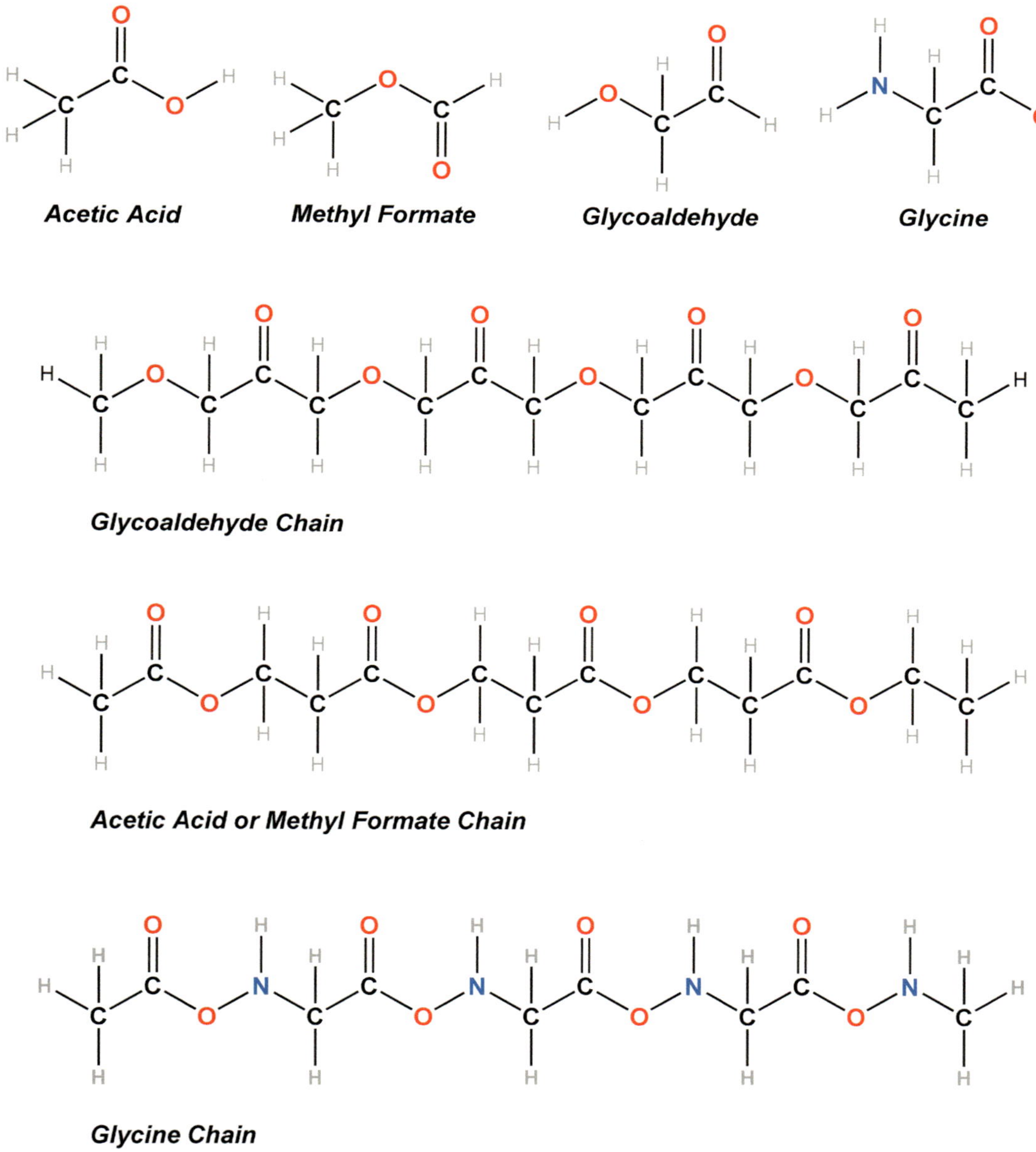

Abiogenesis and the Future of Architecture   367

Acetic Acid   **Methyl Formate**   **Glycoaldehyde**   **Glycine**

Glycoaldehyde Chain

Acetic Acid or Methyl Formate Chain

Glycine Chain

18
4 molecules found in outer space (*top*) can make periodic molecular chains with 4 bends (*bottom*). These chains can be used for building the helical backbones of minimal DNA molecules.

functions of tRNA, will be needed for these molecules to read the molecular code and 3D-print other molecular structures as constructors and assemblers of a higher level system of molecular architecture.

Physical Models

Of the many physical models of minimal DNA-type double helical molecules built by the author, one example is shown in **fig.19** and others can be similarly derived. This one is particularly interesting since the helix is a glycine chain obtained by replacing the outer H atom with CH3 to make an asymmetric repeat unit. The helix can be joined to the triads in any sequence at CH3 locations. Two such complementary chains can be joined by hydrogen bonds into Hex1 or Hex2 configurations and folded into a helix during the building process. Alternatives are possible by connecting at N or sp2 C locations of the glycine chain to provide other minimal DNA-type alternatives using the same backbone. A paired pair comprising A-C and A'-F can be directly derived from the two DNA base pairs.

DNA Periodic Tables

The bottom up progressive order in nature starting from elementary particles to atoms and chemical elements, to molecules and compounds, to the 4 DNA (RNA) alphabets which define the biological genetic code, unfolds with increased physical size and complexity as depicted visually in *Powers of Ten*. The method of mapping chemical elements and compounds in higher dimensions as shown earlier and the extension to abiogenetic space, is here applied to DNA sequences. The dual strategy of organizing-and-generating simultaneously, initiated in [46,47], increases with complexity. As the complexity of information increases, so does the complexity of its representation in n-dimensional spaces. This provides a visual challenge. In DNA, the number of dimensions of genetic sequences having n bases increases in scale ranging from codons (3 bases, $n=3$), gene markers ($n=6$ or more), genes (few thousand to over a million bases), gene clusters (two or more genes; hox genes, for example, is a cluster of 8 genes), and entire genomes. These hierarchical possibilities can be mapped in recursive, self-similar n-dimensional periodic tables that branch from each other in increasing scales. 1-base branches to 2-base sequences, which branches to 3-base sequences, to 4-bases, and so on in an open-ended hyper-tree (higher-dimensional tree) for any number of base sequences.

Representations of Base Sequences

The number of sequences of 4 bases, A, T, C, G, increases in powers of 4 and equals 4^n for n-base sequences. Since 4^n equals $(2^2)^n$ or 2^{2n}, there are two representations of DNA periodic tables. The first is n-dimensional and the 4 bases are arranged in a *linear* row in the sequence A-C-G-T with complementary pairs A-T on the outer ends of a line and C-G in the middle. The second is $2n$-dimensional and the four bases are in a *spatial* arrangement, a square or a tetrahedron, with A-T and C-G on the diagonals of a square. Both representations display

Abiogenesis and the Future of Architecture 369

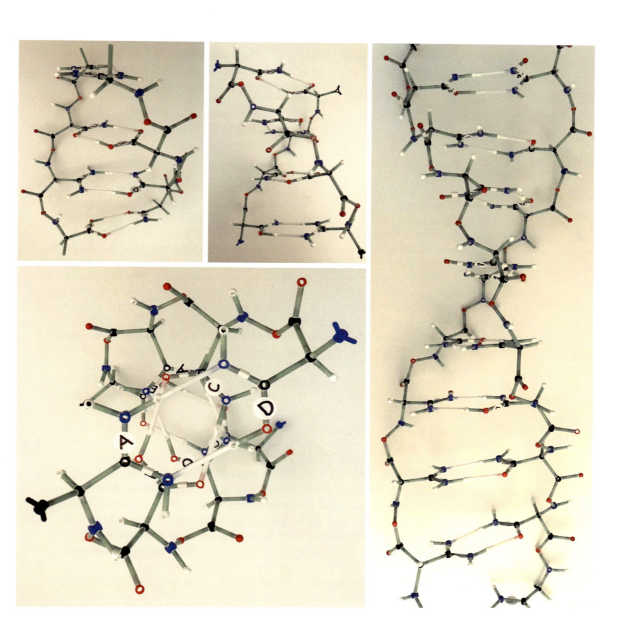

19
Photographs of two models of the minimal double helical DNA-type molecule constructed with a commercial molecular kit and using a pair of glycine chain helices in an anti-parallel arrangement. The model on the left shows complementary pairs A-D and E-C in the sequence AEDC (or DCAE) in three views, perspective view (*top left*), side and top views (*top right, bottom*). The model on the right shows the pairs A-C and B'-E' in the sequence CAAB'CD'D'CB'B' (or ACCD'AB'B'AD'D').

complementarity in DNA sequences located at opposite positions around the center of symmetry. In higher dimensional representations, complementarity, which is fundamental to the Watson-Crick base-pairing of A-T and C-G, extends to base sequences in a visual way due to the complementary location of complementary sequences. Complementarity also applies to how base sequences change from one to another. The change in a base sequence at one vertex to a second vertex has a complementary change across the center of symmetry in the opposite direction from the complementary base sequence at a complementary vertex to a complementary second vertex. This change follows the method first shown for polyhedra in [46].

Linear

The n-base sequences in the linear version (**fig.20**)[19] occupy the vertices of a subdivided n-cube, a finite portion of a n-cubic lattice. The 4 bases are positioned linearly along each edge of the n-cube and mark the n independent axes of the lattice. 2-base sequences ($n=2$, 2-codons) are mapped in 2D space, 3-base sequences ($n=3$, 3-codons) in 3D space, 4-base sequences ($n=4$, 4-codons) in 4D space, and so on [48]. The 2-base sequences yield 16 (4^2) pairs mapped in a 4x4 square periodic table (*top left*). The 3-base sequences[20,21,22] yield the 64 (4^3) triplets arranged in a 4x4x4 cubic lattice (*top right*), a 3D periodic table of amino acids, the building blocks of proteins which make up structural and functional materials in organisms. The 4-base sequences lead to a 4x4x4x4 4D-cubic periodic table of 4^4 possibilities (*bottom right*). The 5-base and 6-base combinations that include gene markers, some used in genealogies and forensics, are part of the 5D and 6D periodic tables of base sequences, and so on. For a large number of bases, say 3 billion as in humans, a 3 billion-dimensional periodic table is possible with similar tables of varying sizes for other organisms.

Spatial

The spatial representations of n-base sequences in $2n$-dimensional tables are shown in **figs. 21-23** for 2-base, 3-base and 4-base sequences. The three diagrams layered together lead to the beginnings of a fractal hyper-tree, a higher dimensional tree. Each point branches to four different spatial directions and leads to a branched hyper-fractal (higher-dimensional fractal) periodic table of base sequences that includes codons, gene markers, genes, gene clusters and genomes of varying sizes.

The usefulness of these two mappings will depend on finding efficient

---

[19] Fig.20 was first published in Lalvani [48] in the section "Biological Genomes and Number" (p.21,22). Author's work on DNA periodic tables and number systems dates back to 2010.
[20] The 16 2-base and 64 3-base combinations were first mentioned in Crick et al, [49].
[21] The 64 codons are arranged in a 4x16 table, the standard table (see, for example, DNA Codon Table, https://en.wikipedia.org/wiki/DNA_codon_table); an alternative 8x8 table using 0,1,0,1 is in Petoukhov [50]; a 3D 4x4x4 cubic array of 64 codons was developed by William Katavolos with Stan Wysocki with his cubic blocks [personal communication, 1975].
[22] The 4-cubic array applies to any 2 pairs of complementary numbers. Of the several variations tried by the author, the 4-base number system is presented in Fig.20. The 0,1,2,3 is suited for inter-conversion with number systems but, when used as a shape parameter, 0 is not always convenient.

Abiogenesis and the Future of Architecture 371

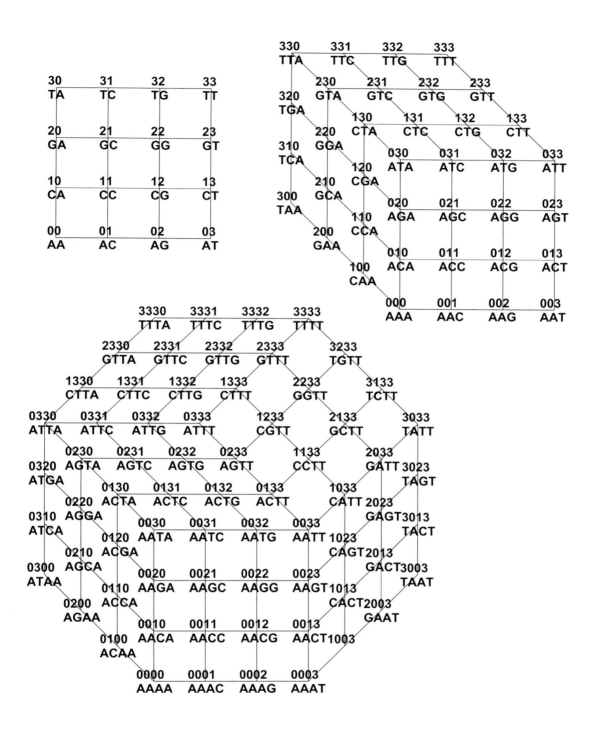

20
DNA codons, 2-base pairs (*top left*), 3-base triplets (*top right*) and 4-base codons (*bottom*) organized in a square, a cube, and a hyper-cube, respectively. (Lalvani, 2018).

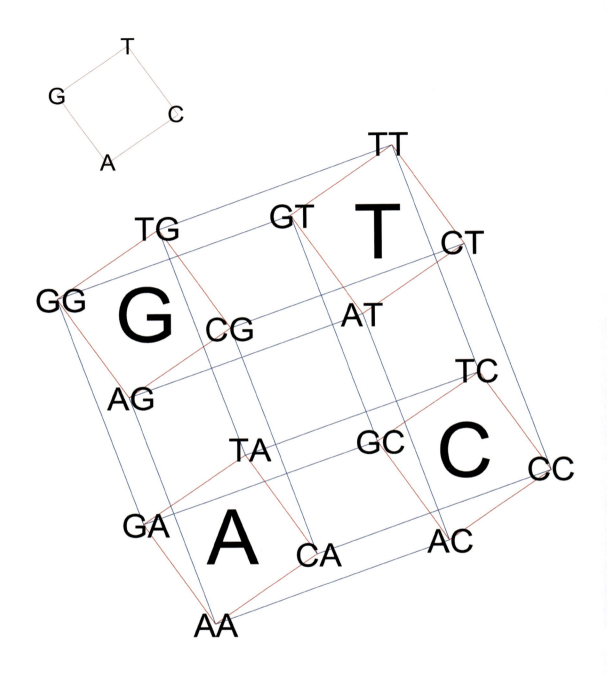

21
4 bases are arranged in a square in generation 1 (gen1; *top left*) which branches into 16 2-base sequences (gen2) comprising all base pairs obtained by adding a third base to each of the 4 original bases. All base pairs are arranged on the vertices of a 4D-cube built of four squares marked by bold letters.

Abiogenesis and the Future of Architecture

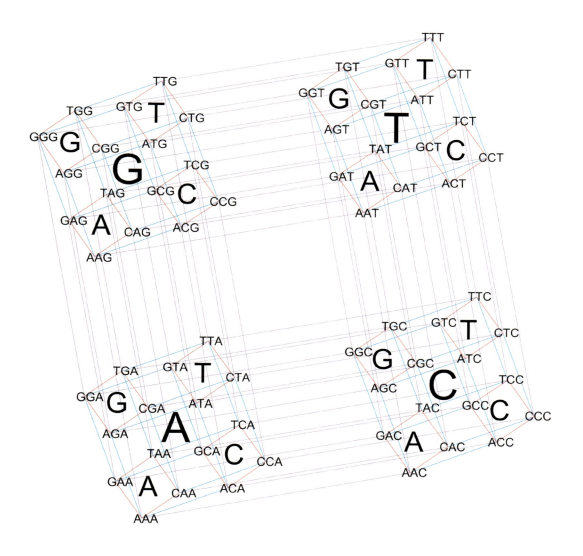

22
64 3-base sequences (gen3) comprising all codons (amino acids which make proteins) obtained by adding a third base to each pair in fig.21 which branches further in the same way gen1 branched to gen2. All 3-base sequences are arranged on the vertices of a 6D-cube built of four 4D cubes marked by larger letters.

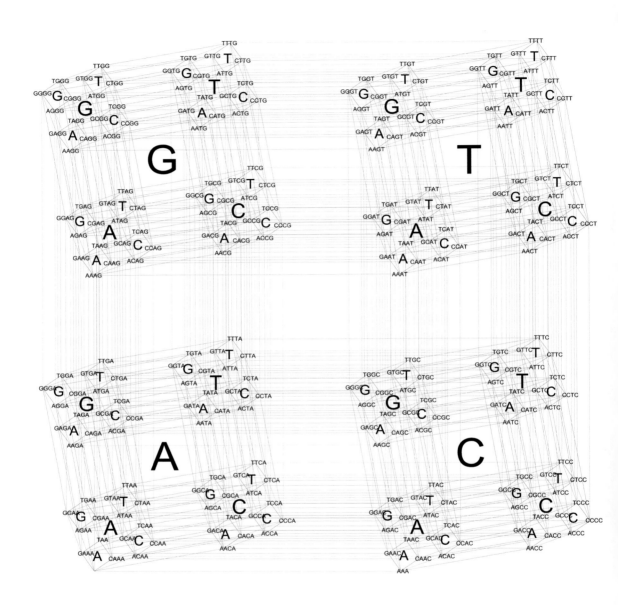

23
256 4-base DNA sequences (gen4) by adding a fourth base to each triplet in fig.22 which branches further in the same way as gen2 branched to gen3. All 4-base sequences are arranged on the vertices of an 8D cube which is built of four 6D cubes marked by the boldest letters.

algorithms to visualize and navigate such vast spaces interactively, and in a user-friendly way, to zoom in on specific parts of these highly structured ultra-large spaces. The unity of living designs is such a space is anchored in genetics while diversity of designs comes from how each individual species, each individual organism, each genome, each gene cluster, each gene, and each group of smaller base sequences, emerges over time within this space. The hierarchical morphological space of possibilities in higher dimensions provides a bridge between unity and diversity of life and adds a representational tool to the unification emerging from Evo Devo, the field of Evolutionary Developmental Biology.

Transforming Sequences (Mutations)

The changes (mutations) in DNA sequences, take place in several ways — adding, deleting or by replacing one or more bases with others in the sequence. Addition and deletion applies to a single base or an entire gene. Replacing a single base with another is a more complex transformation since there are more options. There are 16 such one-base mutations termed SNPs (single nucleotide polymorphisms) which permit changing the genetic code one base at a time. These SNPs can be mapped on the vertices of a 4D cube (**fig.24**). This diagram corresponds to the 16 base pairs in **fig.21** where any pair has been separated into two parts, the initial base and the target base in a mutation. It is easy to imagine a similar conversion of **fig.22** with 3-base sequences by separating into 2 initial bases and 1 target base for a two-nucleotide mutation into a single one, a 2-1 morphing, or vice versa. Similarly, **fig.23** with 4-base sequences into 3-1, 2-2, or 1-3 morphings. The process will apply to all morphings of any number of base-sequences into any number of target sequences. The DNA periodic tables thus are tables of all genomes, but also embedded are tables of all mutations. This will include extinct, extant and future genomes, as in the case of sea-shells and molecular universes mentioned earlier.

DNA And Number

In Chapter 8, Form was tied to Number suggesting a numeric code for form, a morph code for the few classes of forms and structures shown, and beyond. In this section, a relation between Number and DNA is presented so the two systems, number-coding and DNA coding, can become interchangeable. When molecular mechanisms will be established to achieve the transcription between number and DNA coding, this will provide a way to convert number-coded form to DNA-coded form. This will lead to numerically controlled DNA form-making. The linking of the morph code with DNA will require the development of a DNA equivalent of the ASCII code of form and processes. The interconversion between DNA and Number is an important step in linking the number-coded morph coding to DNA scripting of form.

The equivalence of number and DNA was shown in **fig.20** where the 4 DNA bases (A, T, C and G) were illustrated with their associated 4-base numbers

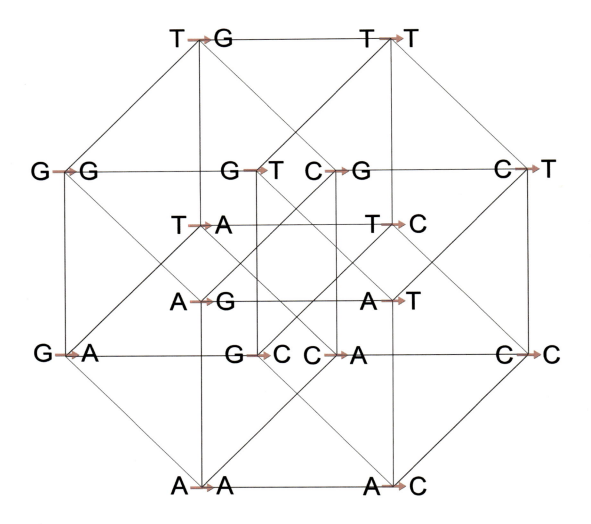

24
The 16 possible mutations (single nucleotide polymorphisms, or SNPs) by changing one base to another.

by assigning A=0, C=1, G=2 and T=3. The sum A+T (0+3) equals C+G (1+2) matches Watson-Crick's complementarity in base-pairing that was central to the discovery of the molecular structure of DNA. This numeric equivalence converts the sequence ATGGCTAAGC to 0322130021, and its complementary sequence TACCGATTCG to 3011203312, each sequence representing one strand of the double helix. The antisymmetry of the two DNA helices is thus embedded in the complementarity of base pairs. Since the number systems are interconvertible, the 16 2-base combinations shown earlier in **fig.21** are shown here in **fig.25** with three different number codes under each base pair: a 4-base code, a binary code and a decimal code (reading from top to down). The equivalence between the 16 DNA base pairs and four number systems – binary (2-base), quaternary (4-base), octal (8-base) and decimal (10-base) — are shown in **fig.26**. A similar mapping can be shown for 3-base combinations, 4-base combinations, and higher DNA base sequences along with their associated number codes in any number-base system. This makes DNA code and Number codes interchangeable. These are presented here as formal relationships and will require a translation into molecular terms so number codes can enable physical construction.

The mappings shown in **fig.25** reveal global and local numeric complementarity between the 2-base sequences. The sum of numbers on any vertex and its opposite vertex located across the center of the 4D cube (a point in the center of the diagram) equals 33 in 4-base system, 1111 in binary system, 17 in the octal system and 15 in the decimal system. Locally, within each square face of the 4D cube, the sum across diagonal locations with each 4D cube is also conserved through the sums are different. The same holds true of the sums across the diagonals of each cube. The concept of local and global complementarities holds in the important 3-base codons, 4-base sequences, and will hold in higher numbers of base sequences including gene markers, genes, gene clusters and entire genomes of organisms. This leads to the idea that each gene marker, or a gene or a gene cluster, has a complementary reversed sequence wherein the complements are located on the opposite vertices of the hypercube or hyper-cubic lattice of base sequences. Such hierarchical complementarity provides a graded link between the Watson-Crick pair on the molecular scale and the Chargaff rule[23] on the organism scale by providing an incremental bridge of order between the two extreme scales of any biological system.

DNA Shape-Scripting

A link between number scripting of shapes (as in Chapter 8 and other parts of the book) and DNA scripting is shown with examples dealing with geometry and topology. This link is a formal one and disregards the molecular basis of DNA transcription required to make physical structures. It is presented as an exploratory form-generation tool for designers based on the interchangeability

---

[23] Chargaff's rule was important in the development of the Watson-Crick DNA model. According to Chargaff, DNA in any species of any organism has the: global rule A+G=T+C. The amounts (percentages) of A and T are equal, and so are the amounts (percentages) of G and C. https://en.wikipedia.org/wiki/Chargaff's_rules

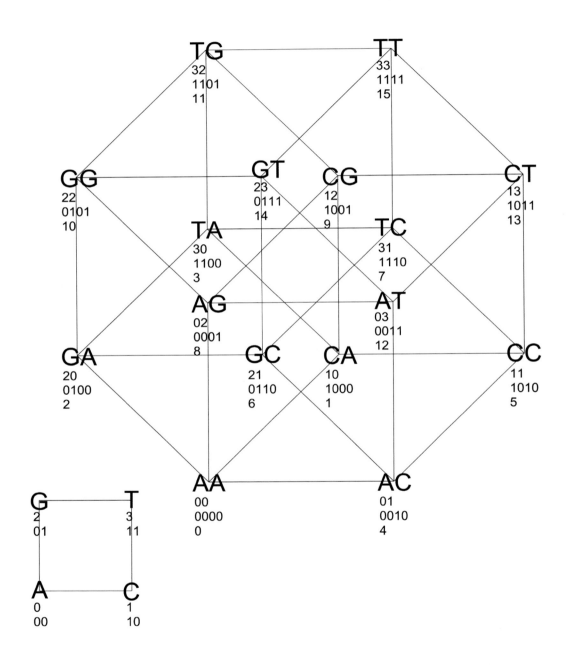

25
16 DNA base pairs and their equivalent number codes in 4-base, binary, and decimal systems (reading top to down). Single bases and their numeric codes are shown at the bottom left.

Abiogenesis and the Future of Architecture

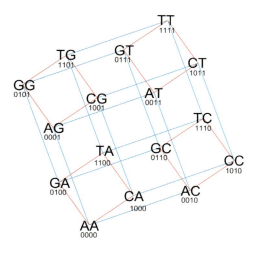

16 Base Pairs (2-Base Codons) on a 4D-cube in binary coordinates

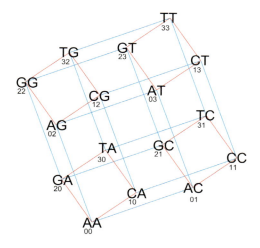

16 Base Pairs on a 4D-cube in 4-base coordinates

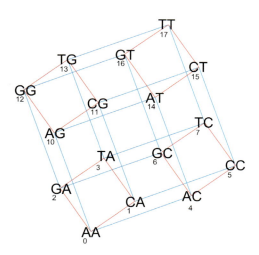

16 Base Pairs on a 4D-cube in octal (8-base) coordinates

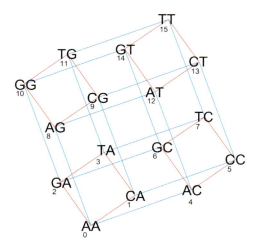

16 Base Pairs on a 4D-cube in decimal (10-base) coordinates

26
16 DNA base pairs and their equivalent number codes in the binary (*top left*), 4-base (*top right*), 8-base octal (*bottom left*) and 10-base decimal systems (*bottom right*).

of number and DNA sequences. When a direct relationship between DNA-coded molecular biology and number-based shape will be found in living structures, it would explain how numbers are used in nature's designs and also provide a design tool for synbio (synthetic biology) form generation. The examples shown here have a 4-base number code and are selected to connect complementarity in topology and geometry to complementarity of DNA bases.

Complementary Angles

The example shown here (**fig.27**) is an extension of shape-coding with angle number sequences described earlier (Chapter 7, p.217) and shows the equivalence between a number code and DNA code. It uses complementarity of angles with the geometric rule that sum of two angles equals 180 degrees (in half-space, i.e. lying on one side of a line) or 360 degrees (around a point). The number codes shown use the 180-degrees sum rule and are based on two pairs of complementary angle numbers, 1-4 and 2-3 derived from $n=5$, the spatial dimension of 5D space from which these angles are 2D-projected. The sum of complementary numbers equals 5. The shapes generated with these angles are built from straight line segments bent at these angles. The pair 1-4 comprises angle numbers 1 ($=36^0$) represented by A, and its complementary angle 4 ($=144^0$) represented by T, to complete the equivalent of A-T pair in DNA. The C-G pair represents the second pair of complementary angles 2 ($=72^0$) and 3 ($=108^0$). These are the angles of the pair of golden rhombii in the Penrose tiling. In the diagram, these angles are used to script a line in 2D space with numeric sequences which are shown with their corresponding DNA scripts. The sequences indicate the movement or growth of a line that grows in equal length segments at the 4 angles in any combinations starting from the start point indicated by a dot and moving in a clockwise direction. 16 examples of 4-angle movements (*top 2 rows*), 5 examples of 10-angle paths (*third row*), and a 50-angle path (*bottom*) are shown in their plan views.

The complementary helix having complementary DNA sequence represents the exterior angles and uses the 360-degree sum rule where this sum equals 2n, 10 in this case. This makes the interior-exterior angle paired complements as 1-9, 2-8, 3-7, 4-6 and 5-5 (the latter 5-5 is not shown in these illustrations). The sequence of interior angles 3342133243 or GGTCAGGCTG represented by a single DNA strand has a complementary sequence 7768977867 or CCAGTCCGAC which is scripted by the complementary strand of the double-stranded DNA. Complementarity of numbers representing complementarity of angles is equated with complementarity of DNA base pairs.

This example of complementary angles has relevance for encoding 2D and 3D structures, movements or growth. More angle number pairs, associated with higher values of the dimension of projection n, lead to a higher N-base number system.[24] Their DNA codes can be derived by converting the N-base number system to a 4-base system ($N=4$) which has a one-to-one correspondence with

---

[24] The base N is twice the number of rhombi which equals $n/2$ (n even) and $(n-1)/2$ (n odd) (see Fig.4, Ch 7). For even n, the square has the same pair of angles and need to be distinguished from each other for pairing.

Abiogenesis and the Future of Architecture

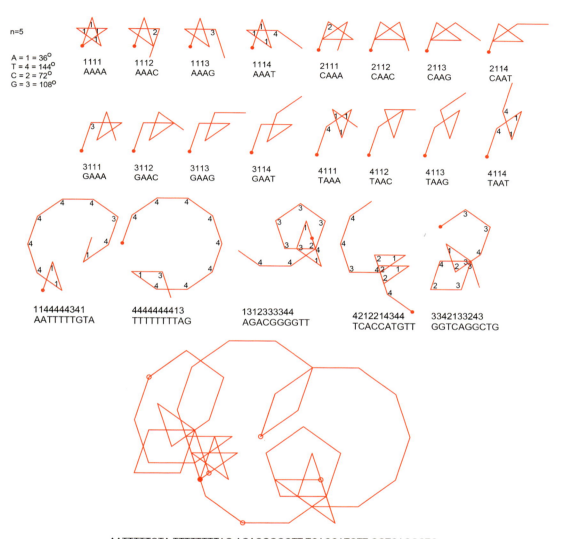

27
Angle number sequence codes and their corresponding DNA codes for angles projected from 5D space where A=360, C=720, G=1080 and T=1440 degrees.

the DNA code. Alternative morphologies for art, design and architecture can be created by using other formal morphological generators like changes in length or curvature including recursive operations and generative procedures as other formal applications of DNA shape-scripting. DNA shape-scripting will lead to personalized DNA art, design and architecture.

Complementary Topological Elements

Topological duality provides cases of 4-base form systems and are amenable to DNA shape-scripting. This applies to the elements of 3D topological structures as well as topologies of entire structures in 3D space. The four basic topological elements comprising 3D structures — vertices (V), edges (E), faces (F) and cells (C) — have sequential dimensions, 0 (vertex), 1 (edge), 2 (face) and 3 (cell), and correspond to the four units of the 4-base number system, 0,1,2,3. The complementarity of topological elements can be paired the same way as the two DNA base pairs. The first pair, vertices and cells (V-C) pair, can correspond to A-T, and the second pair, edges and faces (E-F) pair, to C-G.

Snips of Form

The four topological elements also define a form continuum and can continuously morph from one to another (**fig.28**). These continuous transformations, and their combinations, underlie all 3D structures in the 4D VEFC space described in Chapter 8. Similar to SNPs (pronounced "snip") in biological mutations, the 16 polymorphisms shown earlier in **fig.24** where one DNA base is replaced with another, the topological transformations are an isomorphic set of 16 but are continuous and not discrete as in DNA. These 16 transformations are the *snips of form* which permit any basic formal element of design, commonly described in the design arts as point, line, plane and "volume" or "solid", to morph continuously from one to the other. As in DNA, these transformations include four identities (V>V, E>E, F>F, C>C) with no topological change. The interchangeability of DNA and number systems permits the 16 topological transformations to be number-scripted or DNA-scripted.

One example, F>C transformation, where each 2D face of a 3D cell morphs into an independent 3D cell, is shown in **fig.29**. It is an example of topological cell division with implications for future architecture when 3D spaces and its elements can morph, divide, and replicate.

Combining Elements

Another feature of the 4-base structure of 3D topology, similar to other types of mutations like adding (inserting) or removing (deleting) bases or genes in living systems, relates to the 16 fundamental combinatorial transformations of the four basic elements, V, E, F, C.[25] These transformations lead to the 16 fundamental classes of building systems in nature and architecture. This is one more set of transformations that define the universal morph tool kit. In

---

[25] The 16 combinations of V, E, F, C elements can be represented by the combinations of their dimensional numbers 0, 1, 2, 3, or their binary codes (1,0,0,0), (0,1,0,0), (0,0,1,0), (0,0,0,1), or their DNA codes *a, c, t, g* (p.375). These 16 combinations include the 4 singlets (V, E, F, C), 6 doublets (VE, VF, VC, EF, EC, FC), 4 triplets (VEF, VEC, VFC, EFC), all elements (VEFC) and no element, and fit neatly on the vertices of a 4D cube.

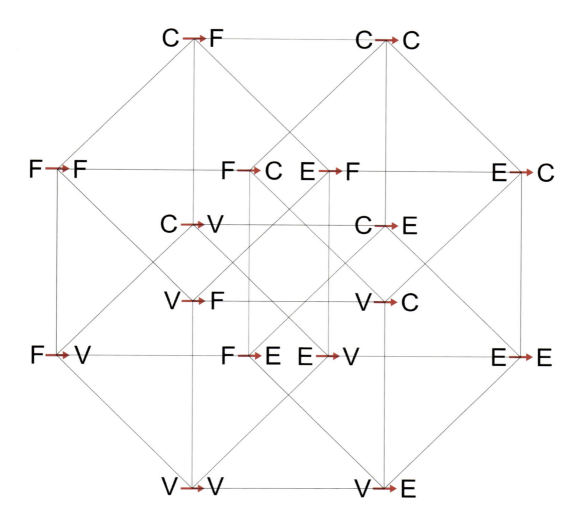

28
16 fundamental inter-dimensional transformations of elements of any 3D topological structure and its components - points (V, vertices), lines (E, edges), planes (F, faces) and 3D cells (C) with the same 4D organization and codes as the point mutations in DNA (fig.24).

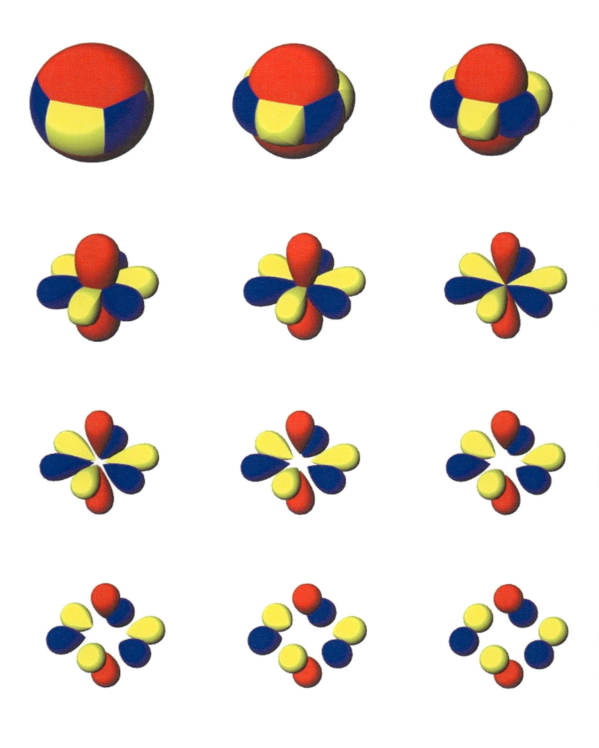

video 37

29
F>C transformation of a hex prism, a cell with 8 faces, morphing to 8 cells in an example of topological cell division (**video 37**).

growing structures, each of the 4 elements exist in 3 states – growing, shrinking or remaining unchanged. This expands the tool kit to 81 states which are mapped on a (3x3x3x3) 4D cube in a combinatorial continuum.

Dual Structures

The complementarity of elements and their continuous transformations extend to 2D surfaces and 3D structures and can be tracked by the element count which occurs in complements in dual topological structures. For 2D surfaces, the pairs of dual pairs present in any surface, can be similarly DNA-scripted.

Shape-Mixing

Artists use three colors to get all the colors they need. Shape-mixing, however, is more complex than color-mixing but, as we saw in Chapter 8, shapes can be mixed in a similar way using graded combinations. Here, it is extended to potential DNA-scripting in **figs.30-32** with an example of the tetrahedral family (3,3). This is just one example from the infinite family of (p,q) structures we saw in the last chapter. In **fig.30**, four fundamental 3D shapes generate the rest. These four come in 2 pairs of duals (complements) similar to the two complementary pairs in DNA. The red tetrahedron (a) located at 1000 is paired with its dual blue tetrahedron (t) located at 0100, the octahedron (c) locate (0100) and its dual cube (g) at (0001). The DNA base symbols A, T, C, G are replaced with a, t, c, g and represent an example of complementarity at a higher level of form. The lower case symbols, in terms of DNA-scripting, represent a higher-level code and must be broken down into sequences of ATCG letters where a composite shape like a tetrahedron is built from DNA base sequences. DNA origami methods could be useful here. For example, a can be built from a prescribed sequence of ATCG letters as a single DNA strand, its complement t from the complementary sequence in the complementary DNA strand, and the other pair of complementary shapes, c and g, from two different pairs of complementary DNA strands. Other shapes in this diagram are combinations of the four basic ones, and all 16 are arranged in 8 pairs of compliments, each pair located on the opposite vertices across the center od the 4D cube. For example, tg (0101) is a combination of t (0100) and g (0001) and its complement ac (1010) is a combination of a (1000) and c (0010) located on the opposite vertex; atc (1110) is a combination of a (1000), t (0100) and c (0100), and is a complement of g located symmetrically across the center of the 4D cube, and so on. In addition, $atcg$ (1111) is obtained by adding g (0001) to atc and is a combination of $a,t,c,g,$ and similarly for other transformations in the binary and DNA codes along the edges, faces and cells of the hypercube.

A graded shape-mixing, where one shape can transform or grow/shrink to another in increments, is shown in **fig.31**. This shows one way how size (scaling), for example, can be added to shape in DNA shape-scripting. This diagram has the same 16 on the vertices of the 4D cube we saw in **fig.30** but an intermediate is introduced on each edge, face and cell of the 4D cube. The red tetrahedron has grown to a larger red tetrahedron aa (location 2000), where the edge of aa is twice the edge of a. This can be seen from the four edges emanating from the origin 0 of

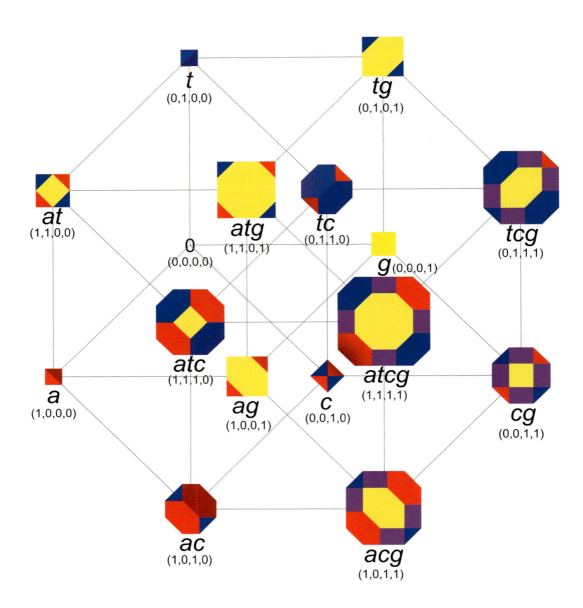

30
An example of shape-mixing using DNA-scripting of 4 shapes in 2 pairs of complements, the *a-t* pair comprising *a* (red tetrahedron, 1000) and *t* (blue tetrahedron, 0100), and *c-g* pair comprising c (octahedron, 0010, red+blue) and g (yellow cube, 0001). The combinations are shown on a 4D cube.

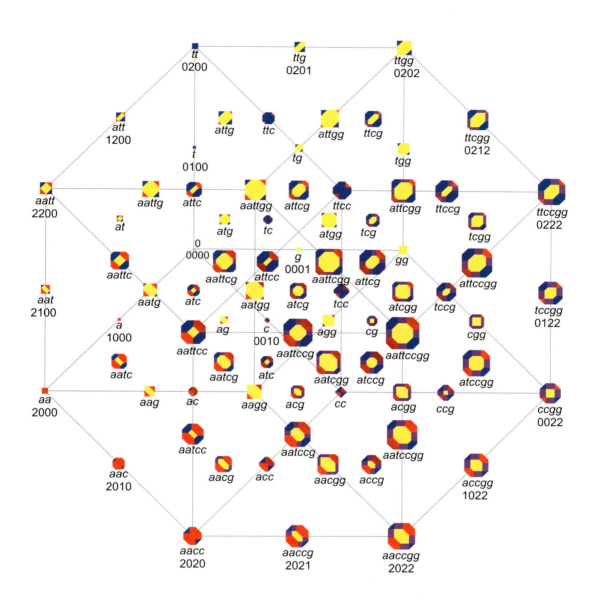

31
The example of shape-mixing is extended with size-mixing shown here with the same 4 shapes in Fig.30 with larger ones twice the original size. This adds *aa* (larger red tetrahedron, 2000), *tt* (larger blue tetrahedron, 0200), *cc* (larger octahedron, 0020), and *gg* (larger yellow cube, 0002). Their 81 combinations are shown.

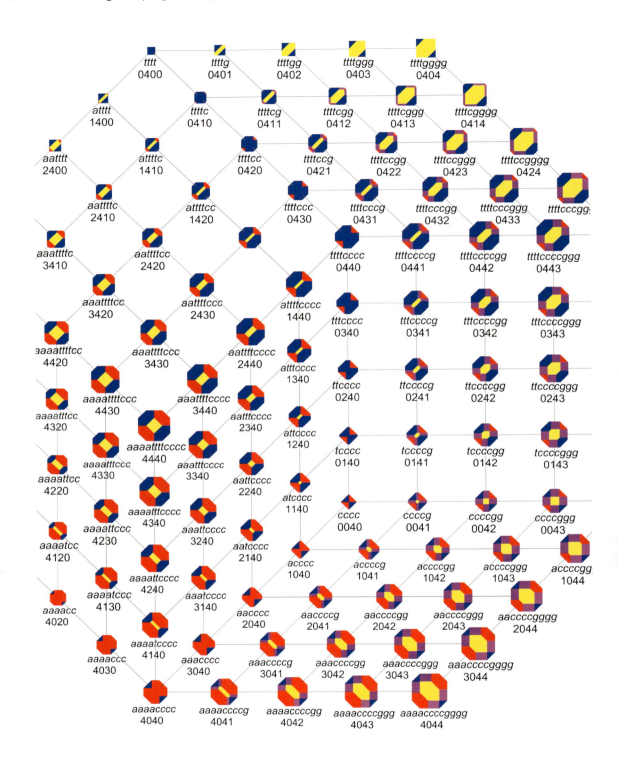

32
Further extension of shape-mixing and size-mixing by adding each of the 4 shapes in Fig.30 in 4 sizes. A close-up of a portion of the 625 combinations with each shape in 4 sizes, for example the red tetrahedron is four sizes DNA-scripted as *a, aa, aaa, aaaa*. Asymmetry will require a non-repeating code.

this diagram. The smaller shapes, *a, t, c, g*, lie closer to the origin and the larger ones, *aa* (2000), *tt* (0200), *cc* (0020), *gg* (0002), at twice the distance from 0 along the same direction. This leads to a hypercube twice the size of the smaller one in **Fig.30**. The four fundamental entities on the vertices of the larger hypercube are (*aa*), (*tt*), (*cc*), (*gg*). All others are derived from their combinations. The transformation pathway along the edge of the hypercube from *aa* (2000) to *aacc* (2020) is through the intermediate *aac* (2010). The other transformations can be reviewed by inspection.

The concept of shape-mixing in incremental sizes is further extended in **fig.32** where the smallest four entities are four times the size of the smallest leading to *aaaa* (4000), *tttt* (0400), *cccc* (0040), *gggg* (0004) as the fundamental shape-mixers. The blue tetrahedron *tttt* morphs to the truncated tetrahedron *ttttcccc* (0440) in the following sequence: *tttt* (0400), *ttttc* (0410), *ttttcc* (0420), *ttttccc* (0430), *ttttcccc* (0440). Other transformations can be similarly read and lead to transformation pathways between DNA-scripted shapes. This is just one example from the morphoverse which extends to other forms, symmetric as well as asymmetric.

Abiogenesis And Our Future

The impact of biology on design will continue to accelerate as it becomes the defining technology for making physical artifacts. The current divide between software and hardware in building technologies will gradually give way to a greater integration between the two, leading to a greater economy of means and materials. Information coding self-replicating molecules like DNA (RNA) are the only technologies where software and hardware are one. Possible non-DNA alternatives suggested by the minimal DNA-type presented here, and other molecular alternatives that encode information and self-replicate that are yet to be discovered, will be an important step in that direction. Such molecules will also provide candidate precursors for DNA and RNA and add to our ongoing active searches in astrochemistry and exobiology which explore our deep past. These lie at the core of our search for the origin of life (abiogenesis) and are driven by a deep humanistic interest in knowing whether we are alone in the universe and whether Earth is unique for the emergence of living systems. The search for our origin is also a search for minimal structures which define new emergent ones.

Prebiotic (abiotic) technologies, molecular as well as cellular, encoded and non-coded, provide a way to tie the future of architecture to the origin of life as mentioned in the opening of this chapter. This tie-in brings closure, a harmony, between us and our deep origin within the universe. On our planet, this harmony has been severely disrupted in our times and, as we rethink and re-position our fundamental relationship with nature to restore a semblance of a balance between us and the rest of nature[26] [51], an overarching philosophical and strategic re-linking is essential.

---

[26] Since the four spatial elements are also common to physical processes, a universal *process tool kit* mirrors the morph tool kit. For example, force can be pointal (0D), linear (1D), planar (2D) or spatial (3D), leading to 16 classes of force-driven processes, same as the 16 V,E,F,C combination, and map on the vertices of a 4D cube (see footnote 25, p.382). These also determine the classes of physical tools we use in making objects.

REFERENCES

1. *Powers of Ten* (1977) Directed by Charles and Ray Eames, narrated by Philip Morrison, Pyramid Films.

2. Sagan, Carl, L.S. Sagan, F. Drake, (1972) Message from Earth, *Science* (AAAS), Feb. 25, 1972, New Series, Vol. 175, No. 4024 (Feb. 25), pp. 881-884.

3. Planck, Max (1989) *Sitzungsber. Dtsch. Akad. Wiss.* Berlin, Math-Phys. Tech. Kl., pp.479-80.

4. Mendeleev, Dmitri (1869) *Zeitschrift für Chemie 12*, 405-406.

5. Jensen, William, B., (2002) *Mendeleev on the Periodic Law, Selected Writings, 1869-1905*, Dover.

6. Dragoset, R.A, A. Musgrove, C.W. Clark, W.C. Martin, and K. Olsen, (2017), *Periodic Table: Atomic Properties of the Elements* (Version 12), NIST SP 966. [Online] Available: http://physics.nist.gov/pt. National Institute of Standards and Technology, Gaithersburg, MD.

7. Scerri, Eric (2013) *Cracks in the Periodic Table, Scientific American*, June, 69-73.

8. Mazurs, E.G. (1957/1974) *Graphic Representations of the Periodic Systems During One Hundred Years*, University of Alabama Press, Alabama, 2nd Edition, 1974.

9. van Spronsen, J.W. (1969) *The Periodic System of Chemical Elements, A History of the First Hundred Years*, Elsevier.

10. Leach, Mark R. (1999-2018) The Internet Database of Periodic Tables, *The Chemogenesis Web Book*. http://www.meta-synthesis.com/webbook/35_pt/pt_database.php

11. Bradley, David (2011) Periodic Debate, *Chemistry Views*, June 09, Wiley-VChapter DOI: 10.1002/chemv.201000093

12. Scerri, Eric (2007) *The Periodic Table, Its Story and Significance*, Oxford University Press.

13. Lalvani, Haresh (2020) 4D-Cubic Lattice of Elements, *Foundations of Chemistry* 22, pp.147-194 (published online, November 14 2019), Springer.

14. Mendeleev 150: 4th International Conference on the Periodic Table endorsed by IUPAC, July 26-28, 2019, ITMO University, St. Petersburg, Russia.

15. Raup, David (1966) *Geometric Analysis of Shell Coiling: General Problems*, Journal of Paleontology, Vol. 40, No. 5 (September), pp. 1178-1190.

16. Darvas, György (2018) *Symmetry: Culture and Science*, Vol. 29, No. 1, pp.5-6, Symmetrion.

17. Watson, James, A. Berry, K. Davies (2017). *DNA, The Story of the Genetic Revolution*, Knopf.

18. Doudna, Jennifer A. and S. H. Sternberg, (2017). *A Crack in Creation, Gene Editing and the Unthinkable Power to Control Evolution*, Mariner Books.

19. Dyson, Freeman (2007) Our Biotech Future, *The New York Review of Books*, July 19.

20. Church, George (2012) *Regenesis, How Synthetic Biology will Reinvent Nature and Ourselves*, Basic Books.

21. Antonelli, Paula with Burckhardt, A (2020) *The Neri Oxman Material Ecology Catalogue*, The Museum of Modern Art.

22. Armstrong, Rachel, Spiller, N., Synthetic biology: Living quarters, *Nature*, 467, 916-918 (2010).

23. Tibbits, Skylar ed. (2017) *Active Matter*, MIT Press.

24. Williams, Adam (2012) They live: Synthetic biology in architecture, *Design Build Network*, Sep 20

2012, https://www.designbuild-network.com/analysis/featureleaf-review-living-buildings-biotechnology/

25 Cumbers, J. (2019) Can We Redesign The Modern City With Synthetic Biology? Could We Grow Our Houses Instead Of Building Them? *Forbes*, Sep 26, 2019.

26 Venter, Craig (2013). *Life at the Speed of Light, From the Double Helix to the Dawn of Digital Life*, Penguin.

27 Callaway, Ewen. (2016) Race to design life heats up, *Nature*, 31 March, Vol. 531, pp.557-558.

28 https://bioethicsarchive.georgetown.edu/pcsbi/sites/default/files/PCSBI-Synthetic-Biology-Report-12.16.10_0.pdf

29 Baltimore, David et al. (2015) A prudent path forward for genomic engineering and germline modification, *Science*, April 3, 348 (6230), pp.36-38.

30 Seeman, Nadrian (1982) Nucleic Acid Junctions and Lattices, *Journal of Theoretical Biology*, 99, pp.237-247. Seeman's website: http://seemanlab4.chem.nyu.edu/

31 Adelman, Leonard M. (1994) Molecular Computation of Solutions to Combinatorial Problems. *Science, New Series*, Vol. 266, No. 5187 (Nov 11), pp.1021-1024.

32 Rothemund, Paul W. K. (2006) Folding DNA to create nanoscale shapes and patterns, *Nature*, Vol. 440 (16 March). doi:10.1038/nature0458

33 Woods, Damien, D. Doty, C. Myhrvold, J. Hui, F. Zhou, P. Yin & E. Winfree (2019) Diverse and robust molecular algorithms using reprogrammable DNA self-assembly, *Nature*, Vol. 567 (21 March), 366.

34 Jung, C., P.B. Allen, A.D. Ellington (2016) A stochastic DNA walker that traverses a microparticle surface, *Nature Nanotechnology*, Vol. 11, February.

35 Katavolos, William (1962) *Organics*, [Hilversum] Steendrukkerij de Jong & Co.

36 Giorgini, Vittorio (1996) Early Experiments in Design Derived from Nature's Morphologies, In: Morphology And Architecture, Guest Ed. Lalvani, H., *International Journal of Space Structures*, Vol. 11, Nos. 1&2, pp.57-67, Multi-Science Publishing, U.K. https://doi.org/10.1177/026635119601-212

37 Watson James D. F.H. Crick (1953) A structure for deoxyribose nucleic acid, *Nature*, 171, pp.737–738. Also available from: http://www.nature.com/genomics/human/watson-crick/

38 Crick, Francis H.C. and J.D. Watson (1954) The complementary structure of deoxyribonucleic acid. *Proceedings of the Royal Society Series A*, 223, pp.80–96.

39 University Of Illinois At Urbana-Champaign. "Formic Acid Found Toward Hot Galactic Molecular Cores". *Science Daily*, 8 July 1999. <www.sciencedaily.com/releases/1999/07/990708080344.htm>.

40 Rubin, R.H., G.W. Swenson Jr., R.C. Benson, H.L. Tigelaar, W.H. Flygare (1971). Microwave detection of interstellar formamide. *Astrophysical Journal*, Vol. 169, L39.

41 Mehringer, David M.; et al. (1997). "Detection and Confirmation of Interstellar Acetic Acid". *Astrophysical Journal Letter,*. 480 (1): L71.

42 Hollis, J.M, F. J. Lovas, A.J. Remijan, P.R. Jewell, V.V. Ilyushin, and I. Kleiner (2006). Detection of Acetamide (CH_3CONH_2): The Largest Interstellar Molecule with a Peptide Bond. *The Astrophysical Journal*, Vol. 643, No. 1, L25.

43 Brown, R.D., J.G. Crofts, F.F. Gardner, et al (1975) Discovery of interstellar methyl formate. *The Astrophysical Journal*, Vol 197, Pt 2, L29.

44 Jørgensen J.K., C. Favre, S.E. Bisschop, T.L. Bourke, E.F. van Dishoeck, and M. Schmalzl (2012) Detection of the simplest sugar, glycolaldehyde, in a solar-type protostar with ALMA. *The Astrophysical Journal Letters*, 757: L4 (6pp), September 20.

45 Elsila, J.E, D. . Glavin, and J.P. Dworkin (2009). Cometary glycine detected in samples returned by Stardust. *Meteoritics & Planetary Science* 44, Nr 9, 1323–1330.

46 Lalvani, Haresh (1981) *Multi-dimensional Periodic Arrangements of Transforming Space Structures*, PhD Dissertation, University of Pennsylvania, Published by University Microfilms, Ann Arbor.

47 Lalvani, Haresh (1982) *Structures on Hyper-Structures* (self-published) Lalvani, New York; a slightly revised version of [46].

48 Lalvani, Haresh (2018) Morphological Universe: Genetics and Epigenetics in Form-Making, *Symmetry: Science and Culture*, Vol. 29, No. 1, 240pp, January.

49 Crick, FrancisH., Griffith, J.S, and Orgel, L.E. (1957) Code without Commas, *Proc Natl Acad Sci USA*, May 15; 43(5):416-21. https://www.ncbi.nlm.nih.gov/pmc/articles/PMC528468/

50 Petoukhov, Sergei V. (2001) Genetic Codes 1: Binary Sub-Alphabets, Bi-Symmetric Matrices and Golden Section, *Symmetry: Culture and Science,* Vol. 12, Nos. 3-4, pp.255-274.

51 Wilson, Edward O. (2016) *Half-Earth, Our Planet's Fight for Life*, Liveright Publishing.

SUPPLEMENT (Chapter 8, p.275)

The Zero Cyclic Sum Property in Graphs

Zaran Lalvani

Consider a graph with weighted vertices and weighted directed edges that has the following properties:

(a) There is a directed edge from every vertex to every other vertex

(b) Every edge's weight is equivalent to the difference between the weight of its origin and destination vertex

The *zero cyclic sum property* comes from the observation that any cycle in such a graph has an edge weight sum of zero. This is an intrinsic property of the representation, as will be proven below.

Proof (by weak induction)

Given a complete directed digraph G with vertices V, corresponding vertex weights W, edges $E(i,j)$ $\forall i,j \in V$ where $i \neq j$, and edge weights $|E(i,j)| = W_j - W_i$, it will be proven that the sum of all edge weights in any directed cycle is equal to zero.

Base Case

A cycle with two vertices i, j has an edge weight sum $|E(i,j)| + |E(j,i)|$. This can be algebraically simplified as follows:

$$|E(i,j)| + |E(j,i)| = (W_j - W_i) + (W_i - W_j)$$
$$= W_i - W_i + W_j - W_j$$
$$= 0$$

∴ A cyclic sum in a cycle of two vertices must always be zero.

Induction

Assuming that all cycles of k vertices have a zero cyclic sum, it will be shown that any cycle of length $k+1$ must also have a zero cyclic sum.

Let us examine a pair of adjacent vertices in an arbitrary cycle of length k, labeled x and y and connected by the edge $E(x,y)$. We can define our cycle as having a set of edges A where $E(x,y) \in A$.

Without loss of generality, we can construct a new cycle of length $k+1$ by adding a vertex z to our existing cycle, removing $E(x,y)$ from its edges, and adding $E(x,z)$ and $E(z,y)$ in that edge's place. Refer to our new cycle's edges as B.

If it can be shown that $|E(x,y)| = |E(x,z)| + |E(z,y)|$, then our new cycle must have the same cyclic sum as the original by applying the algebraic substitution property. Thus,

$$|E(x,y)| = |E(x,z)| + |E(z,y)| \implies \sum_{n=0}^{k} |A_n| = \sum_{n=0}^{k+1} |B_n| \qquad (1)$$

We can algebraically simplify the antecedent (left hand side) as follows:

$$|E(x,y)| = W_y - W_x \qquad (2)$$

$$\begin{aligned} |E(x,z)| + |E(z,y)| &= (W_z - W_x) + (W_y - W_z) \\ &= W_z - W_z + W_y - W_x \\ &= W_y - W_x \end{aligned} \qquad (3)$$

$\therefore$ By (2) and (3), $|E(x,y)| = |E(x,z)| + |E(z,y)|$. Applying this and the induction assumption to (1) implies all cycles of length $k+1$ have a zero cyclic sum.

Proof Conclusion

It must follow by applying the induction to the base case that all cyclic sums in any graph constructed with the constraints of G are zero, *i.e.* all cycles in G have a zero cyclic sum.

Generalization

Without loss of generality, we can extend our proof to graphs with vertices that have weight-vectors in an n-dimensional space. Vector addition and subtraction will have an equivalent result, but instead of the scalar 0, all cycles will have an edge weight-vector sum of $\vec{0}$.

Intuitively, this property maps to a special case of *net displacement* in a physical system. If one considers a two dimensional plane, a cycle can be constructed by traversing a series of points and returning to the origin. Clearly, the net displacement in this scenario would be $\vec{0}$, just as the property dictates.

ACKNOWLEDGMENTS

The work presented in the book spans several decades during which I have benefited from many individuals and organizations. Looking back requires using a telescope with zoom and micro lenses. The acknowledgments include the immediate and the broader, subtle and sustained, interactions which have shaped the work and my thinking. Writing this acknowledgment reminds me of Richard Feynman's "sum of all histories" of an elementary particle.

Gyuri Darvas's invitation to present a keynote as part of the Vienna Lecture Series (*Weiner Vorlesungen*) at the Symmetry Festival 2016, Vienna, led to the "monograph" published as full issue of the journal *Symmetry: Culture and Science* (2018). This set the groundwork for this book for an architecture and design art audience, and was primed by lectures at Pratt, RISD, UPenn and RPI, as well as TED and TEDx.

Individuals, with whom I've had the rare honor and privilege to connect over the years have had a lasting impact on my thinking through brief, sometimes longer, encounters of which they may be unaware. Some are no longer with us. James D. Watson, Aaron Klug, Donald Caspar, Cyril Stanley Smith, and Steve Harrison for their approach to fundamental structure in nature; Stephen Jay Gould for his love of form within and outside paleontology; H.S.M. Coxeter for his receptiveness and encouragement; John Conway; Nicholas De Bruijn and Peter Kramer on many exchanges related to quasi-crystals. Thomas Benjamin for opening the early world of structural biology to me and enabling the meetings with the leading scientists including some mentioned above; Anand Sarabhai and Hildegard Lamfrom for their important role during this period. From all of them I learned that beauty, dear to us in the design arts, is also a driver (an unspoken one) in science and mathematics.

My thanks to Irmgard Bartenieff in dance/movement studies for equating form and movement; Hoshyar Nooshin for opening up space for me within structural morphology, and Martin Mikulas in aerospace structures. Janos Baracs, Michael Burt, Peter Pearce, Koji Miyazaki and Tony Robbin — my colleagues in parallel pursuits — for their inspiring life-long commitment to understanding fundamental structure of form and space; Dean James Parks Morton, The Cathedral of St. John the Divine, New York (see also p.246, 250), for his unwavering support of explorations in geometry within a cathedral setting. To Paul Gerome for the opportunity to work on an international standard for telebiometrics for human interactions with the environment, and seeing it though to adoption.

To many who have played an important role through their support and conversations; Evan Douglis, Karl Chu, Jay Kappraff, Bruce Hannah, Thomas Zung, Joe Clinton, Istvan Hargittai, Eric Scerri, Ted Goranson, Nat Friedman, to name a few; Carole Sirovich and Mark Rosin; Timothy Collins at NASA-Langley; Terence Riley at MoMA, Murray Moss for representing me during a brief but significant period; Peter Franck and Tarik Currimbhoy at OMI; Hans De Castellane, Deborah Buck and Helen Varola; Lydia Bradhsaw at MTA Arts; Barbara Ferry and Ellen Strong at Smithsonian's NMNH; Robert McDermott, Patrick Hanrahan and David

Sturman at Computer Graphics Laboratory (CGL), NYIT. I am indebted to my colleague John Lobell for his continuing support through decades of development of this work; he was instrumental in initiating the collaboration with Milgo and has played a significant role as an advisor and advocate.

At *Milgo-Bufkin*, Bruce Gitlin, for his unwavering support of the physical experiments in "making" at his factory during the period 1997-2014 and since. None of the metal works in the book would have been possible without him. Gitlin's exceptional commitment, vision and knowledge in addition to the open access to his factory enabled my "Florentine moment" to borrow Martin Seligman's words (p.4). Stewart Gitlin, for continuing the trajectory. The extraordinary metal-workers at Milgo for sharing their skills and also teaching me how a hard material like metal is malleable to them as clay is to a potter: Alex Kveton, Wayne Lapierre, Robert Warzen, Tony Pukulinski, Fahan Mohammad, Pauli Mussi, Scott Krissow and Frank Alicea. Additional behind-the-scenes support from: Bill Soghor, Steve Messler, Scott Kranzler, Boris Umansky and Barbara Kanter.

Pratt Institute, my academic home base since 1970, has provided the stability to continue. It has been a privilege to receive the generous support and encouragement from all levels of the administration over the years including the Research Recognition Award, Pratt's highest honors in research. I would like to acknowledge the School of Architecture (SOA) for its continuous support of open-ended explorations and for the Morphology program. To Pratt for establishing the Center for Experimental Structures (CES) and Frances Halsband for initiating CES in the early 1990's. My association with esteemed senior colleagues Vittorio Giorgini, John Johansen and William Katavolos (p.275), all co-founding partners of CES, has been a rare and distinct honor which continued with Katavolos until recently. Exchanges with all three have enriched me considerably. My special thanks to Tom Hanrahan, Peter Barna, Kirk Pillow, Erika Hinrichs, Jason Lee, Allison Druin, and also to Pam Gill; the roles of Frances, Tom and Kirk have been significant. To Harriet Harris and Adam Elstein for the careful photo-documentation of the models at CES, and Theo David for his continuing interest in archiving the work. From earlier years, Steve Kagan, Constantine Karalis, Richard Penton and Michael Trencher. Parts of the book were supported by Pratt's Faculty Development Fund, Seed Grant, SOA, and the sabbatical in Spring 2021.

At *CES*, the unending thirst for knowledge from our students, research assistants and interns at CES reminds us we are all students when we search. For this, they deserve my unreserved thanks. For CES research projects: my colleagues and associates Che-Wei Wang, Ajmal Aqtash, John Gulliford, Peter van Hage, Robinson Strong and Ahmad Tabbakh. To Che-Wei and Ajmal for insightful discussions and feedback, to Ajmal for his counsel and guidance. Recent student assistants for prep work and some digital images related to the book: Emma Chan, Ardon Lee, Anna Oldakowski, Kalliopi Economou, Sharvari Mhatre, Matthew Mitchell, Jonathan Hamilton, Ivan Man Hin Yan and Tron Le. To industry partners, Milgo-Bufkin and LERA+ (Alfonso Oliva and team) for their important roles in the *3D Hypersurface* project (p.262), Alfonso for several exciting discussions on experimental projects.

University of Pennsylvania (1976-81) for providing the environment for doctoral work; my academic advisor, Buckminster Fuller, for the freedom to do "independent" work and his critical comments. Fuller, whose models I re-constructed as an undergraduate architecture student in India, and later learned more of him through his writings and meetings. His influence has been unmeasurable. He alerted me early on to the idea of design science as a solution for solving planetary problems and as a scale-free exploration; his spontaneous personal remarks and unconditional support at the time have given me the strength to continue on. Peter McCleary for his generous support of the work in early stages; Holmes Perkins; Robert LeRicolais, my first connection at UPenn, for many inspiring discussions and guidance in my student years; Anne Tyng for her friendship and sharing her insights.

My gratitude to the *Indian Institute of Technology, Kharagpur* (IIT-KGP), where this search began in 1966-67 as an undergraduate student within an architecture department set in a STEM institution. This provided the spirit of ArtSci integration and the opportunity to "discover" morphology. To architecture Professors M.A. Rege, Himanshu D. Chhaya and Gopal Mitra, special thanks for supporting me during that formative period. To Dr Sahib Ram Mandan (Mathematics) who loaned me his exquisite wooden models of polyhedra and would give me feedback on my drawings, Dr B. Chatterji (Chemistry) on shapes of molecules, Prof A. P. Gupta (Civil Engineering) on structural principles and physical tests of natural fibers and an egg shell, and Dr.D. De (Agriculture) on plant morphologies. J.G. Helmcke's stereo electron micrographs of diatoms received from him in Berlin were eye-openers during that period. More recently, for the honor of their Distinguished Alumnus Award.

At *Lalvani Studio*, I have been privileged to be supported by many talented and brilliant individuals for their dedicated assistance over the years. They have generously shared their exceptional skills in digital modeling, scripting, rendering and animations for most of the digital images in the book. Neil Katz, Ajmal Aqtash, Che-Wei Wang, John Gulliford, Peter van Hage, Patrick Donbeck, Robinson Strong (listed chronologically) stand out for sustained duration and depth of involvement. Practically all early "pre-scripting" computer modeling by Neil Katz for published papers in the mid-80's thru 90's, some images from that period are included in Chapters 7 and 8 in addition to the acknowledgements in Chapters 1-3 and postscripts on AlgoRhythms which are a tribute to his contribution. Special thanks to Che-Wei for introducing animations in the body of work and, more recently, for developing interactive digital tools for navigating portions of the morphoverse. Additional animations by Neil, Brian Bulloch, Robinson and Ahmad Tabbakh. For renderings, Mohamad Al-Khayer, Ajmal, Che-Wei, Neil and Kristof Dubose in Chapters 1-3 and postscripts, followed by John, Peter, Patrick and Robinson in later chapters. Additional digital support in recent years for new images in Chapters 8 and 9: Ahmad Tabbakh, Sharvari Mhatre, Kalliopi Economou, Anna Oldakowski, Nubia Lluvicela, Mari Kroin, Arie Salomon, Matthew Malcom and William Vandenburg. The collective contributions of all have been invaluable. Some images in Chapters 8 and 9 modeled by the author have benefitted from the skills they have taught me.

To *Graham Foundation for Advancement of the Arts* for their support of the unpublished manuscript *Higher Dimensions and Architecture* (1994) and the self-published *Patterns in Hyper-Spaces* (1982) which have contributed to parts of

Chapters 7 and 8, respectively.

For insightful discussions, the biophysicist Loren Day on the molecular biology aspects of the work (Chapter 9); he has been my go-to sounding board over the years in this area. The biochemist Larry Luchsinger, on chemistry aspects of minimal DNA-type molecules (Chapter 9); his informal review over five years of development has been invaluable; the paleontologist Niles Eldridge for his comments and encouragement; the mathematician Egon Schulte, for reviewing the mathematical aspects of the work (Chapter 8) since 2016. Uttara Asha Coorlawala, for her constructive feedback on parts of the manuscript. All have contributed generously with their time which has improved the book greatly.

At *Routledge*, Katharine Maller for her enthusiastic reach out that began the book, Francesca Ford for transitioning support, Krystal (La Duc) Racaniello for her steering me through early manuscript preparation, Christine Bondira for early production feedback, Lydia Kessel for behind-the-scenes guidance, Jake Millicheap and Adam Guppy for carefully seeing it through to publication.

Book production at Lalvani Studio, Safa Mehrjui, Naini Bansal, Chloe Ni for preparing the book for publication. Safa for design and layout, Naini for collating and proof-reading, Matthew Malcolm for the cover, Chloe and Matt for their help in taking it over the finish line. The book would not have taken its physical form without their care and diligence.

Photography: Peter Tannenbaum (p.254) and Alexander Severin (p.226, bottom middle and right), Pratt Institute; Drew Lackey, Robert Wrazen, Bruce Giltin, Ajmal Aqtash and Haresh Lalvani; Patrick Donbeck and Zaran Lalvani at Lalvani Studio.

Video: All video and animation credits appear inside the list of videos on p.viii.

INDEX

2n-dimensional 368, 370
360-degree sum rule 380
Abiogenesis 6, 3, 128, 210, 340, 385, 389
    abiogenetic space 4, 352, 353, 359, 368
abiotic 352, 366, 389
    technologies 366
Adelman 357, 391
affine transformations 313, 316
Alber Einstein 212
Albrecht Dürer 313
AlgoRhythm(s) 2, 17, 18, 19, 44, 64, 66, 70, 72, 74, 81, 83, 97, 159
    Beams 66, 82
    Columns 17, 159
    Glass Panel Systems 81
    Kinetic AlgoRhythms 70
    Skins 66
    Soft AlgoRhythms 70, 74
    Truss System 81
Algo Signage 83
allotropes of carbon 108
"alphabet" of form 74
Amoebas 319
    amoeboid shape 319
angle
    rise angle 322
    central angle 359
    cone angle 322
    departure angle 322
    dihedral angles 301
angle number(s) 301, 319, 380
    sequences 217, 321, 380
Anne Tyng 273
anti-parallel 359, 366, 369
    helices 359, 366
Antonio Gaudi 42, 131, 132, 148, 162, 275, 327, 328, 329, 330, 331, 339
ARCH 07201 205, 206
Archimedean spiral 98, 99
architectural genome 28, 30, 79
architectural taxonomies 28
Aristid Lindenmeyer 130, 157, 158
Arthur C. Clarke 142, 158

artificial
    architecture 28
    biology 29
    genetic code 5, 10
    genomics 41
    intelligence 28, 273
    life 28, 41, 78
ASCII code of form 375
assemblers 4, 43, 366
astrochemistry 389
asymmetry 30, 108, 221, 318, 319, 359, 366
    asymmetric repeat unit 359, 368
atomic number(s) 342, 344, 346, 352
A-T pair 380
automorphic 82
automorphogenesis 41, 78, 130

Baer 43
Balaenoptera musculus .52
base
    1-base 368
    2-base 368, 370, 371, 372, 377,
    3-base 368, 370, 371, 373, 375, 377
    4-base 368, 370, 371, 374, 375, 377, 378, 379, 380, 382,
    5-base 370
    6-base 370
    8-base 377, 379
    10-base 377, 379
binary
    code 327, 377
    continuum 2, 3, 305
    system 377
biomimicry 1, 29, 104, 142, 209
    biomimicry in reverse 205
Bmp4 336, 339
bottom up 6, 74, 76, 80, 155, 368
branched surface 142
Bruce Giltin 17, 148, 164
Buckminster 152, 154, 157-159, 338-340, 396
buckytubes 108

CALLISTO 172
CaM 336
carapaces 115, 124, 127, 142, 168, 241

CARAPACE Series 166, 168
carbon nano-threads 108
Carl Sagan 4, 340, 341
cell-removal 294
cellular automata 74, 76, 80, 114, 155, 168, 275, 327, 332, 333
Center for Experimental Structures 3, 18, 227
centrally symmetric 124, 216, 352
Chargaff rule 377
Charles Darwin 270
chemical architecture 357
chemogenetic pathways 352
Che-Wei Wang 316
Christopher Castelino 159
Cloud Cover 9
Cloud Nine 152
color-mixing 385
Column Museum 10, 13, 18, 25
combinatorial
    continuum 2
    transformations 382
complementarity 4, 125, 275, 346, 352, 366, 368, 375, 377, 380, 382, 385
    of angles 380
    of DNA base pairs 380
    of numbers 380
complementary
    angle numbers 380
    base pairs 358, 361, 366
    sequence 375, 380
    strand 380
Compound curves 63, 64
connection atoms 360, 366
conservation
    of cost 96, 104, 128
    of material 98, 106, 358
    rule 346
continuous
    numbers 97, 98
    or discrete 74, 272
    transformations 2, 30, 44, 97, 130, 155, 270, 273, 274, 275, 382
continuum 2, 3, 4, 29, 30, 31, 44, 45, 70, 74, 78, 80, 130, 131, 132, 154, 155, 156, 191, 209, 272, 274, 281, 284, 289, 294, 299, 301, 305, 316, 319, 322, 327, 332, 336, 382
    morphologic continuum 45
    2D continuum 123
    binary continuum 2, 3, 305
    combinatorial continuum 2
    CA Continuum 332
    flexible-rigid continuum 132
    form continuum 130, 191, 272, 274, 382
    form-process continuum 130
    platter continuum 108
    pattern continuum 96, 97, 108, 332
    process continuum 130, 131
crescents 6
    crescent-shaped 6, 7
CRISPR-Cas9 356
curved-
    curved-edges 301
    curved-faces 301
    curve-folded 2, 17, 81, 82, 159
    curved polygons 301
    curved polyhedra 301
    curved structures 299
    curved hyperstructures 241
CYCLONE 195
cymatics 357

D'Arcy Thompson 41, 79, 106, 114, 124, 130, 131, 209, 313, 322
Darwin's finches 336, 339
David Raup 327
decimal 98, 99, 116, 117, 155, 274, 336, 377, 378, 379
    code 377
    system 377
deltahedra 246, 281
Dendrocalamus giganteus 152
derivative structures 216, 272
developable surfaces 64, 72
digital biomimicry 104
digital emergence 168, 327
digons 285, 301
dihedron 79, 191
Dimension
    0D 316
    1D 98, 316
        1D cellular automata 332, 333,

Index

334
1D Series 300 96

2D
- 2D amoeboid shapes 319
- 2D building blocks 216
- 2D building kit 233
- 2D hypersurfaces 216, 221, 246
- 2D hypertiles 216, 227,
- 2D lattice 298, 299, 301, 304, 305, 346
- 2D periodic table 217, 285, 289, 292, 301, 302, 303, 323, 324, 347, 348
- 2D projection 6, 172, 213, 215, 216, 217, 221, 236, 270, 319, 380
- 2D singularity structures 301
- 2D space 98, 216, 316, 370, 380
- 2D space-filling curves 115
- 2D structures 212, 241, 281, 285, 380
- 2D surface 132, 213, 221, 382, 385
- 2D tiling units 217, 250

3D
- 3D blocks 232, 233, 234, 313
- 3D cell(s) 132, 227, 228, 250, 255, 258, 260, 276, 295, 316, 319, 327, 342, 382, 383
- 3D cube 213, 227, 233, 236, 275, 294
- 3D cubic lattice 288, 289, .322
- 3D Euler space 276
- 3D faceted form 316, 319
- 3D folded structures 159
- *3D Hypersurface* 262, 264
- 3D periodic table 301, 319, 346, 349, 370
- 3D projections 216, 236, 250, 270, 295
- 3D saddle hyper-tile 241
- 3D singularity structures 276
- 3D space 98, 148, 151, 169, 170, 196, 212, 213, 233, 313, 316, 327, 341, 342, 370, 382
- 3D structures 162, 196, 212, 216, 241, 243, 275, 276, 281, 294, 322, 335, 337, 357, 358, 380, 382,
- 3D taxonomy 213, 301
- *3D VEF space* 276, 281, 283
- *3D X-STRUCTURES* 205.

4D
- 4D abiogenetic space 352
- 4D architectonics 227
- 4D Cartesian coordinates 342
- 4D cells 250, 258, 260, 342
- 4D compositions 230, 231, 232, 233
- 4D cube 4, 44, 83, 116, 117, 119, 120, 123, 124, 227, 228, 229, 233, 236, 262, 289, 292, 294, 313, 315, 316, 332, 342, 352, 375, 377, 385
- 4D cube space frame 237
- 4D cubic frame 227
- 4D-cubic lattice 116, 237, 339, 342, 346, 390
- 4D-cubic periodic table 342, 370
- 4D cubic space 327
- 4D cubic super-lattice 346
- *4D Form Solid* 313, 316
- *4D HCNO space* 352, 353
- 4D hyper-cell 316
- 4D lattice 119, 352, 356, 236, 352, 356
- *4D Modulor* 233, 234
- 4D morphological space 124
- 4D periodic table 305, 306, 342, 343, 344, 345, 346, 347, 363
- 4D polytopes 285, 288, 289, 293, 294, 295,
- 4D quantum coordinates 346
- 4D scaling 314, 316
- 4D space 116, 213, 232, 233, 275, 227, 228, 285, 351, 352, 370
- 4D space frame 227, 228, 237,
- 4D space-time 213
- 4D structures 276, 289, 294, 342
- 4D suspended structures 227
- 4D tetrahedron 294
- *4D VEFC space* 276, 281, 283,

382
4th dimension 213, 231, 233, 342
5D
    5D cells 258, 260
    5D cube(s) 4,
    5D Euclidean space 6
    5D Form Solid 316
    5D lattice 285, 289, 292
    5D periodic table 370
    5D space 221, 262, 276, 285, 380, 381
    5D space-time 213
6D
    6D cell 258, 260
    6D cube 35, 216, 235, 236, 299, 327, 331, 373, 374
    6D Form Solid 316
    6D lattice 289, 299
    6D microverse 292
    6D non-periodic infinite polyhedron 258
    6D periodic tables 370
    6D Schwarz surface 258
    6D space 35, 236
7D
    7D cube 213, 215, 216, 217
    7D lattice 293, 294
    7D projections .217
    7D space .116, 294, 296
8D
    8D cube .332, 334, 374
    8D space .262, 332
    8D table .347, 363
9D
    9D coordinates .299
    9D cube .299
    9D lattice .299
    9D microverse .296
    9D space .285, 294
10D Periodic Table .220
11D
    11D model .322
    11D space .322
12D model .326
13D space .298, 299
14D space .350, 352
15D
    15D lattice .285
    15D microverse .299
    15D space 285
118D
    118D cubic lattice 346
    lattice 347, 363
    118D periodic table 346
    118D space 346, 351
discrete
    and continuous 74, 274, 305
    numbers 97
DIY
    genetics 356
    microverses 338
Dmitri Mendeleev 270
DNA
    DNA alternatives 4, 358, 389
    DNA and Number 375
    DNA art 382
    DNA code 377, 380
    DNA-coded form 375
    DNA coding 375
    DNA computer 357
    DNA-computing 357
    DNA periodic tables 368, 370, 375
    DNA scripting 375, 377
    DNA scripts 380
    DNA shape scripting 114, 377, 382, 385
    DNA technology 356
double helical molecules 357, 358, 366, 368
doubly-emergent 98
Doxiadis 134, 158
Drexler 41, 43, 76, 80
dumbbell plan 195

economy of scale 2, 98, 104, 128
edge
    edge-preserving transformations 281
    edge-removal 294, 299
    transformations 289
electronic blocks 342
emergence 3, 74, 80, 97, 155, 162, 168, 176, 262, 270, 275, 322, 327, 332, 338, 389
    of order 97

emergent 2, 3, 74, 132, 155, 165, 168, 174, 191, 195, 270, 313, 327, 332, 358, 389
   designs 2, 96, 97, 98, 108, 115, 327, 358
   folds 165
   morphologies 270, 313, 332
   EMERGENT OPENINGS 168, 170
   EMERGENT RIPPLES 168
emergency shelter 3, 191, 195
English wheel 64
epigenetic(s) 30, 72, 131, 155, 156, 157, 209, 210, 269, 339, 392
   continuum 155
   form-making 131
   phenomena 131, 155, 209
Escher 31, 44, 148, 158, 222, 267
Euclidean 6, 30, 191, 192, 213, 241, 266, 273, 288, 289, 319
   space 6, 30, 273
Euler
   Euler Characteristic 276
   Euler count 275, 276, 281
   Eulerian structures 79, 276
   Euler-Schlafli 342
Evo Devo 128, 272, 335, 375
exobiology 389
expandable minimal surface 142
expanded structures 155
explosion 305
exponential scale 341

fabrication genome 66
face-addition 299
face-removal 294, 299, 300
   of curved structures 301
   of forms 273
   of structures 299
flat-faced 301
form
   continuum 130, 191, 272, 274, 382
   form-process continuum 130
   form-process-material triad 335
   Form Solid 313, 316
   is process 66, 180
   form follows number 217

fractal 108, 236, 237, 238, 249
   Fractal AlgoRhythm Ripples 83
   Fractal Fold Roof 34
   Fractal High-Rise 10, 15, 79
   hyper-tree 370
Fractured Folds 169
Frank Drake 341
Freeman Dyson 340, 356, 390
Frei Otto 132, 148, 339
frequency of subdivision 294, 296
fullerenes 35, 42, 359
   buckminsterfullerenes 338
function follows form 174

Garrett Gourlay 82
gene-editing 356
generative
   morphology 35
   principles 125
   taxonomies 108
   taxonomy 346
genetic(s) 2, 16, 79, 128, 130, 131, 209, 335, 336, 356, 370
   Genetic Architecture 130, 157
   code 5, 10, 11, 16, 30, 42, 131, 273, 336, 359, 366, 368, 375
      of architecture 16, 30
      of form 30, 273
   engineering 5, 16, 72, 356
   tool kit 128
gene tool kit 272, 335
genomic
   alphabet 83
   Genomic Architecture 2, 6, 28, 41, 64, 79, 130, 157, 209, 210
genus 30, 116, 121, 195, 276, 278, 336
geodesic spheres 42, 285, 294
geometric automaton 327
Geospiza 336
Giorgini 16, 18, 41, 80, 130, 134, 157, 357, 391, 394
golden rhombii 246, 380
GPS 112, 114
GPS of form 273
graded shape-mixing 385
gravitational rotational forming 148
Gravitational Waves 148

GR FLORA 176, 178, 179
grid transformations 313
growing morphologies 270
Gyorgi Darvas 269, 390, 395

HCO molecular chain 366
helix
    2-bend 366
    3-bend 366
    4-bend 366
    helical
        axis 359
        backbone 359, 360
        chain molecule 366
        molecular chain 359
        Molecules 366
higher dimensional
    space 6, 30, 35, 212, 273
    tree 370
high technology 10
H.S.M. Coxeter 158, 213, 266, 285, 339, 395
human proportions 313
hyper 6, 44, 70, 273, 274, 275, 316, 337, 342, 368, 370, 377
    Hyper Architecture 6
    hypercube 44, 337, 342, 371, 377, 385
    hyper-Escher 44
        pattern 31
    hyper-fractal periodic table 370
    hyperfullerenes 42
    hyper-networks 274
    hyperspace 17, 30, 125, 159, 212, 213, 233, 236, 265
        Hyperspace Architecture 6
    hyperspheres 35
    hyperstructures 3, 6, 16, 17, 35, 129, 159, 266
    hypersurface(s) 5, 6, 7, 17, 33, 157, 159, 252, 255
    hyper-tiles 6, 216, 241
    hyper-tree 368, 370
    Hyperuniverse 29, 31, 44
        of Form 29, 44
hyperbolic space 288, 289
hyperbolic tessellations 285, 292, 301
hyperboloid(s) 195, 196, 205, 241, 246

HYPERION 174
Windows into 31
HyperWall 68, 70, 81, 83
    system 66

Ilene Shaw 262
inclusion-exclusion 294
industrial production 98, 104, 128, 131, 337, 358
Industriall Revolution 128
infinite
    2D periodic table 285, 346, 348
    emergent designs 97
    infinities 72, 272, 338
    polyhedra 246, 258, 262, 297, 299, 304
information
    coding 4, 358, 366
    encoding 358
instant
    architecture 3, 155, 191, 196
    Cubism 172
    structures 185, 195
InterRipples 45, 54
inter-transforming 31, 72, 76, 272
intra-transformations 272
irregular geometries 270
irregular morphologies 270, 294
isomer space 352
isomorphism 337
    isomorphic 335, 382
    isomorphic set 382

Jackson Pollack 134
James Joyce 357
James Watson 63, 340
John Conway 395
John Holland 63, 130, 157
John Johansen 16, 80, 130, 157, 394
John Lobell 63
Jung 357, 391
JVCI-syn3.0 357

Karl Chu 130, 393
Koji Miyazaki 43, 236, 268, 395
Kristof Fenyvesi 227

Lautomaton 98, 104, 106, 108, 112, 115

laws of nature 29
Leach 41, 390
Le Corbusier 134, 158, 233, 234, 267, 313, 339
Leonhard Euler 276, 357
Leopold Kronecker 270
lightweight structures 358
L-systems 155, 168, 327, 332
Ludwig Schläfli 285
Lymnaea 336

MD150 342
magic carpet 152
Mandelbrot 134, 158
man-nature dichotomy 10
Mario Carpo 2, 18
Martin Seligman 4, 396
Martyn Poliakoff 341
mass-active 195
mass customization 2, 18, 28, 37, 44, 66, 97, 104, 128, 185, 213, 222, 316
    in Nature .125
Mendeleev 270, 341, 342, 390
meta
    meta-algorithm 72
    Meta Architecture 2, 5, 6, 19, 74, 79, 130, 154, 157, 209
    meta-dimension 6
    meta-morphology 273
    meta space 5
    meta-structure(s) 3, 30, 35, 265, 273, 316
    meta-symmetry 272
metamorphosis 30, 273
Michael Burt 246, 268, 305, 339, 395
microverse 272, 292, 296, 299
Milgo-Bufkin 5, 131, 162
    Milgo Experiments 2, 17, 35, 63, 159, 162, 211, 250
    Milgo Gallery 40, 45, 60
        Gallery 1 45, 60
        Gallery 2 45, 60
        Gallery 3 45, 60
minimal
    DNA molecule 366
    DNA-type double helical molecules 357, 358, 368
    DNA-type molecular structures 358
    DNA-Type Molecules 358
    double-helical molecules 360
    helices 359, 366
    molecules 358, 366
    processes 358
    structures 358, 389
    surfaces 258, 262, 297, 299, 301, 305, 327, 331, 358
    saddle surfaces 305
        saddle prisms 305
minimum
    architecture 191
    genome 357
    triad pair 359
Mobius strip 262, 265
Modulor 134, 158, 233, 234, 267, 313, 339
molecular
    architecture 368
    evolution 352
    scale 377
    structure 377
monohedron 191
morph
    chip 76
    morph code 30, 42, 273, 274, 375
    morph gene(s) 72, 80, 274, 337
    morph genome 66, 74, 79, 115, 274
    morph tool kit 6, 270, 272, 274, 285, 294, 301, 313, 335, 336, 338, 382
morphing(s) 30, 42, 98, 104, 105, 106, 107, 124, 217, 221, 226, 281, 283, 375
    MORPHING88 114
    MORPHING PLATTERS 172, 358
    states 33
morphogenesis 41, 78, 114, 155, 162, 205, 211, 275, 327, 335, 336, 352
    of molecules 352
    morphogenetic pathway 37, 274, 336, 337
morphogenomics 17, 28, 30, 74, 159
morphological
    code 5, 11, 30, 31, 130
    genes 72, 76, 191, 209

genome 28, 30, 37, 41, 154, 274
Genome 63, 70, 72, 76, 79, 157, 339
hyperuniverse 30, 41, 42
informatics 28
space 123, 124, 352, 375
universe 5, 16, 17, 29, 41, 44, 64, 70, 72, 130, 155, 156, 157, 159, 209, 210, 269, 270, 271, 339, 392
morphology 4, 35, 42, 64, 68, 76, 273, 316, 322, 339, 352
morphology of morphology 273
Morphoverse 6, 2, 3, 209, 270, 271, 276, 336, 337
multi-cellular 281
structures 294
multi-directional expansion 138, 162, 164, 174
Municipal Arts Society 17, 18, 66, 68, 159
Murray Moss 96, 104, 128, 129, 393
mutations 273, 375, 376, 382, 383
Mycoplasma mycoides JVCI-syn1.0 357
nano-structures 357

NASA-Langley 66, 276
n-base sequences 368, 370
NCN 359, 360, 362
NCO 359, 360, 362
n-cube 370
n-cubic lattice 370
N. De Bruijn 236, 395
n-dimensional 172, 214, 215, 216, 217, 273, 368
n-dimensional cubes 172, 273
n-dimensional periodic tables 368
n-dimensional spaces 368
Neil Katz 66, 77, 332
nodal-Pitx gene 336
non-deformational bending 35, 64
non-Euclidean
geometries 191
space(s) 30, 273
non-periodic 6, 30, 42, 134, 172, 270
n-simplexes 273
number code 217, 319, 375, 377, 380

number-coded form 375
number scripting 377
numerically controlled DNA form-making 375
numeric blobs 319
numeric code for form 375
numeric complementarity 377
NYSTAR 70, 80, 81, 82, 205

octal 377, 379
octet truss 82
ontogenic design 274
organism scale 377
origin of form 97, 216, 276
origin of life 4, 216, 389
orthographic view 352

paired complements 380
paired pairs 360, 366
PANEL Series 162, 165, 166
parallel expansion 174
Patrick Donbeck 98
Patrick Hanrahan vi 236
pattern-code 31
pattern-generation 31
Penrose tiling(s) 43, 82, 172, 217, 246, 258, 380
perimeter-preserving 104, 105
periodic molecular chains 366, 367
periodic table
periodic table of elements 341, 342, 346
periodic table of form 41
periodic tables of form 273, 341
periodic tables of periodic tables 273
Peri-Tube 152
Perrella 6
"perspective" space 352
perspective transformation 316
Peter Pearce 269, 305, 339, 395
Peter van Hage 98, 115, 322
Phalium 116
phylogenetic design 274
physical emergence 162, 168, 322, 327
pigmentation patterns 125, 116, 115, 322
Planck scale 132

Planck units 341
plane-faced 299, 305
pneumatic morphologies 322
polymorphisms 335, 375, 376, 382
polytopes 246, 273, 285, 288, 289, 293, 294, 295
Portico A 327, 329, 330
Powers of Ten 340, 368, 390
Prague Biennale 17, 18, 19, 159
Prague Column 17, 159
prebiotic 389
prebiotic information coding 366
precursor molecules 360
principle of similitude 106, 142
process universe 335
Project X 66, 68

quantum numbers 342, 346, 352
quasi-crystals 35, 42
quaternary 377

Rene Thom 63
repeat unit 359, 366, 368
representations 274, 332, 368, 370
rheology 3
    rheologic process 148, 176, 210
Richard Feynman 397
Roald Hoffmann 2, 129
Robert Le Ricolais 1, 157, 396
Robert McDermott 42
Robert Oppenheimer 357
Roger Penrose 172, 236
Rothemund 357, 391
ROUND WHORL 169, 170, 172
Rudolph Doernach 130

saddle
    polygons 241, 244, 245, 246, 258, 305, 306, 307
    polyhedral 262, 305
    prisms 305, 306, 310
    rings 305, 307, 308, 309
Santiago Calatrava 42
Scerri 341, 342, 390
Schectman 42
Schläfli symbol 285
Schwarz surface 258, 260, 261, 327
SEED54 356

Seeman 357, 391
self-
    self-adjusting 154
    self-architecture 2, 6, 154
    self-crumpling 176
    self-folding 3, 176, 179
    self-replicating 357, 358, 366, 389
    self-replicating molecules 357, 389
    self-rigidization 152, 180
    self-rigidization to self-shaping 152
    self-rigidized 132, 152, 154, 195
    self-shaping 3, 131, 152, 154, 155, 156, 157, 162, 164, 165, 166, 168, 169, 170, 174, 176, 185, 209, 210
    self-similar 82, 98, 114, 246, 273, 322, 327, 332, 368
    self-similarity 35, 114, 273, 274, 316
    self-undulations 168, 195
    self-wrinkling 176
Sequioa sempervirens 152
SEQUOIA 180, 182, 185
SEQUOIA 2 180, 182
shape-coding 380
shape-mixing 385
shape-scripting 377
single-cell structures 276
single nucleotide polymorphisms (SNPs) 335, 375, 376, 382
singularities 30, 276, 301
    singularity structures 276, 281, 283, 301
skewed subdivision 294
snips of form 335, 336, 382
space-time 152, 213, 270, 272, 274, 322, 335, 336
space-time-mass 272, 322, 335
space-time transformations 270
Spheroids 33, 35
Stanley Perelman 66
Stanley Wysocki 370
stellated polygons 289, 290
Stephen Jay Gould 41, 395
stitching 17
St. Petersburg 342, 390

straight-edged 299
strength of form 64
strength of material 64
Structures on Hyper-Structures 16, 42, 66, 79, 157, 338, 392
Stuart Kauffman 96
subdivided n-cube 370
subdivisions 285, 294
surface-active 195
Synergetics 158, 339
synthetic
    biology 128, 162, 265, 356, 391
    DNA 357
    life 357
    organism 357
symmetry
    2-fold rotational 360
    meta-symmetry 272
    mirror-symmetry 360
    symmetry-breaking 124, 272, 313, 316, 327
    symmetry-preservation 272

TED2004 4, 17, 19, 21, 159
Ted Goranson 273
TENT PYRAMID 205, 206, 208
Termespheres 148
tesseract 342
The Little Prince 152
titanium 17, 66, 81, 159
top down 6, 74, 76, 80, 155
    emergence 155
topology
    topological
        cell division 382, 384
        duality 382
        transformations 30, 191, 281, 382
    topology-changing 272
    topology-preserving 222, 272, 313, 346
    topoverse 337
transformation
    pathways 30, 43
    transforming columns 35, 37, 44
    step-by-step transformations 327
Transitions 10, 45, 60
tree of life 41, 43, 337, 357

triad pairs 359, 360, 364, 365, 366, 368
triangular grid 289, 290

Umbrellas 10, 15
universal scale 81, 341
universe
    of columnar forms 66
    of continuous transformations 30
    of form 2, 66, 115, 272, 337
    of number 337
Uri Shiran 159
Uttara Asha Coorlawala 17

Variable
    Geometries 275, 313
    Positions 316
    Rotations 316
    topologies 275, 313
vector(s) 275, 342, 346, 352
    vector-star(s) 274, 346, 352
        2-vector-star 346
        3-vector-star 346
        118-vector-star 346
VEFC Space 275, 276, 281, 283, 382
VEF Space 276
vertex
    vertex-removal 294
    symbols 298, 299
v-grooving 17
Vincent DeSimone 81
Vittorio Giorgini 16, 18, 41, 80, 130, 157, 394
Voronoi cells 98

Watson-Crick
    base-pairing 368
    DNA structure 359
WaveKnot 10, 15
wave space labyrinth 37
Wayne LaPierre 168
William Katavolos 357, 396
Wolf Hilbertz 130, 157
Wolfram 115, 129, 333, 334
    Wolfram's Rule 30 332
    Wolfram's rules 327, 332
Woods 357, 391
WORM HOLE 205, 208

X-POD 138 191, 192
X-STRUCTURES Series 2, 3, 162, 174,
    180, 191, 192, 205, 209, 322
X-TOWER(s) 3, 174, 180, 185, 189, 195
    X-TOWER 54.4 180, 183, 185
    X-TOWER 88.2 185, 187, 188
X-TUBES 195
XURF Series 162, 164
    XURF Portrait(s) 172

Zaran Lalvani 275, 393-394
zero-cyclic sum 275, 346
zig-zag spirals 112
zonogons 217, 241, 313